RESEARCH IN ORGANIZATIONAL CHANGE AND DEVELOPMENT

Volume 12 • 1999

RESEARCH IN ORGANIZATIONAL CHANGE AND DEVELOPMENT

An Annual Series Featuring Advances in Theory, Methodology, and Research

Editors: WILLIAM A. PASMORE
　　　　　Organization Development Practitioner
　　　　　New York

　　　　　RICHARD W. WOODMAN
　　　　　Department of Management
　　　　　Texas A&M University

VOLUME 12 • 1999

JAI PRESS INC.
Stamford, Connecticut

Copyright © 1999 by JAI PRESS INC.
100 Prospect Street
Stamford, Connecticut 06901-1640

All rights reserved. No part of this publication may be reproduced, stored on a retrieval system, or transmitted in any form or by any means, electronic, mechanical, photocopying, filming, recording, or otherwise, without prior permission in writing from the publisher.

ISBN: 0-7623-0627-0
ISSN: 0897-3016

Manufactured in the United States of America

CONTENTS

LIST OF CONTRIBUTORS — vii

PREFACE
William A. Pasmore and Richard W. Woodman — ix

CONVERSATIONS AND THE EPIDEMIOLOGY OF CHANGE
Jeffrey D. Ford — 1

ORGANIZATIONAL DEVELOPMENT AS FACILITATING THE SURFACING AND MODIFICATION OF SOCIAL RULES
Craig Lundberg — 41

COLLABORATION AND ALLEGORY: EXTENDING THE METAPHOR OF ORGANIZATIONAL CULTURE IN THE CONTEXT OF INTERORGANIZATIONAL CHANGE
Joseph W. Grubbs and Robert B. Denhardt — 59

MAKING CHANGE PERMANENT: A MODEL FOR INSTITUTIONALIZING CHANGE INTERVENTIONS
Achilles A. Armenakis, Stanley G. Harris, and Hubert S. Feild — 97

TQM AND ORGANIZATIONAL CHANGE: A LONGITUDINAL STUDY OF THE IMPACT OF A TQM INTERVENTION ON WORK ATTITUDES
Jacqueline A-M. Coyle-Shapiro — 129

IMPLEMENTING EFFECTIVE CROSS-FUNCTIONAL TEAMS: A MULTILEVEL FRAMEWORK FOR ANALYSIS
Long W. Lam, Sheri J. Bischoff, La Verne H. Higgins, and D. Lynne Persing — 171

COLLABORATIVE ORGANIZING: AN "IDEAL TYPE"
FOR A NEW PARADIGM
  Peter J. Robertson   205

THE RELATIONAL HEALING DIMENSION OF
ORGANIZATIONAL DEVELOPMENT:
TRANSFORMATIVE STORIES AND DIALOGUE IN
LIFE-CYCLE TRANSITIONS
  Gurudev S. Khalsa and David S. Steingard   269

THE PROFESSIONALIZATION OF ORGANIZATION
DEVELOPMENT: A STATUS REPORT AND LOOK TO
THE FUTURE
  C. Ken Weidner, II and Orisha A. Kulick   319

ABOUT THE CONTRIBUTORS   373

LIST OF CONTRIBUTORS

Achilles A. Armenakis   Department of Management
                                Auburn University

Sheri J. Bischoff   Department of Organizational Leadership and
                                  Strategy
                                Brigham Young University

Jacqueline A-M. Coyle-Shapiro Industrial Relations Department
                                London School of Economics and Political
                                  Science

Robert B. Denhardt   School of Public Affairs
                                Arizona State University

Joseph W. Grubbs   School of Public and Nonprofit
                                Administration
                                Grand Valley State University

Hubert S. Feild   Department of Management
                                Auburn University

Jeffrey D. Ford   Max M. Fisher College of Business
                                The Ohio State University

La Verne H. Higgins   Department of Industrial Relations and
                                Human Resource Management
                                Le Moyne College

Stanley G. Harris   Department of Management
                                Auburn University

Gurudev S. Khalsa   Department of Organizational Behavior
                                Case Western Reserve University

Orisha A. Kulick	Organization Development Practitioner Westmont, Illinois
Long W. Lam	School of Business and Public Administration University of Houston at Clear Lake
Craig Lundberg	School of Hotel Administration Cornell University
D. Lynne Persing	ESC Recherche Ecole Superieure de Commerce de Toulouse, France
Peter J. Robertson	School of Policy, Planning, and Development University of Southern California, Los Angeles
David S. Steingard	Erivan K. Haub School of Business St. Joseph's University
C. Ken Weidner, II	Center for Organization Development Loyola University, Chicago

PREFACE

Volume 12 of this series is arguably the most challenging and creative collection of chapters we have yet assembled. As editors, we have urged our colleagues to be more daring in their thinking. Much of what one reads in the literature of our field, we feel, is a rehash of well-known and shop-worn ideas. Apparently, our colleagues have heard our message and accepted the challenge. Anyone who reads this volume carefully will probably find that at some point, it tests the limits of their thinking. As one of our authors writes, it is important to approach the subject matter with an open and inquisitive mind.

Ideas which are comfortable and familiar are not likely to challenge or transform our thinking. As human beings, our need to reduce cognitive dissonance causes us to seek the familiar and reject the unfamiliar, often without careful reflection. Yet as scholars, we must overcome such natural tendencies in order to look beyond the reaches of well accepted doctrine. We must explore less-understood and less-accepted explanations of the way things are and consider instead the possibilities that alternative futures could hold. Then, with courage and persistence, we can add our small contribution to the advancement of science and society. We applaud the authors of Volume 12 for their creativity, hard work, and courage.

Jeffrey Ford's chapter on conversations as an instrument for organizational change sets the tone for what is to follow. Using the science of epidemiology as a backdrop, Ford helps us to understand how ideas catch on, spread, and eventually become institutionalized in the form of institutionalized organizational changes. In a stimulating and well-documented discussion, Ford explains how the tools and frameworks of epidemiology have been used to understand patterns of proliferation of such diverse phenomena as violence, alcoholism, suicides, and smoking. With rich metaphors, Ford describes how the epidemiology of change conversations can be analyzed by examining characteristics of infective agents, the organizational environment, and host susceptibility. He then goes on to make recommendations about how to enhance the spread of beneficial change-enhancing conversations by increasing the infectivity of agents and increasing host exposure and susceptibility. Students of dialogue, structuration theory, and the social creation of reality will find much of interest here, while practitioners will value this fresh look at what is required to make change visible, powerful, and permanent.

Craig Lundberg begins his piece on social rules theory by first reviewing shortcomings in traditional approaches to thinking about organizational change. He notes that OD has been dominated by models which posit unidirectional causality and are blind to the pluralism of contexts, structural systems differences, inherent conflict, and politics. Social rules theory, rooted in ideas put forth by Marx, Weber, Chomsky, Goffman, March, Weick, and others has several advantages over these traditional models.

Social rules theory views organizations as social systems, composed of members who are agents with varying degrees of influence. Rule systems govern the transactions among members of social systems by shaping the nature, impact, content, and outcomes of conversations. The basic questions addressed by social rules theorists are how social rules are formulated and later transformed. Rules are created, learned, maintained, and modified by agents. Agents develop distinct identities and associated capacities to influence social rules in different contexts. Gaining insight into how identities develop, are maintained, and change is fundamental to understanding efforts to bring about organizational change. Because rules are a combination of generic societal patterns of interaction, like the way people behave in hierarchies, and specific contextual patterns or the way people treat a particular individual because of their reputation in an organization, scientists and practitioners need to engage the help of local actors in conducting inquiries into how social rules function.

Rules are necessary in systems to make transactions easier and more predictable but they also serve as barriers to change. By taking a holistic view of how rules are formed and function, members of social systems can be more choiceful about the ways in which rules function and more knowledgeable about the processes that must be followed to change rules when the time comes to react to internal or external events. Lundberg provides us with a new perspective on organizational change and the seeds of a new intervention methodology in his ground-breaking work for this volume.

Continuing the theme of new ways of thinking about the nature of the field, Joseph Grubbs and Robert Denhardt demonstrate the application of the literary tool of allegory to enable a deeper and richer view of organizational dynamics. Allegory, like metaphor, is a story about one thing that means another. Allegory takes the form of a story or extended dialogue, in which major characters symbolize points of view, hidden meanings, unconscious tensions, or other social dynamics that would normally go unnoticed or not be discussed.

Using a case involving a network of public organizations, Grubbs and Denhardt demonstrate the application of allegory to organization analysis and change. Characters such as Charity, Thrift, Knowledge, and Suspicion engage in dialogue that the actors involved in the real organizations were thinking but not saying. The power of the allegory is enhanced by utilization of Argyris' two column method of describing the difference between espoused and real sentiments. This method reveals important tensions and motives among the actors that might otherwise go unexplored and, therefore, weaken the working relationships among the network members. While the actors may debate the interpretation of the allegory, the purpose of stimulating more frank and open dialogue is served. Defensive routines are overcome and a deeper level of meaning is attained. The significance of this addition to the change agent's tool kit should be obvious to all who have felt stymied by interparty politics in complex systems.

Achilles Armenakis, Stanley Harris, and Hubert Feild begin their chapter by telling the truth, which is that many efforts to change organizations come and go like the "program of the month." Certainly, this is not the intention of either the consultant or the client, yet the situation persists. Through their analysis of change efforts in organizations like Whirlpool, General Motors, Xerox, and GE, the authors construct a model of the factors that affect the longevity of interventions.

The four major phases of intervention in the model are readiness, adoption, commitment, and institutionalization. Within these phases, many specific actions can be undertaken to enhance intervention longevity. These include active participation, persuasive communication, managing internal and external information, human resource management practices, diffusion practices, rites and ceremonies, and formalization. What becomes clear upon reading the article is that institutionalizing change is not a simple matter and certainly not the result of chance circumstances. Understanding the model offered by Armenakis, Harris, and Feild should be made part of every organization development professional training program because the investment made in change by organizations warrants competent efforts to secure the intended return.

Following the discussion of intervention longevity by Armenakis, Harris, and Feild, Jaqueline Coyle-Shapiro investigates the long-term effects of one of the most powerful forms of intervention to date: Total Quality Management (TQM). Employing a longitudinal methodology, Coyle-Shapiro collected both quantitative and qualitative data from an organization in the United Kingdom, which undertook TQM as an adjunct to modular manufacturing.

Coyle-Shapiro describes the intervention, which involved typical steps such as education, training, and communication. She concludes that the intervention had little impact in the long run and reviews the data to explain why this was the case. Two factors emerged prominently in her analysis. First, the cascading process of education and training met with resistance from middle level managers. Second, the intervention took place in an organizational unit that was already regarded as high performing, making the need for change less apparent. Armenakis, Harris, and Feild would probably agree heartily with her conclusion that TQM will continue to fall short of its promise as long as it is viewed as a program, rather than as an intervention, into a complex system in which multiple conditions must be addressed to create powerful and lasting change.

In many organizations today, focus has shifted from TQM to interventions designed to help manage complexity and change. One commonly pursued course of action is the implementation of cross-functional teams, which are intended to provide integration across boundaries that would otherwise stand as barriers to teamwork. Unfortunately, cross-functional teams often function poorly for a number of reasons. These issues are explored by Long Lam, Sheri Bischoff, La Verne Higgins, and D. Lynne Persing in a conceptual piece providing a multilevel framework for analysis of cross-functional team effectiveness.

Noting that cross-functional teams have been employed to speed product development, create coordination among divisions, and increase competitiveness, the authors put forth a model that explains why the outcomes of these and other applications of cross-functional teams vary so greatly in practice. The components of the model include industry characteristics, the characteristics of the cross-functional teams, effectiveness criteria, benefit-enhancing interventions, and detriment-reducing interventions. Based on the framework, implications for interventions that may increase cross-functional team effectiveness are straightforward, including: selecting and training members for technical and interpersonal competencies; training in conflict resolution; increasing geographic proximity; and using performance feedback tools. Given the increasing application of cross-functional teams in industry, this chapter should be of interest to a wide audience of readers.

If collaboration is the underlying goal of cross-functional teams and indeed of all forms of organizing, we should understand more about what makes collaboration possible. To say we desire collaboration is simple enough; but to produce it upon demand is another matter. In his chapter, Peter Robertson helps us think about replacing competition with collaboration as the paradigm for a new ideal type of organization. Drawing heavily upon new paradigm literature, Robertson challenges us to accept the fact that our existing ideas about organizing are out of touch with the world in which we live. To the extent that our work continues to support outdated notions of organizing based on competition and survival of the fittest, we will add to the perils we face in the next century rather than helping to mitigate them. If we can accept a new set of beliefs regarding the purpose for our existence, we can begin to entertain new ways of organizing and new methodologies for intervention. Using logic and evidence, Robertson systematically knocks the underpinnings of our current approach to organizing out from beneath us and suggests new and better ways for us to exist. Some readers will find Robertson's arguments disturbing if not threatening. Others will find hope for a different future which, if we can create it, will arrive none to soon.

Large transformations like that suggested by Robertson will create difficult tensions between the old and new state of existence. Managing these tensions productively will be essential if change is to occur. In fact, significant change at every level involves the management of such tensions and this is the topic of the chapter by Gurudev Khalsa and David Steingard, who offer relational healing as one approach to meeting the emotional challenges of transformation.

Khalsa and Steingard note that transitions create polarization within individuals, groups, functions, and organizations. Inevitably, there are pieces of us that prefer the past to the future. We must reconcile our desire to change with our need for security. To this end, Khalsa and Steingard show how appreciative inquiry and dialogue can be used at different levels of systems undergoing change to hasten and improve the effectiveness of interventions. Using an example from their work with a nonprofit organization, Khalsa and Steingard demonstrate that change progresses through a sequence of psychological phases: splitting, engagement, appreciation, release, and re-integration. Change agents who are alert to these phases can plan interventions using appreciative inquiry and dialogue to surface and deal with issues that would otherwise create resistance to change.

In the final chapter of this volume, C. Ken Weidner and Orisha Kulick tackle the thorny issue of professionalization of organization development. Using the medical profession as a model, the authors explore the meanings attributed to professionalization and the advantages of professionalization to practitioners, clients, organizations, and society.

The authors note that the issue of professionalization is important to the course of evolution of the field and that more rigorous debate of the topic should be undertaken. In favor of professionalization are the autonomy that is gained by being a true professional with respected credentials, the prevention of professional abuses, and quality control of practice. Against professionalization are the intervention of external entities into individual practice, the difficulty of certifying practitioners, and the challenge of conducting rigorous reviews.

Weidner and Kulick argue that the field has suffered from the lack of a unified direction and voice, from fragmentation and a self-perpetuating identity crisis, which professionalization would address. They go on to propose different road maps for the future and suggest that the most likely scenario to find support will be one in which professional liability insurance becomes required to practice. Under this scenario, standards for being insured could be set which would influence the training of practitioners and the quality of practice. By stating their views, Weidner and Kulick have opened the invitation to rejoin the debate about an issue which has confronted the field for some time.

Collectively, the chapters that make up Volume 12 are a statement of the vibrancy and ever changing nature of the field of organizational change and development. If these chapters are any indication, the new millennium will be an exciting time for all of us.

William A. Pasmore
Richard W. Woodman
Series Editors

CONVERSATIONS AND THE EPIDEMIOLOGY OF CHANGE

Jeffrey D. Ford

ABSTRACT

In the network of conversations that constitute the realities called organizations, the focus and unit of work in producing and managing change is conversation. This means that change agents work with, through, and on conversations to generate, sustain, and complete new conversations in order to bring about an altered network of conversations that results in the accomplishment of specific commitments. This chapter proposes that bringing about this alteration is an infective process in which change agents "infect" organizations with new conversations. Drawing on the field of epidemiology, it explores the nature of that infective process and the roles infective agents, susceptible hosts, and environmental factors play in it. These factors are then put into a conversational context and their implications for organizational change explored.

CONVERSATIONS

At the most basic level, conversations are "what is said and listened to" between people. A broader view of conversations as "a complex, information-rich mix of auditory, visual, olfactory and tactile events" (Cappella & Street, 1985), includes not only what is spoken, but the full conversational apparatus of symbols, artifacts, theatrics, and so forth that are used in conjunction with or as substitutes for what is spoken (Berger & Luckmann, 1966). In this respect, conversations are the sum total of communicative relations in action in which language, body, and emotion are inextricably linked (Broekstra, 1998). The speaking and listening that goes on between and among people and their many forms of expression in talking, singing, dancing, and so on, may all be understood as "conversation." Facial expressions and body movements, with or without the use of instruments or tools, constitute speaking. Similarly, listening is more than hearing, and includes all the ways in which people become aware and conscious of, or present to the world.

Conversations can range from a single speech act, for example, "Do it," to an extensive network of speech acts which constitute arguments (Reike & Sillars, 1984), narratives (Fisher, 1987), and other forms of discourse (e.g., Boje, 1991; Thachankary, 1992). Conversations may be monologues or dialogues and may occur in the few seconds it takes to complete an utterance, or may unfold over an extended period of time lasting centuries (e.g., religion). A single conversation may also include different people over time, as is the case with the socialization of new entry people in an organization (Wanous, 1992).

Although conversations are themselves explicit utterances, much of the way in which they support the apparent continuity of a reality is implicit, by virtue of a network of background conversations similar to Harré's (1980) latent structures and Wittgenstein's (1958) form of life. A background conversation is an implicit, unspoken "backdrop" within which explicit conversations occur and on which they rely for grounding and understanding. Background conversations are manifest in our everyday dealings as a taken-for-granted familiarity or obviousness that pervades our situation and is presupposed in our every conversation. A conversation between a female manager and male worker, for example, may occur against a background for gender, manager and worker, oppression or exploitation, human rights, business, organization culture, family relations, or the singles' dating market.

Background conversations are already and always there (Harré, 1980), comprising the intertextual links on which current conversations build and rely. As Bakhtin (1986, p. 86) points out, "our speech is filled with others' words ... which we assimilate, rework, and reaccentuate." When we speak, our conversations are populated and constituted to varying degrees by what others have said before us and by our own sayings and ways of saying (Bakhtin, 1986). Through intertextuality (Spivey, 1977), conversations bring both history and future into the present utterance by responding to, reaccentuating, and reworking past conversations while anticipating and shaping subsequent conversations. When we are asked to justify or explain our linguistic characterizations, we respond with other linguistic characterizations which are themselves based in still other linguistic characterizations and so on (Searle, 1969).

It is the accumulated mass of continuity and consistency in the intertextuality of these conversations that maintains and objectifies organizational reality (Berger & Luckmann, 1966; Fairclough, 1992; Watzlawick, 1990). Objects exist for us as independent tangible "things" located in space and time and impose constraints we cannot ignore (e.g., brute force; Searle, 1995); they are manipulable, and we can do something with and to them (Holzner, 1972; Watzlawick, 1990). When conversations become objectified, we grant them the same permanence as objects by assuming that they exist as some "thing" independent of our speaking them. But this is not the case. Conversations are ephemeral and have no existence or permanence other than when they are being spoken (Berquist, 1993).

Not only are conversations the process through which we construct organizations, they are also the product of that construction: conversations become the organization (Berquist, 1993). What we construct when we construct the reality of organizations are linguistic products, that is, conversations that are interconnected with other linguistic products to form an intertextuality or network of conversations. Organizations exist in the words, phrases, and sentences that have been combined to create descriptions, reports, explanations, understandings and so forth, that in turn create what is described, reported, explained, understood, and so forth. When we describe, we create what is being described in the description. Whether the characterization is taken for granted or is a basis for argument, we have nevertheless created the objects and their properties in our conversations (Winograd & Flores, 1987).

Conversations, therefore, are not only the process of social transmission, they are also the product of that transmission, that is, the "what" that is transmitted. Conversations occur in and constitute the order and pattern of discourse that we understand as "the organization." Order of discourse refers to the ordered set of discursive practices between individuals and groups within a particular organization such as informal conversations, one-on-one meetings, formal presentations, and so on (Fairclough, 1992).

ORGANIZATIONS AS NETWORKS OF CONVERSATIONS

Organizations do not simply have conversations, they *are* conversations. More specifically, they are networks of conversations. For example, planning, budgeting, hiring, firing, promoting, managing, rewarding, and so forth are all "macro-conversations" that are interconnected and constitutive of organizations and which are themselves constituted by different "micro-conversations." As networks of conversations, organizations are not discursively monolithic, but pluralistic and polyphonic with many conversations occurring simultaneously, sequentially, and recurrently (Fairclough, 1992; Hazen, 1993) within and as the context of other conversations. These conversations, in turn, establish the context in which people act and, thereby, set the stage for what will and will not be done (Berquist, 1993; Schrage, 1989).

Organizations, therefore, exist neither as objective entities nor as meanings people carry around in their heads, but in the conversations for, about, and around a limited number of matters in a few physical places with the particular people usually encountered there. Some of these conversations constitute the "social talk" in which people are related and share their lives, wants, likes, dislikes, and so on. But other of these conversations engender commitments that are fulfilled through special networks of recurrent conversations in which only certain details of content differentiate one conversation from another (Winograd & Flores, 1987).

Recurrent conversations are interesting because they become embodied in the offices and departments that specialize in fulfilling some part of the engendered commitments and become background conversations for other departments that are not part of the fulfillment, but simply utilize the recurrent conversations (Winograd & Flores, 1987). For example, recurrent requests for travel reimbursements create a relatively predictable pattern of recurrent conversations called "travel reimbursement" which include all attendant forms and protocols. Although other departments may not be engaged in fulfilling travel reimbursements, they nevertheless may refer to, use, or in some other way rely on or refer to these conversations in their own conversations. Recurrent conversations are also embodied in and constitute the orders of discourse (Fairclough, 1989, 1992) that are variously described as formalized, centralized, hierarchical, and so forth.

Recurrent conversations contribute to structural coupling between organizational participants in which people become habituated to (Berger & Luckmann, 1966) and naturalized in (Fairclough, 1995) the conversations that connect them. Indeed, if we employ Gergen's (Gergen & Thatchenkery, 1996) concept of a distributed self in which identities rest not so much "in" the individual as "in" the conversations in which they are socially engaged and embedded, recurrent conversations give individuals their identities. This means that people will work to maintain the coupling in the face of environmental perturbation (Maturana & Varela, 1987) so as to maintain their identities. For this reason, structural coupling holds conversations in place and contributes to the persistence of existing conversations and orders of discourse.

What we come to know as an "organization," therefore, is a network of conversations that is the result of who has conversations with whom, about what, when, and where (Broekstra, 1998). Some conversations, because they occur between certain people (e.g., CEO and CFO), about particular subjects (e.g., downsizing), in certain places (e.g., board room), and at specific times (e.g., all day meeting), precipitate a particular set of conversations and lead to something happening (e.g., change). Other conversations, however, even though they may be about the same subject (e.g., downsizing), lead nowhere because they are between the "wrong" people, at the "wrong" time, or in the "wrong" place (Hardy, Lawrence, & Phillips, 1998). But all of these conversations constitute the network of conversations we call an "organization" and provide the context in which change occurs.

Given this terminology, we can define the state of an organization at any point by its network of conversations and the actions, behaviors, and practices associated with those conversations. Accordingly, the goal of organizational change is to bring about an alteration in the network of conversations such that there is a correlated alteration in the distribution of actions, behaviors, and practices within the organization. Changing the network of conversations, including orders of discourse, however, implies that some conversations grow and spread while others retract or disappear altogether. In this context, change can be seen as an infective process and epidemiology as a framework for understanding that process.

THE EPIDEMIOLOGICAL TRIANGLE

Epidemiology is the systematic study of the patterns and frequencies of disease or injury within a population and the factors that influence those patterns and frequencies (Ewald, 1994; Mausner & Kramer, 1985). Patterns include the rate of disease or injury within a population and differences in that rate between groups within the population. At the heart of epidemiology is the assumption that the pattern of disease and injury is not randomly distributed throughout a community, but is the result of systematic differences among subgroups that effect their exposure and susceptibility to infectious or harmful agents. Furthermore, it is assumed that these factors and differences can be identified and the knowledge gained used to establish programs of prevention and control.

Although historically considered in conjunction with the acute spread of infectious disease, epidemiology has been used more recently as a framework for understanding excessive changes in the patterns of other social phenomena. The Center for Disease Control, for example, has studied changes in the frequency and patterns of violence within an epidemiological context (Jason, 1984). Epidemiology has also been used as a basis for understanding such social phenomena as alcohol problems (Weisner & Schmidt, 1995), drug use and abuse (Duncan, 1997), mental health (Turner & Marino, 1994), smoking trends (Guba & McDonald, 1993), stress (Hoyt, O'Donnell, & Mach, 1995), depression (Bromberger & Costello, 1992), political violence (Zwi & Ugalde, 1989), affective disorders (Turns, 1978), teenage and college student suicides (Schwartz, 1990; Shaffer, 1988), marital unhappiness (Singh, Adams, & Jorgenson, 1978), childhood injuries (Rivara & Mueller, 1987), and the alienation of college professors (Sandhu, 1972). By treating the spread of social phenomena in a manner analogous to the way infectious diseases move through populations, it is possible to gain new insights into the dynamics of these phenomena.

Epidemiologists contend that the pattern, frequency, and severity of disease and injury within a community cannot be attributed to the presence of only one factor. Rather, disease and injury is a function of the interaction among three factors: agent, host, and environment (Ewald, 1994; Mausner & Kramer, 1985). Together, these three factors comprise an "epidemiological triangle." The significance of this triangle is that it implies an alteration in any one of the three components can disturb the equilibrium among all of them such that there is an increase or decrease in the frequency of disease or injury.

Infective Agents

Agents are typically considered to be the biological, chemical, or physical entities that *must* be present for disease or injury to occur. For example, HIV is the biological agent of AIDS, lead is the chemical agent of lead poisoning, and a gun is the physical agent of a shooting. In the spread of infectious diseases, agents are the entities that are transmitted and "infect" their hosts.

Infect comes from the Latin word "in facere" meaning to put in. To infect someone, therefore, means to induce or insert something into their system. This something is an "infective agent" and where its induction is successful *and* the agent establishes itself within the host, infection results.

Infectivity and Disease

The ability of an agent to successfully enter and establish itself within a host, that is, to produce an infection, defines the agent's infectivity (see Table 1; Mausner & Kramer, 1985). The easier it is for an agent to "get in," to overcome host immunity, the more infective the agent. One way to think of infectivity is in terms of the quantity of agent that is required to produce an infection in a host. The smaller the quantity required, the higher the agent's infectivity.

In a conversational context, infectivity is evidenced both by the number of times something has to be said and the amount that has to be said before someone "gets" the conversation. Conversations that are relatively short and only have to be said once (e.g., "fire" in a crowded theater) have high infectivity, whereas, those that are

Table 1. Infectivity and Disease

Infectivity	Ability of agent to successfully enter a host.
Pathogenicity	Ability of agent to produce an effect (e.g., disease) in host.
Virulence	Severity or magnitude of effect produced by agent in host.
Infection	Presence of infective agent in host.
Disease	Reactions in host produced by infection.

more drawn out or have to be repeated numerous times have low infectivity. High emotion conversations, for example, are more infective and more contagious than low emotion ones (Hatfield, Cacioppo, & Rapson, 1994).

Just because an agent "gets in," however, does not mean that it will have any effect on its host. Infection and disease (illness) are not the same. Infection refers to the presence of an infective agent in a host. Disease, on the other hand, refers to the effects or symptoms produced by that agent. Fevers, chills, vomiting, headaches, crying, upsets, and so on are all symptoms of an infection that is sufficient to bring about physiological or emotional changes in the host (Ewald, 1994; Hatfield et al., 1994).

Whether an infective agent produces disease, and the severity of the disease it produces, depends on its pathogenicity and virulence. Pathogenicity is the ability of an agent to produce symptoms once it has entered a host. Nonpathogenic agents have no symptomatic effect on their hosts, whereas pathogenic agents produce some alteration in their host's normal functioning. Within a given population, agent pathogenicity is determined by the number of detectable cases of disease produced within that population. The greater the number of cases, the more pathogenic the agent and the greater the agent's *pathogenic effect*.

Whether pathogenic effects are mild or severe, however, depends on the agent's virulence. Virulence determines the degree of alteration an agent produces in the host with more severe effects being indicative of more virulent infections (Ewald, 1994). Low virulent infections produce little or no alteration in their hosts' performance. Day-to-day social or small-talk conversations, for example, although highly infectious and pathogenic, are generally non-virulent in terms of the alteration they produce in the actions, behaviors, and practices of participants. Highly virulent infections, on the other hand, produce significant alterations in host performance.

Infectivity, pathogenicity, and virulence determine whether or not an infective agent gets into a host (infectivity), whether the agent is able to produce an effect (pathogenicity), and, if so, the magnitude of the effect produced (virulence). Together, these three factors could explain why some changes seem to "take off" or produce dramatic results (high infectivity, pathogenicity, and virulence) while others never seem to "get off the ground" or produce relatively minor results (low infectivity, pathogenicity, and virulence). Indeed, revolutionary, rapidly moving changes may be the result of inducing infective conversations that are highly pathogenic and virulent, whereas slower moving, incremental changes are the result of infective conversations with lower pathogenic effects and virulence. If this is the case, then one way change managers can alter both the speed and degree of change is to find ways to increase the infectivity, pathogenicity, and virulence of their conversations while neutralizing the infectivity, pathogenicity, and virulence of alternative conversations.

Stages of Disease

Differences in the magnitude or severity of the effects produced by infectious agents makes it possible to distinguish different levels or stages of disease (Mausner & Kramer, 1985). These stages are shown in Table 2. At the lowest level is the *exposure stage* in which no disease has developed, but the groundwork for it has been laid. Susceptible hosts may or may not have been exposed to the agent, and hosts may even be infected, but the infection has not reached a stage where it is detectable or symptomatic. In other words, there is no detectable evidence of an infection, if in fact one even exists, but there is an opportunity for exposure to an infective agent. From a conversational standpoint, people in organizations are always in the exposure stage because there is always the opportunity that they will be exposed to some "new" conversation. In this sense, organizations are rich reservoirs of infective conversations.

In the *preclinical stage*, there are no manifest reactions or symptoms in the host, but the presence of the infection can be detected through screening tests. With biological diseases, such as AIDS, there are laboratory tests that can be used to determine whether the disease is present before any symptoms are evident. In the case of conversations, the corollary tests are questionnaires that may reveal "private conversations," for example, what people think, although there is no corresponding public manifestation of that thinking.

The *clinical stage* occurs when there are recognizable and overt symptoms or evidence of disease. Some of these symptoms reflect direct alterations in the appearance (e.g., a rash) and functioning of the organism (e.g., reduced mobility). Other symptoms reflect the interaction of the organism's immune system with the infection. Fever, for example, is indicative of an organism's interaction with an infection. In the case of conversations, the clinical stage is evident in all instances where people publicly resist, endorse, or question what is being introduced. Within the clinical stage, there may be multiple levels of disease progression. For example,

Table 2. Stages of Disease

Stage	Characteristics
Exposure	Host is being or has been exposed to agent; conditions are such that exposure is likely; may be infection, but it is not yet detectable.
Preclinical	Host is infected and infection has progressed to the point that it can be detected through some form of screening test. There are no overt signs of infection.
Clinical	Infection has progressed to the point that the host has manifest symptoms.
Disability	Infection has run its course and host is left permanently with the consequences of the infection.

in diffusions of innovation, movement from acknowledged awareness to adoption (including rejection) would all be seen as occurring within what is called the clinical stage. Similarly, much of what constitutes the transition period in change (Beckhard & Harris, 1977) occurs within this stage.

Finally, the *disability stage* refers to any altered or diminished capacity suffered by the host as a result of the disease. This disability may be relatively minor, as in retarded mobility, or severe, as in paralysis or death. Within the context of organizational change, disability is more appropriately replaced with adoption (as in innovations), habituation, or institutionalization. Given this replacement, organizational change can be seen as a process of moving a conversation through the stages of disease such that it becomes part of the network of conversations that constitute the organization.

Disease, therefore, refers to the outcomes produced by a given infection and the severity of disease refers to the magnitude of those outcomes. As applied to organizational change, disease, at its most general level, refers to the effects, intended and emergent, that result from introducing new conversations into an organization. In this respect, the exposure stage implies that the opportunity for infection occurs the moment someone comes into contact with an infected host or any other carrier of an infectious agent (e.g., email). The preclinical stage suggests that it is possible for someone to be infected with a conversation and to be unaware they are infected. As Weick (1979, 1995) points out, people may not know what they think until they see what they say. Because there are no observable symptoms, the preclinical stage can give the appearance that nothing has happened and that the change is "not taking" when in fact this is not the case.

At the clinical stage, however, there is manifest evidence that people are engaged with the conversation. This engagement may range from open skepticism, challenges, and complaining to the emergence of new language, actions, and practices consistent with the conversation (Barrett, Thomas, & Hocevar, 1995). At this stage, overt resistance is evidence that the new conversation is engaged and that it is being contested by other conversations. The outcome of this contest will determine whether the disease progresses and thus its severity.

Finally, the disability stage implies that a change has "institutionalized" itself by becoming incorporated into the organization's network of conversations and generating new actions, behaviors, and practices within the organization. The pervasiveness of this disability, however, may be systemic or local. That is, the disability may affect the entire organization (e.g., a new compensation program that covers everyone), or be limited to some subset of the organization (e.g., use of total quality management principles in manufacturing only).

As a disease progresses through the different stages, there is always the possibility that the host will recover. Recovery may be the result of treatments intended to counter the disease (e.g., chemotherapy, surgery), or the host's immune system overcoming the disease. When recovery occurs, the disease goes away and the host appears to revert to its original healthy state. But this appearance is misleading in those cases where the infection results in increased host immunity, making subsequent infections both less likely and more difficult to produce. When and whether subsequent infection is possible depends on whether the immunity produced from prior infections is local or systemic and whether it is temporary or permanent. Systemic permanent immunity makes subsequent infection impossible, whereas local temporary immunity makes subsequent infection both more likely and more frequent.

The ability of infections to produce immunity means that changes, especially those that fail, can increase resistance to subsequent change (Reichers, Wanous, & Austin, 1997). Indeed, people can become cynical about change and dismiss all efforts, regardless of their utility or benefit. As a result, subsequent changes will require higher infectivity, pathogenicity, and virulence if they are to have a chance of succeeding. This cycle offers one explanation for why managers find it increasingly difficult to produce change: prior changes and change attempts have increased immunity and the level of resistance subsequent changes must transcend if they are to take in an organization.

Memes: Conversational Agents of Infection

In his work on the relationship between genes, culture, and human diversity, Durham (1991) proposes that cultural evolution and change is symmetrical to organic change. That is, culture is a "paragenetic" transmission system whose influence on individuals is symmetrical to that of genes. Among other things, this means that culture is an inheritance system that is transmitted socially through "units of transmission" and that changes in these transmission units can cause changes in the actions, behaviors, and practices of a given population.

According to Durham (1991, p. 188), any unit of cultural transmission must meet three conditions: "(1) consist of information that actually or potentially guides behavior, (2) accommodate highly variable kinds, quantities, and ways of organizing information (that is, with variable amounts of hierarchy and integration), and (3) demarcate bodies of information that are, in fact, differentially transmitted as coherent, functional units." In organic evolution, these units of transmission are genes. In cultural evolution, these units of transmission are memes.

Memes. Drawing on the work of Dawkins (1989), Durham, among others (Dennett, 1991, 1995; Lynch, 1996), proposes that the unit of cultural transmission that is symmetrical to the gene is a "meme" (rhymes with gene). Dawkins (1989, p. 192) provides the following explanation for the derivation of this term:

> We need a name ..., a noun that conveys the idea of a unit of cultural transmission, or a unit of *imitation.* "Mimeme" comes from a suitable Greek root, but I want a monosyllable that sounds a bit like "gene." I hope my classicist friends will forgive me if I abbreviate mimeme to *meme.*

According to Dennett (1991, p. 201), memes are, roughly speaking, ideas: "Not the 'simple ideas' of Locke or Hume (the idea of red, or the idea of round or hot or cold), but the sort of complex ideas that form themselves into distinct memorable units." Examples of memes are clichés, tunes, catch-phrases, clothes fashions, ways of making pots or of building arches, right triangle, wheel, chess, evolution by natural selection, and so on (Dawkins, 1989; Dennett, 1991).

Memes are the smallest functional units of cultural transmission. They are also variable units of transmission in that they can vary in size, form, and internal organization (Durham, 1991). The opening five notes of Beethoven's 5th symphony, for example, constitutes a meme just as "symphony" itself is a meme. Yet, the two memes differ significantly in their size, form, and organization. What is critical to memes, therefore, is not size, form, or organization, but coherence and the ability to replicate with reliability and fecundity (Dawkins, 1989).

Longevity, fecundity, and accuracy of replication influence the propagation of memes (Dawkins, 1989). If meme "C" exists longer than meme "D," all other things being the same, there will be more of meme "C" in a given population than meme "D." But longevity, by itself, is of little value in the absence of replication because the meme could be located in only one vehicle. A meme published in one book that sits on the shelf in someone's library and is never read may exist a long time, but it will not propagate itself in the population. On the other hand, a meme that is published in a best seller has the opportunity to propagate widely throughout the population. The existence of a meme, therefore, depends on its ability to propagate through a continuous chain of vehicles that are able to persist (Dennett, 1991, 1995).

Fecundity refers to how prolific memes are in reproducing themselves. If meme "A" makes copies on the average of once a week, while meme "B" makes copies on the average of once an hour, "B" will outnumber "A" even if "A" has greater longevity. Fecundity means that the "success of a meme depends critically on how much time people spend in actively transmitting it to other people" (Dawkins, 1989, p. 198). Quite simply, a meme will spread only if people are engaged in transmitting it through any available meme vehicle. Even speaking against a meme helps its spread. Why? Because in order to speak against something, you must bring that something into the conversation, thereby spreading it. For example, one cannot speak against sexism without mentioning sexism in some way and thereby replicating the "sexism" meme.

Accuracy of replication has to do with the amount of variation produced in a meme (these variations constitute allomemes). If "X" makes a mistake every 10th time and "Y" makes a mistake every 100th, there will be more Y than X (Dawkins, 1989). But accuracy in this case does not mean an exact replication of the meme. As Dawkins points out, there need not be an exact copy of each meme, written in some memetic code in each person's brain. Memes are susceptible to variation and mutation during replication. This susceptibility is known to everyone who has ever played the "telephone" game in which one person repeats what they are told to the next person and so on in a sequential pattern. At the end of the sequence, there is often considerable difference between what the first person said and the last person heard.

During replication there are errors which are themselves replicated. In the preceding example, "X" is the source of more variants than "Y," each of which replicates with a particular accuracy, fecundity, and longevity. If any of these variants are more vivid and gripping than the original, they will eventually out propagate it (Lynch, 1996). This phenomenon of being "out propagated" occurs during organizational change when there are several understandings (i.e., misunderstandings and rumors) about why a change is being made and what its consequences might be. Some of these variants will be favorable toward the change and others will not. If, for example, one of the unfavorable variants propagates faster than the original, change agents can end up explaining themselves or addressing variant issues which are different from the original change.

Allomemes: Variations of a meme. A meme is like a template for which there are many possible variations. For example, there are many different forms of the "handshake" meme. Consistent with this idea of variation, Durham (1991, p. 189) subdivides memes into two categories: holomemes and allomemes. Holomemes are the more inclusive of the two categories and "represent the entire cultural repertoire of variation for a given meme, including any latent or unexpressed forms." A holomeme, therefore, is the set of all existing and possible variations of a particular meme. For example, the holomeme of the meme "number" includes all the existing forms and variations of numbers (e.g., whole, rational, imaginary), as well as those forms and variations that have not yet been invented but which will be classified as numbers. Likewise, the holomeme for the meme "management" includes all the various forms and expressions of management such as strategic management, human resource management, financial management, and so forth. Holomemes, therefore, are the set of all existent and possible allomemes (variants) of a meme.

An allomeme is one particular form or variant (i.e., a specific element) of a meme that is actually used by some members of a population in some circumstances (Durham, 1991). If a meme is like a theme, then an allomeme is a variation of that theme. Examples of allomemes include alternative corporate strategies, different forms and programs of compensation, alternative styles of leadership and motivation, different conceptions about "resistance to change," and alternative definitions of words such as change.

Instructive effects. What is particularly significant for us about memes and allomemes is that they have "instructive effects." What this means is that memes influence the overt and latent properties of their carriers (Durham, 1991). Instructive effect implies a correlation between the distribution of allomemes and the distribution of properties within a given population such that an alteration in the distribution of allomemes within a population results in an alteration in the distribution of corresponding properties within that population. Some of these alterations may be relatively small, incremental, and virtually undetectable, whereas others may be quantum-like and readily observable (Durham, 1991; Tushman & Romanelli, 1985).

The idea that allomemes have instructive effect is highly significant when considered in the context of organizations as networks of conversations because it implies that a change in the distribution of those conversations can produce corresponding alterations in organizational properties. For example, if an organization can alter the distribution of conversations in its network such that the conversation "complain to someone who can do something about it" occurs more frequently and in more places, there will be an alteration (increase) in the number of complaints resolved. This implies that the properties of organizations, such as their structures, operating practices, orders of discourse, and so on are all a function of the content and distribution of conversations (allomemes) that constitute the organization's network of conversations and that alterations in these conversations will have correlative instructive effects. Under these conditions, organizational change becomes a function of the differential transmission and selection of conversations in which the frequency of a conversation or set of conversations changes through time.

Viral sentences. If memes, as root conversations, are the foundational agents of transmission in organizations, how are they able to propagate? Hofstadter (1985) provides some insight into answering this question in his discussion of "viral sentences." Viral sentences are self-replicating conversations that operate in a manner similar to viruses. Viruses are small "objects" that enslave larger and more self-sufficient hosts (e.g., cells), getting them by hook or crook to carry out a complex sequence of replicating operations that bring new copies into being, which are then free to go off and infect other hosts (Hofstadter, 1985).

For example, consider Sentence A: "It is your duty [job, obligation, calling, mission, etc.] to convince [inform, tell, persuade, etc.] others that 'this' is true [false, real, etc.]." Whenever you are in Sentence A, you execute it and engage in convincing others that "this" is true (real). If others have the same sentence, they will do the same, and so on. In other words, Sentence A gets you to engage in transmission, that is, it is self-replicating.

Not all conversations, however, are self-replicating. So how do they spread? Hofstadter proposes that they spread by combining with self-replicating sentences in a symbiotic relationship in which each plays a complementary and mutually supportive role in the survival of the sentence system they together comprise. In other words, conversations that are not self-replicating must combine with or drag along self-replicating conversations to ensure their own replication. Because conversations can be fragmented, combined, and recombined, it is possible for any sentence to become viral. How infective, pathogenic, and virulent the resulting sentence is, however, will be a function of the infectivity, pathogenicity, and virulence of the combining elements.

For example, let us assume that the entire system of sentences (e.g., a body of theory, research, and beliefs) which comprise "resistance to change" is not self-replicating. This means that this system of sentences, by itself, will not get you to engage in transmission, that is, it will not reproduce itself. But consider what happens when the word "this" in Sentence A (which is a self-replicating sentence) is replaced with the phrase "resistance to change," giving us "It is your duty to convince others that resistance to change is real." The result of this combination is a new self-replicating sentence in which people transmit to others that "resistance to change" is real.

In the preceding example, Sentence A and "resistance to change" form a symbiotic relationship in which each plays a complementary role in the survival of the new system they together constitute. In such cases, sentences like Sentence A are *hooks* and the sentence or system of sentences with which they combine (e.g., "resistance to change") are the *bait* (Hofstadter, 1985). Once the bait is taken, the recipient is "hooked" and will engage in the transmission of the new sentence combination.

Another, more general version of this "bait-and-hook" combination is found when sentences such as "The *villain* is *wronging* the *victim*" (Hofstadter, 1985) combine with Sentence A giving us the sentence "It is your duty to inform others that the *villain* is *wronging* the *victim*." In this example, villain, wronging, and victim are themselves place markers (like "this" in Sentence A) that can be replaced with other sentences or sentence systems. For example, "The President is lying to the people," "Management is cheating the union," and "The Dean is not leading the faculty," are all variations (i.e., allomemes) of this sentence.

Hofstadter's work implies that in order for conversations to replicate, they must either be self-replicating or combine with ones that are. In other words, conversations that are not self-replicating (i.e., "bait") must combine with or drag along self-replicating conversations in order to ensure their own replication. Although Hofstadter never explicitly addresses the characteristics of "hooks," other than their self-replicatability, his examples suggest that normative conversations, such as "should," "ought," "must," and "have to," are central.

According to Dennett (1991, 1995), normative conversations are among the most entrenched in our culture and among the conversations that constitute us, they play a central role. If we accept Hofstadter's viral sentences meme, the central role of normative conversation is one of propagation not only of themselves, but of all the conversations with which they combine. Thus, for example, in the combination "The whales are in danger of extinction and we should save the whales," the phrase "we should save" is the normative part that has us engage in getting others to help save the whales. Notice that when the normative part is removed, "The whales are in danger of extinction" is just an assertion and does not "hook" one as easily.

Environmental Exposure to Agents

Infective agents produce no contagion in the absence of exposure to susceptible hosts. Environmental factors, such as population density, availability of transmission vehicles, and so forth, are the biological, social, and physical conditions in which hosts reside (Mausner & Kramer, 1985) and determine the opportunities a host has for exposure to infective agents. One environmental factor that influences exposure is prevalence.

Prevalence

Prevalence is the number of existing cases of a specific infection in a population at a given point in time (Mausner & Kramer, 1985). Within a given population (N), there are people who are infected (I), people who are susceptible (S), and people who are immune or have recovered and cannot be reinfected (R) (Cavalli-Sforza & Feldman, 1981), such that $N = S + I + R$. Prevalence, therefore, is the ratio I/N at a given point in time and changes in this ratio over time determine whether an infection is spreading. For any host population, there is a prevalence that is considered normal, or endemic, for that population.

The higher the prevalence, the higher the likelihood a susceptible host (S) will come into contact with an infected host (I). The higher the infectivity of the agent, the higher the likelihood of infection resulting from that exposure. Research on emotional contagion, for example, shows that the likelihood of being infected with the moods of others is a function of the number of people around you in that mood and the severity of the mood (Hatfield et al., 1994). Prevalence, therefore, tells us the extent to which a particular conversation or set of conversations exists within a population and whether those conversations are spreading, remaining the same, or diminishing (see Table 3).

Prevalence depends on incidence rates and duration of infection. An incidence rate establishes the probability a susceptible person will become infected and is defined as the number of new infections that occur within a given period of time. In general, the higher the incidence rate, the higher the rate at which susceptible people are becoming infected, and the greater the prevalence.

Table 3. Prevalence and Determinants

Prevalence	The total number of infected people in a given population.
Incidence	The number of new infections within a population.
Contact rate	The average number of people someone comes in contact with in a given period of time.

Two factors that influence incidence rates are agent infectivity and contact rates. Infectivity refers to the ability of an agent to invade and multiply in a host and is a function of agent–host interaction. In general, where there is high (low) infectivity, there will be high (low) incidence. But levels of infectivity can be offset or amplified by changes in the contact rate. The contact rate, which is a function of social dynamics, refers to the average number of people someone comes in contact with in a given period of time. For a given level of infectivity, increases (decreases) in contact rates will produce higher (lower) incidence rates as infected hosts increasingly come into contact with susceptible hosts.

The level of prevalence is also influenced by duration or how long the disease lasts before the infected host recovers. Where the disease is chronic, such as diabetes, duration can be for the life of the host. For other diseases, such as the 24-hour flu, duration is very short. In general, the longer the duration the higher the prevalence. Once someone has recovered, they are either susceptible (e.g., common colds), and can become reinfected, or they are immune and are removed from the susceptible host pool.

Because conversations are ephemeral, they will be of short duration unless ways are found to increase their existence. Indeed, part of the job of infecting an organization with a new conversation is finding ways to keep the conversation in existence. In this respect, we are all familiar with occasions where great ideas were generated and agreements made only to be forgotten shortly thereafter. They were forgotten not because people did not care, were malicious, or any number of other reasons, but because the conversations had no existence independent of when they were spoken. As will be discussed below, one way to increase existence is through the use of different meme vehicles (e.g., media), which increase the duration of the conversation.

Where organizations are networks of conversations, there is a prevalence for each and every conversation. Indeed, it is this prevalence that determines the distribution of conversations within the organization. Accordingly, alterations in prevalence reflect an alteration in the distribution of conversations and, to the extent these conversations have instructive effects, the actions, behaviors, and practices within the organization. Prevalence, therefore, offers one way to determine the extent to which a new conversation has been successfully introduced into an organization. By increasing a conversation's infectivity (e.g., its attractiveness), the frequency of contact between infected and susceptible hosts, and the duration of the infection, change managers can increase prevalence.

Epidemics. Shifts in prevalence are the basis for declaring whether or not an epidemic exists, with the pattern of prevalence determining the type of epidemic. In a common or point source epidemic, incidence and prevalence increase from the exposure of susceptible hosts to a common source of infective agents. For example, an epidemic of food poisoning is the result of people being exposed to a common food source containing the infective agent. This exposure may involve a single, one time occurrence, several occurrences over time, or be continuous. Common source epidemics are characterized by an explosive onset of infection (very high incident rate) that is limited and localized in time, place, or persons and then diminishes (see Figure 1A).

Propagated or progressive epidemics are caused by the transmission of an infectious agent from infected host to susceptible host on a one-to-one basis. In propagated epidemics, there are many sources of the infection and the number of sources increases until the infection runs out of a sufficient number of susceptible hosts, the sources are contained or eliminated, or immunity sets in. Propagated epidemics are characterized by a slower increase in prevalence up to some limit and then they decline (see Figure 1B).

The difference between point source and propagated epidemics provides a framework for how one goes about introducing a new conversation. If a change agent is interested in a "rapid start," then broadcasting from an infected host to many susceptible hosts is appropriate. Indeed, this particular strategy is used in organizational change when there is an interest in getting the message out as quickly as possible. Alternatively, if the change agent is interested in a slower start that builds momentum, then a progressive or propagated epidemic is appropriate. In this case, the change agent relies on numerous emissaries to spread the word and for those they infect to do the same. One of the nice things about organizational change is that both strategies can be used so that it is possible to infect a large number of people quickly and then have them propagate the infection throughout the organization.

It is worth pointing out that there is a self-correcting character to epidemics in that the rising risk of infection can prompt susceptible hosts to take self-protective measures (Philipson, 1996). These measures tend to reduce their exposure (e.g., lower the contact rate) or increase their immunity (e.g., lower infectivity). As a result, an epidemic will moderate as the number of susceptible hosts effectively decreases relative to infected hosts. In the case of organizational change, this tendency could be evidenced by people avoiding meetings, failing to read e-mail or other information, and hanging out with other noninfected susceptibles so as to "protect" themselves from what is happening.

Transmission

For an infection to propagate through a population, there must be mechanisms for transmitting the infective agent from its source (e.g., infected host) to susceptible hosts. Although speaking and listening, reading and writing are the underlying technologies of cultural transmission (Dennett, 1995), there are many variations in how and in what forms these are used. Indeed, the proliferation in transmission vehicles (e.g., the internet) means that conversations are virtually unquarantineable as they leap from vehicle to vehicle, medium to medium. Epidemiologists distinguish between direct and indirect forms for the transmission of infective agents.

A. Common Source

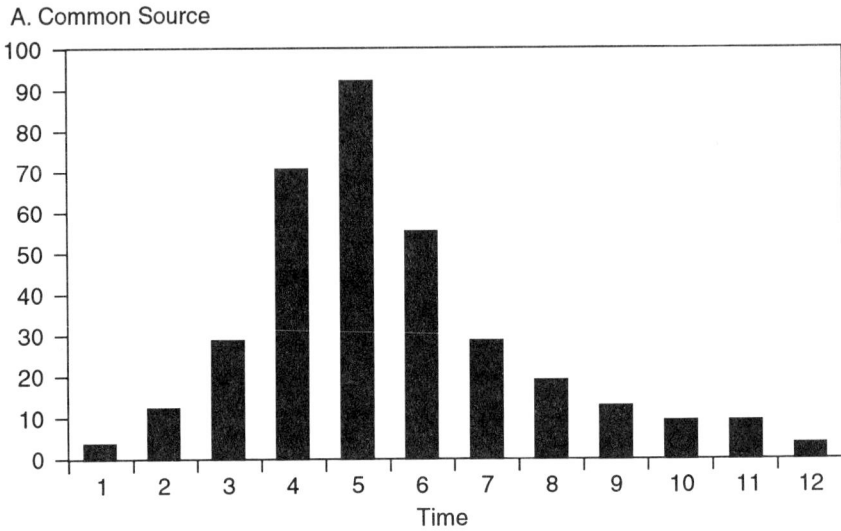

B. Propagated

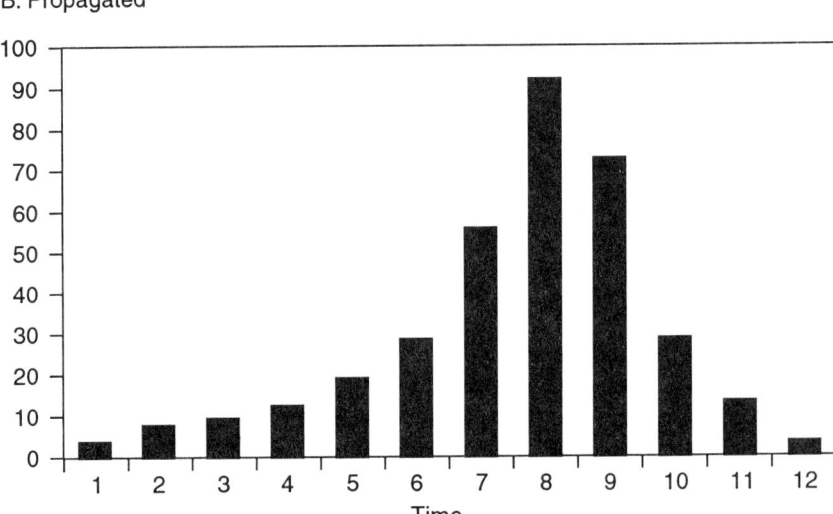

Figure 1. Types of Epidemics

Direct. Direct transmission refers to an essentially immediate transfer of an infective agent from its source (e.g., an infected host) to a susceptible host through direct contact (e.g., touching). In the case of conversations, direct transmission refers only to those forms of direct personal contact that are available in face-to-face interactions. This does not mean, however, that direct transmission requires verbal communication. Research on emotional contagion, for example, shows that emotional states (e.g., upset, anger, irritation, etc.) propagate from individual to individual even in the absence of verbal communication (Hatfield et al., 1994).

Direct transmission can occur whenever there are mobile infectious agents or mobile susceptible hosts. Mobility increases the opportunity for the host and agent to come in contact with each other, thereby increasing the contact rate, incidence, and prevalence. Where the agent is mobile, infections occur when the agent moves into a population of susceptible hosts. For example, someone who is in a bad mood can infect an entire group of people who were otherwise feeling fine. It is this ability to dampen others' moods that gives rise to the attribution that someone is a "wet blanket."

Alternatively, infectious agents may be located in stationary sources. Under these conditions, the infectious agent must "sit-and-wait" for susceptible hosts to come to it (Ewald, 1994). Pathogenic *E.coli* bacteria, for example, require that the susceptible host eat something contaminated with the bacteria. In this method of direct transmission, infection occurs when susceptible hosts come into contact with the source, and prevalence is a function of how many hosts come into contact with the source.

With direct transmission, mobility determines whether it is the infectious agent or the susceptible host that is responsible for exposure. Determining which is mobile is of significance because it dictates both what is needed to propagate an infection (e.g., increase mobility) and how to stop its propagation (e.g., decrease mobility through quarantine).

Indirect. Indirect transmission refers to the transmission of an infective agent from a source to a susceptible host through some form of intermediary vehicle. In the case of conversations, these vehicles include all the artifacts and expressions of human culture including books, pictures, music, papers, sayings, machines, charts, dance, and so on (Dennett, 1991). For example, this chapter and the volume it is in are a form of indirect transmission, as are training videos and journal articles. With indirect transmission, the likelihood of infection increases with the number and form of vehicles and with the number of susceptible hosts who come into contact with those vehicles. This means that as long as an infective conversation is embodied in any vehicle, and there are susceptible hosts, there is a potential for infection. Only when all the vehicles in which a conversation is embodied are destroyed is the potential for infection eradicated.

An advantage of indirect transmission is that it expands the reach of infectious agents by eliminating the need for source or host mobility. By propagating the number and variety of vehicles, infective agents effectively increase their contact rate, thereby increasing incidence and prevalence. As a result, indirect transmission can substantially increase the speed and distance with which an infection can spread.

Direct transmission means that the opportunity for infection exists only as long as the source of that infection survives in the host's environment and its infectivity is preserved. Once the source or its infectivity is removed, there is no opportunity for infection. Hence, removal of the source eliminates the threat of infection while persistence of the source continues the threat.

With indirect transmission, propagation depends on the availability and persistence of different vehicles, and can continue even if the original source no longer exists. The works of Plato, for example, continue to propagate even though Plato is no longer alive. Because conversations are ephemeral, they depend on their embodiment in humans or other vehicles for their continued existence. If all such embodiments disappear or are destroyed, infectious agent's have no way to move from source to host. For this reason, the fate of any conversation depends on the variety of forces that act on the many vehicles that embody them (Dennett, 1991). The elimination of all vehicles will bring about extinction of the conversation much in the same way the extinction of a species eliminates its genes.

The environment, therefore, determines the opportunity that exists for an infectious agent to come into contact with a susceptible host. The more prevalent a disease within a population, the greater the opportunity for exposure and infection. The relative ease with which both sources of infectious agents and susceptible hosts can move through a population also increases the opportunity for exposure and infection. This is one reason why isolation and quarantine is an effective way for fighting the spread of disease. Finally, the availability of alternative transmission vehicles also increases the opportunity for exposure and infection.

Host Susceptibility

At first glance, it would seem that the introduction of an infectious agent into a community where there is a high likelihood of exposure to that agent would be sufficient to produce disease. But this is not the case because host susceptibility influences the likelihood of infection. Susceptibility refers to the responsiveness of a host to a *specific* infective agent. In a sense, susceptibility is like a "readiness for" or "receptiveness to" infection. Highly susceptible hosts are very responsive and easily infected by a particular agent, whereas immune hosts are unresponsive to that agent.

What is significant about susceptibility is that it is always specific. That is, susceptibility is always with respect to specific infective agents. Someone who is susceptible to the measles, for example, is not necessarily susceptible to cholera. This means that when we talk about susceptibility, we must do it relative to a particular agent, not as a general condition.

That susceptibility is specific suggests that "readiness for change" should be treated as a specific condition that can only be established in relation to the induction of a particular change. An organization that is ready for reengineering, for example, is not necessarily ready for cultural transformation even though both may be "ready

Conversations and the Epidemiology of Change 21

for change." For this reason, it may be inappropriate to talk about readiness for change as if it is a generalizable condition independent of the specific change being considered.

Immunity

Immunity refers to a host's capacity to counteract the effects of an infective agent and results from natural endowment, immunization, or prior infections with the same or related agents. Some immunities last a lifetime, whereas other immunities can be lost through continued exposure to infective agents or due to other infections. AIDS, for example, reduces immunity to forms of infectious diseases to which a host was previously immune.

Durham's (1991) work suggests that host susceptibility to conversations is in part the result of preference based on secondary (i.e., learned) values (e.g., "rules of thumb, wise proverbs, social conventions, moral or ethical principles"). Secondary values are conversations from prior infections and immunizations that serve as the very criteria for the evaluation of other cultural phenomena (i.e., other conversations), thereby creating susceptibility differentials that would otherwise not obtain. For example, we all have conversations of the form "Ignore everything that appears in or is said by X."

Existing conversations, therefore, become the filters for subsequent conversations, making it hard for new conversations to invade. If new conversations "pass" the evaluation, they get in. These filters, which are themselves simply other conversations, constitute a type of immunological barrier that responds to the introduction of new challenges. The success of these barriers determines whether or not a host is susceptible and the nature of that susceptibility.

An example of this filter process can be seen in scientific processes (Durham, 1991), such as that found in academic research. The conversations of science—theories, hypotheses, research methods, for example—are transmitted differentially depending on their consequences in explaining phenomena. In this respect, empirical research that explains a significant amount of variance is viewed more favorably and transmitted more widely than research that does not. Indeed, science has even created an entire set of conversations for establishing and communicating if what is discovered is "statistically significant." Conversations that meet or exceed these standards for significance are transmitted with far greater success than those that do not. In fact, it is a rare piece of empirical research that is published without meeting these standards. But "statistical significance" is itself a conversation that is used as the basis for selecting other conversations, that is, statistical significance is a secondary value; and, as a secondary value, statistical significance becomes a limit on what can be published.

Another example of this filtering is found in a study of the influence of mismatches in corporate cultures on the interorganizational relations between two organizations (Wilkof, Brown, & Selsky, 1995). The two organizations were unable to resolve problems, although technical and structural solutions existed, because they did not realize that their perceptions and understandings of the other were filtered through their cultural lenses. When they talked about critical incidents, personnel from the two organizations observed that "not only did the stories not match most of the time, it did not even sound like people were talking about the same incident or the same companies" (Wilkof et al., 1995, p. 377). Managers in both organizations were unable to acknowledge and understand each other's culture and continued to misinterpret each other and to blame each other for the problems. They were, quire simply, unable to see the other's point of view. Failures were attributed to the other's "way of doing things" consistent with their own culture. It was not until the consultants were able to get the managers to consider alternative interpretations (engage in different conversations) that things began to improve.

The preceding examples suggest that the dynamic between infective agent and susceptible host is similar to the active–attractive dynamics found in trialectics. Trialectics proposes that the movement from one state (e.g., uninfected) to another state (e.g., infected) involves two interdependent components: attractives and actives (Ford & Ford, 1994). Attractives are like magnets in that they attract, draw, or pull things toward them. We have all had the experience of being drawn or attracted to something or someone, for example, food, another person, a vision, and so forth. They were attractive.

But attractives are only attractive to things that are "active," that is, that are looking for, listening for, or open to what is being offered, made available, or given off by the attractive. An entity is considered active in that it "acts on," is receptive to, or is susceptible to the attractive entity. Food is not attractive to someone who has just eaten, but it is to someone who is hungry; hunger is active, food is attractive. Things are attractive to us because we are "active" with respect to them; there is no attractive without an active.

Host susceptibility and agent infectivity, therefore, can be seen as an active attractive dynamic in which an agent is infective (active) because a host is susceptible (attractive). For example, if someone is "cynically active," cynical comments will be attractive and they will notice cynicism more than someone who is "empowerment active." In this sense, the relationship between agent and host is like velcro in which the agent has one piece and the host has another piece, and without the complementary piece, nothing sticks.

Herd Immunity

One consequence of individual host immunity is a condition referred to as herd immunity. Herd immunity is the resistance of an entire group to the propagation of an infectious agent based on the immunity of a proportion of individual members within the group. The higher the proportion of immune hosts, the higher the "herd immunity" and the lower the likelihood of a contagion within the group. Herd immunity decreases the probability a group or community will experience an epidemic after the introduction of an infectious agent even though group members may be susceptible.

Herd immunity is of particular significance with infections spread through direct transmission because once the infection encounters an immune host, it stops spreading. Because of herd immunity, it is not necessary to achieve 100 percent immunity in a population in order to halt the spread of directly transmitted infections. Indeed, several well placed, immune hosts can effectively stop the spread of an infection even in a large population of susceptible hosts. Herd immunity explains why some groups never seem to get infected with particular ideas and why it is that only a few people can apparently "shut down" a change. Herd immunity is less effective where infections propagate through indirect transmissions because there are more opportunities for agents to come into contact with susceptible hosts.

DISCUSSION AND IMPLICATIONS

Where organizations are considered networks of conversations, change managers are engaged in altering the network of conversations that constitute the organization through the introduction of new conversations and patterns of discourse. For an alteration in the distribution of conversations to have a sufficient instructive effect in the organization, the change per unit of time in the distribution of conversations must either be large or cumulative (Durham, 1991). This means that change managers must be able to successfully introduce and sustain conversations that spread quickly, as with point source epidemics, or that spread cumulatively, as with propagated epidemics. In this regard, organizational change can be understood as an infective process in which the work of change managers is to purposively infect an organization with new conversations that spread and become part of the organization's network of conversations.

Managing the Triangle

According to the epidemiological triangle, the spread of any infectious disease is a function of agent, host, and environment factors. Increasing agent infectivity, opportunities for exposure, and host susceptibility increases the likelihood of infection and disease. For change managers, increasing the likelihood of infection means finding ways to make the conversations more infective, increasing opportunities for susceptible hosts to come into contact with infectious agents through direct and indirect transmission, and increasing host susceptibility (reducing immunity).

Increasing Agent Infectivity

Ford and Ford (1995) propose change managers can increase conversational infectivity by being more selective in the type of conversations they use and when they use them. Conversations for understanding, for example, are appropriate and useful when understanding is wanted, but not when action is called for. Where change managers are interested in action, they will find conversations for performance to be far more effective. The difference in these conversations implies that conversations for performance are more infective in the domain of action, whereas conversations for understanding are more infective in the domain of understanding. Furthermore, with conversations for performance, unreasonable requests and promises appear to be more pathogenic and virulent than ordinary requests (Goss, 1996).

Clearly there are likely to be other factors that contribute to the infectivity of particular conversations and additional research is needed to discover what these factors might be. For example, conversations that are high in emotional content, particularly of an unsettling nature, are highly infective and spread rapidly as people attempt to cope with their reaction (Kubey & Peluso, 1990). This, along with other research on emotion contagion (Hatfield et al., 1994), suggests that change managers can alter the infectivity of their conversations by the amount and type of emotion they display. Excitement and enthusiasm, for example, are more likely to support a change than are despair and hopelessness. However, if the use of emotion is seen as inauthentic, it could reduce credibility, making the conversation less infective while contributing to distrust and cynicism (Kouzes & Posner, 1993; Reichers et al., 1997).

Increasing Host Exposure

In addition to making conversations more infective, change managers can also find ways to increase the exposure of susceptible hosts to sources of infective agents. Prevalence implies that change managers can spread change by "surrounding" susceptible hosts with infected hosts and other indirect conveyance methods that contain infectious agents. If this is the case, then change managers would do well to keep infected hosts (e.g., advocates and supporters) mobile so as to increase prevalence within different areas of an organization. Any new conversation introduced into an organization has, by definition, little or no prevalence in that organization. Although this means that the infected host has a large pool of susceptible hosts, it also means that the host risks being reinfected with the more prevalent conversations of the organization, that is, a relapse. If a relapse occurs, the more prevalent conversations will persist, giving the appearance of inertia or resistance, when in fact what has happened is that the host has been reinfected with the prevalent conversations.

Increasing exposure also raises issues of existence. Conversations are ephemeral and have no existence independent of the vehicles that embody them. Based on the principles of transmission, existence is a function of the number of infected hosts and the number, variety, and longevity of the vehicles in which infective agents are embodied. If there are few infected hosts and few, if any vehicles, a conversation will have less existence than where there are many hosts and vehicles. Low levels of existence are evident in people "forgetting" or "not knowing." As existence increases, people become "habituated" to the conversation and it becomes part of the network of conversations.

Recognizing that prevalence is a function of existence means that change managers do not have to attribute low incidence rates to personal characteristics or attributes, for example, personality. Rather, change managers can look for ways to add existence to conversations by increasing the number of vehicles that embody the conversation and the frequency with which the conversation is spoken. For example, change managers can increase the contact rate and the duration of each contact, thereby increasing exposure.

Increasing Host Susceptibility

Host susceptibility to a specific conversation is a function of prior experiences with that or related conversations. Where those experiences have been favorable, susceptibility is likely to be higher than where experiences have been unfavorable. In this respect, research has shown that unfavorable experiences result not only in cynicism toward change, but those who initiate it (Reichers et al., 1997). From a procedural fairness standpoint, employees are likely to expect an explanation for a change decision regardless of whether outcomes are positive or negative. Furthermore, if they are not given an explanation, they are likely to feel the procedures for making the decision are unfair, leading to resentment against the decision as well as the decision makers (Daly, 1995).

The impact of prior experiences on susceptibility implies that change managers may be able to increase host susceptibility (reduce immunity) by bringing closure to prior experiences (Albert, 1983; Ford & Ford, 1995). It also implies that how managers conduct changes now can and will have implications for how people react to future changes. Accordingly, managers who are interested in creating organizations susceptible to future changes will want to learn how the changes they are currently introducing and managing are raising or lowering host susceptibility. Research on cynicism toward change, for example, proposes that managers can reduce the likelihood of cynicism if they openly and completely communicate the results of changes, no matter what they are (Reichers et al., 1997). Similarly, research on leadership suggests that host susceptibility to change will increase along with the change manager's credibility which is a function of authentic and straight communication (Kouzes & Posner, 1993).

The realization that host susceptibility is a function of prior conversations suggests the possibility that hosts can intentionally alter their own susceptibility by altering the conversations to which they are exposed. Individuals, for example, can expose themselves to new cultures, thereby increasing their likelihood of infection to the conversations of those cultures. Alternatively, individuals may alter their susceptibility through ontological change processes in which they reveal the background conversations that operate as filters (Marzano, Zaffron, Zraik, Robbins,

& Yoon, 1995). Indeed, such processes seem to underlie some forms of therapy (Watzlawick, 1978; Watzlawick, Bavelas, & Jackson, 1967).

Incubation Periods

One factor that influences the dynamics of infectious disease is the incubation period. The incubation period is the time interval between infection and the onset of disease (Mausner & Kramer, 1985) and is a function of host–agent interaction. The longer the incubation period, the longer the time between exposure to an infectious agent and evidence of illness.

What is significant about the incubation period is that each infection has a different incubation period and change managers can misread the absence of symptoms as "nothing happening." There may be no symptoms not because there is no infection, but because the infection has not progressed to the point where symptoms are detectable or manifest. Where incubation periods are particularly long, managers run the risk of giving up on good ideas not because people have rejected the idea, but because it has not worked its way through the incubation period.

Differences in incubation periods could explain why some changes appear to happen suddenly, in a revolutionary, quantum-like, or punctuated fashion (Gersick, 1991; Romanelli & Tushman, 1994). For example, assume an infection has an incubation period of six months and that a change manager immediately infects, through direct and indirect conveyance, 50 percent of the organization. During these six months, nothing much would appear to be happening. Then, in the sixth month, there would be an explosive onset of new actions, behaviors, and practices resulting from the new conversation.

The incubation period also has implications for managers who assume that once something is introduced that it will (or should) stick. Such an assumption is based on the premise that all introductions have an immediate incubation period. Where this is not the case, managers could become frustrated and attribute the absence of symptoms to resistance. Although host immunity does contribute to the incubation period, the incubation period is not resistance. It is simply the time it takes for the infection to progress to a level where symptoms are detectable. If managers are concerned with successfully infecting susceptible hosts, they should use multiple and frequent conveyance methods over an extended period of time so as to allow for any cumulative effects to overcome host immunity.

Given the importance of incubation periods, research is needed to identify the conversational factors that influence incubation periods. Because incubation is in part a function of host immunity and agent infectivity, altering either of these two factors would be important to altering the incubation periods. Giving managers access to altering this period could bring about a substantial increase in the speed of change without any loss in the outcomes produced by the change.

Research is also needed to help change managers differentiate incubation periods from the absence of infection. Where change managers introduce conversations with low infectivity, there will be no infection and no incubation period. However, if managers confuse the absence of infection with an incubation period, they may take the wrong next steps. For example, if there is no infection, an appropriate next step is to increase exposure of susceptible hosts to the agent with the intent of producing an infection. However, if an infection exists, an appropriate next step is to hasten the incubation period. Given the inability to differentiate these two states, change managers should probably assume low infectivity and continue to expose susceptible hosts through multiple conveyance vehicles.

Disease Stages

Where organizational change is seen as an infective process, change managers will want to recognize and use progression through the stages of disease as a guide for where things are. Barrett et al. (1995) provide an excellent example of how change moves through stages of disease in their study of the introduction of Total Quality Leadership into a Navy Command. As Barrett et al. (1995) point out, the introduction of any change involves the introduction of new language (conversations) and that this new language is introduced through existing orders of discourse. Exposure occurs when the new language is first spoken into the organization. Although everyone in the organization was exposed to the same new language of TQL, their responses to it were different. Initially, some people questioned the new language, pointing out inconsistencies, questioning the authenticity of those proposing it, and complaining. As both the number of people speaking the new language and the degree to which the new language was spoken increased, however, people began to add new language (what Barrett et al., 1995, call "nascent scripts"). The increase in TQL related vocabulary provided a new background set of conversations that provided a basis both for sensemaking and for taking new, novel forms of action. As the new vocabulary expanded, it replaced and transformed the "older" vocabulary, made new actions that were previously unimaginable possible, and altered underlying assumptions and beliefs.

What is significant about the Barrett et al. (1995) study is that it clearly shows the progression of a conversational infection within an organization consistent with the stages of disease. Immediately after exposure, there is a contest between the new language and the existing language, much as there is a contest between a host's immune system and an infectious agent. This contest is evidenced in the challenges to and complaining about the new language and whether it will take or if it will be something that passes. This is like the incubation period in which the infection is working to establish itself in the host. Once the TQL language begins to spread and events are now interpreted using the new vocabulary, it is clear that the infection has gone beyond the incubation period and into the preclinical and clinical stages. In the clinical stage, the issue becomes one of how severe (extensive) the disease will be. In Barrett et al.'s (1995) example, the infection can be seen as increasing in severity as "nascent scripts" are added (the TQL allomemes are combining and recombining with other allomemes), creating new actions that were previously unimaginable. The disability stage is reached as the new conversations of TQL begin to replace and alter the background assumptions and beliefs of the organization. At this point, the new vocabulary has successfully established itself within the network of conversations.

Countering unwanted infections. Understanding the stages of disease not only allows the change manager to see what is needed to accelerate the spread of an infectious conversation, it also provides information for how to counter the spread of unwanted infections. For example, at the exposure stage the prevention strategy is to reduce the susceptibility of the host or reduce host exposure to infective agents and their vehicles. This reduction is accomplished through education and immunization of the host (e.g., giving susceptible hosts conversations which counter the infective conversation) or through the eradication or confinement of the infective agent. Once an infection has moved to the preclinical and clinical stages, the change manager can work to slow the progression of the disease by limiting communicability (e.g., contact rates and access to vehicles) or treating the disease through the use of countering conversations with the intent to cure the disease or limit its severity.

Nonlinear Dynamics

The epidemiological triangle implies that the ability to successfully infect an organization depends in part on understanding that epidemics are a nonlinear function of the dynamics among agent, host, and environmental factors. Linearity assumes that an X percent increase in one factor will produce a corresponding increase or decrease in another factor. But in epidemics, a small (large) variation in the conditions of one factor, such as the contact rate, can result in a disproportionate change in the prevalence of disease. This nonlinearity is similar to that observed in complexity theory and stems from the interconnectivity that constitutes all networks, including an organization's network of conversations (Waldrop, 1992).

Indeed, what is significant about epidemics is that there is a threshold point that they must pass to start (or end) and that until this point is passed, the triangle remains stable at the endemic level of disease. In fact, this threshold point is established by the ratio of the removal rate (effectively the rate at which hosts become immune) to the incidence rate. If this ratio increases, people are being removed from the susceptible host pool at a faster rate than they are entering the infected pool. This means that the number of people who can become infected is getting smaller at a faster rate than people who are becoming infected. But if this ratio goes down, people are being infected at a faster rate than they are being removed, and prevalence increases. The ratio of the removal rate to the incidence rate, therefore, is responsible for setting a threshold to the number of susceptible hosts and prevalence (Cavalli-Sforza & Feldman, 1981).

Consider, for example, a situation in which 1,000 residents in a city are infected with an untreatable strain of 24-hour flu (Gladwell, 1996). The virus has a 2 percent incidence rate, which means that 1 out of every 50 people who come in contact with it become infected. Furthermore, assume that city residents have an average contact rate of 50 people per day. Under these conditions, the 1,000 infected residents will infect another 1,000 residents who in turn will infect another 1,000 residents and so on. But, because the duration of the disease is only 24 hours, 1,000 people will recover each day. As a result, the prevalence of flu will remain endemic at 1,000 cases per day.

But consider what happens if for some reason the contact rate were to rise from 50 people per day to 55 people per day. Although this does not sound like much of a change, the net effect is that prevalence would rise. In fact, within one week, the prevalence would increase to nearly 2,000 cases and the city would have a progressive epidemic on its hands.

This particular example illustrates that the equilibrium among agent, host, and environmental factors can be highly sensitive to fluctuations in those factors. For example, the flu epidemic could have been worse with the introduction of a more pathogenic and virulent virus with an incidence rate of 5 percent and duration of 48 hours. Alternatively, the prevalence level could be significantly reduced if the contact rate were to drop or host susceptibility improved and the incidence rate decreased to 1 percent. For example, if the contact rate went from 50 to 45 per day, prevalence would fall to approximately 500 cases within a week (Gladwell, 1996).

The threshold point at which any subsequent variation in one or more of the factors precipitates a shift up or down in the prevalence of some occurrence is referred to as its "tipping point" (Gladwell, 1996). The tipping point is that point at which ordinary and stable phenomenon become unstable. In the case of epidemics, the tipping point is that point beyond which an alteration in factors influencing the incident rate (e.g., contact rate), duration (e.g., a new strain of infective agent), or removal results in a sudden "outbreak" or "rash" of cases, that is, an increase in prevalence. As long as the relation among these factors remains at or below the tipping point, nothing changes. However, once the tipping point is exceeded, there is a sudden outbreak or rash of events beyond what is normal or endemic for the population. Every epidemic has its tipping point.

One of the things that makes tipping points interesting is that nothing appears to happen until they are crossed and then there is a dramatic and sudden change in incidence. This phenomena is illustrated by the "broken window theory" (Wilson & Kelling, 1982). The essence of this theory is found in the metaphor that if a window in a building is broken and left unrepaired, other windows will soon be broken. This is true in "good" as well as "bad" neighborhoods. Window breaking does not occur because there is a group of window breakers within the community, but rather because the broken window is a single or invitation to others that it is

"OK" to break windows. In other words, a broken window exceeds the tipping point and there is a sudden increase in the number of broken windows.

A classic example of the broken window theory is found in Zimbardo's research in which he put an abandoned car without license plates on a street in the Bronx and a comparable car on a street in Palo Alto (as reported in Gladwell, 1996; Wilson & Kelling, 1982). Within 24 hours after a family of three removed the radiator and battery, virtually everything of value had been removed from the car. The windows were smashed, upholstery ripped, and parts torn off. Most of the "vandals" were well-dressed, apparently clean-cut whites. The car in Palo Alto sat untouched for a week until Zimbardo smashed part of it with a sledgehammer. Within a few hours, the car had been destroyed. The "vandals" again appeared to be clean-cut whites.

Tipping points have also been found for neighborhood transitions in which attainment of a certain minority percentage leads to an accelerated increase in the rate at which white residents leave (Steinnes, 1977). A study by Crane (1991) found that there was no effect on teenage pregnancy rates as long as the percentage of high status people (e.g., professionals) in the neighborhood was between 5 percent and 40 percent. However, once this percentage fell below 5 percent, teenage pregnancy increased like an epidemic. Other studies have found that once the number of homicides reached between 50 and 70, gang-related drive-by shootings became epidemic (Hutson, Anglin, Kyriacou, Hart, & Spears, 1995).

A second thing that is interesting about tipping points is that a series of small steps in one area can produce extraordinary results in apparently nonrelated areas. For example, stopping more suspicious cars, confiscating more guns, chasing away more loiters, and shutting down more drug markets may result in a dramatic drop in major crime ("Defeating the Bad Guys," 1998). Rudy Giuliani, mayor of New York, has used the principle of tipping points as the basis for crime prevention in New York with considerable success ("Defeating the Bad Guys," 1998). By cleaning up environmental factors such as graffiti, and similar allomemes of misconduct (e.g., turnstile jumping in subways), law enforcement officials substantially lowered the incident of major crime in New York. Although there may be other explanations for the drop in major crimes, the drop is consistent with the "broken window" theory and tipping points ("Defeating the Bad Guys," 1998; Gladwell, 1996).

Tipping points offer one explanation for why, even after considerable investment, some change efforts produce no significant alterations in actions, behaviors, or practices while other change efforts, with considerably less investment, produce dramatic shifts. If the tipping point is highly sensitive to variation, then very slight alterations in agent, host, or environmental factors can precipitate dramatic changes. Alternatively, if the tipping point is very insensitive, change managers can invest considerably and still realize no real benefits. Until the tipping point is crossed, nothing happens. But once it is crossed, there is a sudden shift in incidence.

The broken window theory and the principle of tipping points suggests that there may be no such thing as an insignificant or inconsequential conversation during a change effort. Indeed, because it is not possible to know what conversations will lead to what results, every conversation counts and it may be a mistake for a change manager to "step over" or "ignore" some conversations with the justification "it really doesn't matter." For example, the absence of trust is considered to be one of the factors that contribute to resistance during organizational changes. Trust, as well as credibility, have both been found to be a function of the extent to which leaders keep agreements or authentically clean them up when they are broken (Kouzes & Posner, 1993; Lewicki & Bunker, 1995). Tipping points and the broken window theory propose that there is a finite number of agreements that can be broken and not authentically cleaned up before there will be an epidemic of broken agreements and a widespread loss of trust. This means that if change managers "step over" a broken agreement, no matter how small or insignificant, they run the risk of an epidemic of distrust and subsequent resistance to change.

Rethinking "Change"

The adoption of change as an infectious process within a network of conversations calls for an alteration in our understanding of what constitutes "change" in general and "a change" in particular. Traditional, structural–functionalist perspectives talk about change as if it were a clearly definable and identifiable object or thing that is put in place (e.g., a computer system). Even if it is acknowledged that there are many parts, stages, or components, the change is nevertheless represented as if it has object-like properties and clearly defined parameters that exist independent of the conversations in which they are embedded. At best, conversations are simply a tool that are used to put the change in place (Ford & Ford, 1995). Within the context considered here, however, such a monolithic view of change is problematic.

Change, like the organizations in which it occurs, is not monolithic discursively. Rather, it is more appropriately seen as a polyphonic phenomenon (Hazen, 1993) within which many conversations are introduced, maintained, and deleted (Barrett et al., 1995; Czarniawska, 1997). This perspective is evident in Czarniawska's (1997) study of Swedish government agencies in which particular changes were comprised of a series of conversational episodes organized around particular themes (e.g., "decentralization" or "computerization"). It is also evident in Barrett et al.'s (1995) study and Elden's (1994) observation that the transformation of Magma Copper occurred in a "myriad of many, mostly small, local activities" initiated on a local level within a common commitment to a possible future. According to Elden, the key to Magma's transformation was the introduction of a metalanguage that allowed for creating possibilities for action that were outside of what would be predicted based on historical and current operating practices. This new language helped people to break through their habitual ways of thinking, envision futures not possible inside these ways of thinking, and enact that future.

Within an infectious conversation perspective, there is no **the** change, like a single conversation, that is being produced. Rather, change is an unfolding of many conversations within a general theme (Czarniawska, 1997), new vocabulary (Barrett et al., 1995), or metalanguage (Elden, 1994), most of which cannot be anticipated and must be generated "in the moment." Indeed, every time change managers introduce a conversation to a susceptible host, they will need to engage in a variety of conversations depending on agent, host, and environmental factors. In this sense, change as an infectious process is like experimental theater or improvisational jazz where the script (music) is being written while it is being performed (Boje, 1995; Czarniawska, 1997). Although there is a theme or context to the change, the specific conversations that are needed, with whom, and when have to be generated on a moment-to-moment basis. Indeed, as Barrett et al. (1995) have found, entirely new language, the specifics of which cannot be anticipated, is generated as a change grows and spreads within an organization.

Because the conversations required for spreading a change cannot be anticipated, the production of any change involves the generation and unfolding of many conversations, the specifics of which are unforeseeable at the time. The failure to recognize change as this unfolding (spreading) of micro-conversations within a macro-conversation robs change agents of their power in conversations. Indeed, it is the very inconspicuousness of individual transmission events that generate and sustain epidemics, as well as all other conversations endemic to organizations that makes their actual relevance little noticed and underestimated (Lynch, 1996). Yet, as can be seen with epidemics, transmission need not proceed conspicuously to amass an enormous host population or produce dramatic conversational shifts.

Conversational Shifts

Change, as an infectious process, produces a language shift in organizations (Holmes, 1992). When Holmes (1992) refers to a language shift, she is talking about a gradual process whereby the language of a wider community displaces the language of a smaller community. For example, immigrants to a country shift from their native tongue to the language spoken by the wider community. As people speak in one language, the vocabulary in another language diminishes and there is a loss of fluency and competence by its speakers. There is a gradual erosion of the prior language and the minority language retreats in terms of the places in which it is used, who uses it, when, and for what purposes. In the terminology of this chapter, the minority becomes infected with the more prevalent conversations of the majority.

But in the case of organizational change, we are actually proposing something akin to a "reverse" language shift in which a nascent conversation infects and becomes part of the network of conversations that constitute the organization. That is, we are talking about bringing forth new conversations into an existing community and having those conversations prosper such that they become naturalized and habituated within the network of conversations that constitute an organization.

During language shifts, both the majority and minority languages are present in the organization, although the extent of their usage is altering. In particular, if the new conversations are infectious and spreading, the minority language will be growing and expanding within the organization, whereas the majority language will be contracting. Because it is not possible to keep the two languages separate, people within the organization will have to be bilingual (or even multilingual) and there can be a period of "reduced competence" in which people are unlearning the old, but have not yet mastered the new (Gilmore, Shea, & Useem, 1997). Although such bilingualism could be considered a potential source of friction and frustration for some (Barrett et al., 1995), it need not be a limitation. Czarniawska (1997), for example, found that a government agency continued using both the new and old accounting system and ways of acting to the appreciation of all. In a sense, people were bilingual and were able to distinguish which language to use, when, and with whom.

Change as conversational shifts can also be seen in Rappaport's (1993) finding that the power of mutual help organizations in supporting individual transformation stems from their ability to provide shared narratives that are incorporated by the individual into their life narratives. Mutual help organizations "tell stories" which become reflected in the autobiographical stories people tell. Moreover, the stories told tend to have a different focus and flavor than those told by people under professional care, suggesting that different organizations impart different stories. Because identity formation and change takes place within a social context of community narratives (e.g., Gergen & Thatchenkery, 1996), joining a community has consequences for identity development and change. Mutual help organizations provide narratives that interpenetrate with personal narratives bringing about changes in the autobiographical stories people tell.

Rappaport's work suggests that conversational shifts are also found in the interpenetration of organizational stories with individual autobiographical stories. If organizational change cannot occur without individual change (Gergen & Thatchenkery, 1996; Hazen, 1994), then we would expect to see the narratives of organizational change to manifest themselves in the personal stories people tell. More specifically, we would expect to see the autobiographical stories individuals tell about themselves in organizations to be interpenetrated with organizational stories. In fact, interpenetration of organizational and personal stories might be one way to determine the pathogenicity and virulence of an infectious conversation. Extrapolation of Rappaport's suggests that organization change can be seen as the interpenetration of a new conversation (or network of conversations) with the existing network of conversations such that the new conversation is modified and expanded as it becomes part of the organization's autobiography. Such an extrapolation is consistent with the findings of Barrett et al. (1995).

Decentering the Individual

The possibility that organizational change can be understood as an infectious process in conversations moves conversations from a background, support role into a foreground, main character role. Rather than simply a tool in the process of change, conversations are the medium, message, target, and product of change. Indeed, organizations are themselves seen as constituted in and by conversations such that alterations in these conversations provide opportunities for new actions that were previously unimaginable. Moving conversations onto center stage makes it possible to decenter our attention on the characteristics and attributes of individuals in the change process (Barrett et al., 1995). Thus, rather than look at individual factors as key to the change process (Dirks, Cummings, & Pierce, 1996; Neumann, 1989), we begin to look at conversational dynamics. From a conversational perspective, it is the conversations that are being spoken and listened that are of interest, not the characteristics of those who speak and listen.

This shift in the centrality of individuals and conversations implies a shift in the theoretical questions one might ask. For example, where individuals are central and conversations are secondary, we might ask, "What are the individual characteristics that determine why someone would select (resist) that conversation?" However, where conversations are primary, the question is more like "What are the conversations that allow for the inclusion or exclusion of other conversations?" In a sense, the question shifts from one of "Why do people select the conversations they do?" to "Why do conversations select the people they do?" (Lynch, 1996).

Decentering begins to make sense if we accept Durham's (1991) proposition that current conversations become the filters through which all subsequent conversations must pass. Under this proposition, people are not monolithic identities with relatively fixed characteristics that consciously choose which conversations "get in" or "stay out." Rather, individuals are networks of conversations (memes) that have accumulated over time (and which continue to accumulate) and that establish the barriers which subsequent memes must pass if they are to enter the network. As Dennett (1991, pp. 207–208) points out, "It cannot be 'memes vs. us,' because earlier infestations of memes have already played a major role in determining who or what we are. The 'independent' mind struggling to protect itself from alien and dangerous memes is a myth. Our existence as us, as what we as thinkers are—not as what we as organisms are—is not independent of these memes."

Infection by memes, therefore, may not be a conscious choice any more than infection by any virus is a conscious choice. Research into emotional contagion, for example, shows that being infected by another's emotional state occurs unconsciously (Hatfield et al., 1994). Indeed, one important point about infective agents in general and infective conversations in particular is that their capacity to propagate within a population has little to do with their epistemological virtue. As Dennett (1991, p. 203) points out, "The first rule of memes is that replication is not necessarily for the good of anything: replicators flourish that are good at ...

replicating!—for whatever reason." Conversation "X" might spread in spite of its perniciousness and conversation "Y" might go extinct in spite of its virtue. There is no necessary condition between a conversation's replicative power, its "fitness" from its point of view, and its contribution (positive or negative) to its host's fitness (Dennett, 1991).

Tipping points and the nonlinear dynamics of the epidemiological triangle also make it clear that the spread of a particular conversation is a population phenomenon, not an individual one. Although host susceptibility is one of the factors that influence whether a particular host is infected, it does not determine the dynamics of the population as a whole. Thus, rather than arguing that broken windows are the result of people who are "window breakers," the nonlinear dynamics of epidemics implies that "window breakers" occur as a function of the interplay among host, agent, and environment. Such dynamics allow for the possibility that those who are "resistant" with one change will be "receptive" to other changes based not only on their personal susceptibility, but on the dynamics of contact, infectivity of conversations, and so on.

The Power of Conversing

Memes propagate themselves in the "meme pool" by leaping from one vehicle to another vehicle of the same or different form through replication (Dawkins, 1989). If a scientist hears, or reads about an idea, theory, or research finding, and passes it on to colleagues and friends, writes about it in articles, or talks about it in lectures, the idea can be said to propagate or replicate itself. Memes, therefore, are not limited to person-to-person replication. Rather, memes can propagate from person to person, person to book, book to person, paper to person, person to computer, and so on (Dawkins, 1996). But, in all forms of meme replication, human agency is involved. A meme located in a book for example, cannot, without human intervention, move to a computer on its own. It is because of their reliance on human intervention that memes are dependent on human conservators for their continued existence.

The realization that memes rely on human conservators for their existence implies that although infection by a particular conversation is not necessarily a conscious choice, exposure to particular conversations may be. We can, for example, decide not to put ourselves in situations in which we will be exposed to conversations that are unattractive to us. By the same token, we can be aware of the conversations to which we expose others. Bohm (1996), for example, proposes that we can be aware of the conversations that constitute us and choose whether or not we will speak them. Indeed, the idea behind both Bohm's (1996) concept of dialogue and ontological approaches to change (Marzano et al., 1995) is that hosts can discover the conversations that constitute them and then choose whether to persist or alter those conversations.

If we are aware that each time we speak (in all its forms of direct and indirect transmission) we expose the listener to a potentially infectious conversation, then we can look to see if we want to infect them with what we are about to say. For example, if we elect to speak complaints, and complaining spreads, then we should not be surprised when we find ourselves in an organization of complainers. In this respect, Bohm (1996) talks about the pervasive "nowness" of pollution in which we are continually pouring pollution into the stream of conversations. The source of this pollution is not in time—not back in ancient time, when it may have started—but rather the source is always *now*. Each time we speak, we have a choice about what we say. We can say something that is "pollution," or we can say something else. The choice is always ours.

ACKNOWLEDGMENTS

I would like to thank Jay Wilkens and Christie Walters for their helpful comments in the preparation of this chapter.

REFERENCES

Albert, S. (1983). The sense of closure. In K. Gergen & M. Gergen (Eds.), *Historical social psychology* (pp. 159–172). Hillsdale, NJ: Erlbaum.

Bakhtin, M. (1986). *Speech genres and other late essays* (V.W. McGee, trans.). Austin: University of Texas Press.

Barrett, F., Thomas, G., & Hocevar, S. (1995). The central role of discourse in large-scale change: A social construction perspective. *Journal of Applied Behavioral Science, 31*(3), 352–372.

Beckhard, R., & Harris, R. (1977). *Organizational transitions: Managing complex change*. Ready, MA: Addison-Wesley.

Berger, P., & Luckmann, T. (1966). *The social construction of reality*. New York: Anchor Books.

Berquist, W. (1993). *The postmodern organization*. San Francisco, CA: Jossey-Bass.

Bohm, D. (1996). *On dialogue*. London: Routledge.

Boje, D. (1991). The storytelling organization: A study of story performance in an office-supply firm. *Administrative Science Quarterly, 36,* 106–126.

Boje, D. (1995). Stories of the storytelling organization: A postmodern analysis of Disney as "Tamara-land." *Academy of Management Journal, 38*(4), 997–1035.

Broekstra, G. (1988). An organization is a conversation. In D. Grant, T. Kennoy, & C. Oswick (Eds.), *Discourse and organization* (pp. 152–176). London: Sage.

Bromberger, J., & Costello, E. (1992). Epidemiology of depression for clinicians. *Social Works, 37*(2), 120–125.

Capella, J., & Street, R. (1985). Introduction: A functional approach to the structure of communicative behavior. In R. Street & J. Cappella (Eds.), *Sequence and pattern in communicative behavior* (pp. 1–29). London: Edward Arnold.

Cavalli-Sforza, L., & Feldman, M. (1981). *Cultural transmission and evolution: A quantitative approach*. Princeton, NJ: Princeton University Press.

Crane, J. (1991). The epidemic theory of ghettos and neighborhood effects on dropping out and teenage childbearing. *The American Journal of Sociology, 96*(5), 1226–1259.

Czarniawska, B. (1997). *Narrating the organization: Dramas of institutional identity*. Chicago, IL: The University of Chicago Press.

Daly, J. (1995). Explaining changes to employees: The influence of justifications and change outcomes on employees' fairness judgements. *Journal of Applied Behavioral Science, 31*(4), 415–428.

Dawkins, R. (1989). *The selfish gene.* Oxford: Oxford University Press.

Dawkins, R. (1996). *The blind watchmaker.* New York: W.W. Norton.

Defeating the bad guys. (1998, Oct. 3). *Economist,* pp. 35–38.

Dennett, D. (1991). *Consciousness explained.* Boston, MA: Little, Brown.

Dennett, D. (1995). *Darwin's dangerous idea: Evolution and the meanings of life.* New York: Touchstone.

Dirks, K., Cummings, L., & Pierce, J. (1996). Psychological ownership in organizations: Conditions under which individuals promote and resist change. In R. Woodman & W. Pasmore (Eds.), *Research in organizational change and development* (Vol. 9, pp. 1–23). Greenwich, CT: JAI Press.

Duncan, D. (1997). Uses and misuses of epidemiology in shaping and assessing drug policy. *The Journal of Primary Prevention, 17*(4), 375–382.

Durham, W. (1991). *Coevolution: Genes, culture, and human diversity.* Stanford, CA: Stanford University Press.

Elden, M. (1994). Beyond teams: Self-managing processes for inventing organization. In M. Beyerlein & D. Johnson (Eds.), *Advances in interdisciplinary studies of work teams* (Vol. 1, pp. 263–289). Greenwich, CT: JAI Press.

Ewald, P. (1994). *Evolution of infectious disease.* Oxford: Oxford University Press.

Fairclough, N. (1989). *Language and power.* London: Longman.

Fairclough, N. (1992). *Discourse and social change.* Cambridge, UK: Polity Press.

Fairclough, N. (1995). *Critical discourse analysis: The critical study of language.* London: Longman.

Fisher, W. (1987). *Human communication as narrative: Toward a philosophy of reason, value, and action.* Columbia: University of South Carolina Press.

Ford, J.D., & Ford, L.W. (1994). Logics of identity, contradiction, and attraction in change. *The Academy of Management Review, 19,* 756–785.

Ford, J.D., & Ford, L.W. (1995). The role of conversations in producing intentional change in organizations. *The Academy of Management Review, 20,* 541–570.

Gergen, K., & Thatchenkery, T. (1996). Organization science as social construction: Postmodern potentials. *Journal of Applied Behavioral Analysis, 32,* 356–377.

Gersick, C. (1991). Revolutionary change theories: A multilevel exploration of the punctuated equilibrium paradigm. *Academy of Management Review, 16,* 10–36.

Gilmore, T., Shea, G., & Useem, M. (1997). Side effects of corporate cultural transformations. *Journal of Applied Behavioral Science, 33*(2), 174–189.

Gladwell, M. (1996, June 3). The tipping point. *The New Yorker,* pp. 32–38.

Goss, T. (1996). *The last word on power.* New York: Currency Doubleday.

Guba, C., & McDonald, J. (1993). Epidemiology of smoking. *Health Values, 17*(2), 4–11.

Hardy, C., Lawrence, T., & Phillips, N. (1998). Talk and action: Conversations and narrative in interorganizational collaboration. In D. Grant, T. Keenoy, & C. Oswick (Eds.), *Discourse and organization* (pp. 65–83). London: Sage.

Harré, R. (1980). *Social being: A theory for social psychology.* Totowa, NJ: Littlefield, Adams & Co.

Hatfield, E. Cacioppo, J., & Rapson, R. (1994). *Emotional contagion.* Paris: Cambridge University Press.

Hazen, M. (1993). Towards polyphonic organization. *Journal of Organizational Change Management, 6*(5), 15–26.

Hazen, M. (1994). Multiplicity and change in persons and organizations. *Journal of Organizational Change Management, 7*(6), 72–81.

Hofstadter, D. (1985). *Metamagical themes: Questing for the essence of mind and pattern.* Toronto, Canada: Bantam.

Holmes, J. (1992). *An introduction to sociolinguistics.* London: Longman.

Holzner, B. (1972). *Reality construction in society.* Cambridge, MA: Schenkman Publishing.

Hoyt, D., O'Donnell, D., & Mack, K. (1995). Psychological distress and size of place: The epidemiology of rural economic stress. *Rural Sociology, 60*(4), 707–720.

Hutson, H., Anglin, D., Kyriacou, D., Hart, J., & Spears, K. (1995). The epidemic of gang-related homicides in Los Angeles county from 1979 through 1994. *Journal of the American Medical Association, 274*(13), 1031–1036.

Jason, J. (1984). Centers for Disease Control and the epidemiology of violence. *Child Abuse and Neglect, 8*(3), 279–283.

Kouzes, J., & Posner, B. (1993). *Credibility: How leaders gain and lose it, why people demand it.* San Francisco, CA: Jossey-Bass.

Kubey, R., & Peluso, T. (1990). Emotional response as a cause of interpersonal news diffusion: The case of the space shuttle tragedy. *Journal of Broadcasting & Electronic Media, 34*(1), 69–76.

Lewicki, R., & Bunker, B. (1995). Trust in relationships: A model of development and decline. In B. Bunker, J. Rubin, & Associates (Eds.), *Conflict, cooperation, and justice: Essays inspired by the work of Morton Deutsch.* San Francisco, CA: Jossey-Bass.

Lynch, A. (1996). *Thought contagion: How belief spreads through society.* New York: Basic Books.

Marzano, R., Zaffron, S., Zraik, L., Robbins, S., & Yoon, L. (1995). A new paradigm for educational change. *Education, 116*(2), 162–173.

Maturana, H., & Varela, F. (1987). *The tree of knowledge: The biological roots of human understanding.* Boston, MA: New Science Library.

Mausner, J., & Kramer, S. (1985). *Epidemiology: An introductory text.* Philadelphia, PA: W.B. Saunders.

Neumann, J. (1989). Why people don't participate in organizational change. In R. Woodman & W. Pasmore (Eds.), *Research in organizational change and development* (Vol. 3, pp. 191–212). Greenwich, CT: JAI Press.

Philipson, T. (1996). The economic epidemiology of crime. *Journal of Law & Economics, 39*(2), 405–433.

Rappaport, J. (1993). Narrative studies, personal studies, and identity transformation in the mutual help context. *Journal of Applied Behavioral Analysis, 29*(2), 239–256.

Reichers, A., Wanous, J., & Austin, J. (1997). Understanding and managing cynicism about organizational change. *Academy of Management Executive, 11*(1), 48–59.

Reike, R., & Sillars, M. (1984). *Argumentation and the decision making process* (2nd ed.). Glenview, IL: Scott, Foresman.

Rivara, F., & Mueller, B. (1987). The epidemiology and causes of childhood injuries. *The Journal of Social Issues, 43*(2), 13–31.

Romanelli, E., & Tushman, M. (1994). Organizational transformation as punctuated equilibrium: An empirical test. *Academy of Management Journal, 37*(5), 1141–1166.

Sandhu, S. (1972). The epidemiology of alienation: A study of college professors. *International Journal of Contemporary Sociology, 9,* 100–107.

Schrage, M. (1989). *No more teams! Mastering the dynamics of creative collaboration.* New York: Currency Paperbacks.

Schwartz, A. (1990). The epidemiology of suicide among students at colleges and universities in the United States. *Journal of College Student Psychotherapy, 4*(3–4), 25–44.

Searle, J. (1969). *Speech acts: An essay in the philosophy of language.* Cambridge, England: Cambridge University Press.

Searle, J. (1995). *The construction of social reality.* New York: Free Press.

Shaffer, D. (1988). The epidemiology of teen suicide: An examination of risk factors. *The Journal of Clinical Psychiatry, 49,* 36–41.

Singh, B., Adams, L., & Jorgenson, D. (1978). Epidemiology of marital unhappiness. *International Journal of Sociology of the Family, 8*(2), 207–218.

Spivey, N. (1997). *The constructivist metaphor.* San Diego, CA: Academic Press.

Stiennes, D. (1977). Alternative models of neighborhood change. *Social Forces, 55*(4), 1043–1057.

Thachankary, T. (1992). Organizations as "texts": Hermeneutics as a model for understanding organizational change. In R. Woodman & W. Pasmore (Eds.), *Research in organizational change and development* (Vol. 6, pp. 197–233). Greenwich, CT: JAI Press.

Turner, R., & Marino, F. (1994). Social support and social structure: A descriptive epidemiology. *Journal of Health and Social Behavior, 35*(3), 193–212.

Turns, D. (1978). The epidemiology of major affective disorders. *American Journal of Psychotherapy, 32*(1), 5–19.

Tushman, M., & Romanelli, E. (1985). Organization evolution: A metamorphosis model of convergence and reorientation. In B.M. Staw & L.L. Cummings (Eds.), *Research in organizational behavior* (pp. 333–375). Greenwich, CT: JAI Press.

Waldrop, M. (1992). *Complexity: The emerging science at the edge of order and chaos.* New York: Touchstone.

Wanous, J. (1992). *Organization entry: Recruitment, selection, orientation, and socialization of newcomers* (2nd ed.). Reading, MA: Addison-Wesley.

Watzlawick, P. (1978). *The language of change: Elements of therapeutic communication.* New York: Basic Books.

Watzlawick, P. (Ed.). (1960). Reality adaptation or adapted "reality"? Constructivism and psychotherapy. In *Münchhausen's pigtail: Or psychotherapy & "reality"—essays and lectures.* New York: W.W. Norton.

Watzlawick, P., Bavelas, J., & Jackson, D. (1967). *Pragmatics of human communication: A study of interpersonal patterns, pathologies, and paradoxes.* New York: W.W. Norton.

Weick, K. (1979). *The social psychology of organization.* Reading, MA: Addison-Wesley.

Weick, K. (1995). *Sensemaking in organizations.* Beverly Hills, CA: Sage.

Weisner, C., & Schmidt, L. (1995). The community epidemiology laboratory: Studying alcohol problems in community and agency-based populations. *Addiction, 90*(3), 329–341.

Wilkof, M., Brown, D., & Selsky, J. (1995). When the stories are different: The influence of corporate culture mismatches on interorganizational relations. *Journal of Applied Behavioral Science, 31*(3), 373–388.

Wilson, J., & Kelling, G. (1982, March). Broken windows. *The Atlantic,* pp. 29–38.

Winograd, T., & Flores, F. (1987). *Understanding computers and cognition: A new foundation for design.* Reading, MA: Addison-Wesley.

Wittgenstein, L. (1958). *Philosophical investigations* (G.E.M. Anscombe, trans.). Englewood Cliffs, NJ: Prentice-Hall.

Zwi, A., & Ugalde, A. (1989). Toward an epidemiology of political violence in the Third World. *Social Science and Medicine, 28*(7), 633–642.

ORGANIZATIONAL DEVELOPMENT AS FACILITATING THE SURFACING AND MODIFICATION OF SOCIAL RULES

Craig Lundberg

ABSTRACT

This chapter employs an emerging theoretical perspective—social rules theory—to reframe and redefine organizational change and development in ways potentially overcoming some of their conceptual inadequacies. After elaborating social rules theory, it is first utilized to reframe organizational change, and second, to redefine OD as a special type of project system in organizations which facilitates change by surfacing, assessing, and modifying as needed, the rules and rule systems within or linking organizational components.

Organizational change and development are surely the leading candidates in recent years for the phenomena that have most fascinated both organizational scholars and management practitioners. Managing change and enhancing organizations are widely common aspirations included in the claims and justifications of organizationally responsible persons as well as scholars worldwide. Change competencies are widely believed to be crucial for "thriving on chaos" (Peters, 1987), "managing in permanent whitewater" (Vaill, 1989), "strategic restructuring" (Kochan & Useem, 1992), and "redesigning the corporate future" (O'Toole, 1985) in the "new realities" (Drucker, 1989) of the "age of unreason" (Handy, 1989) as we move into a "post industrial society" (Trist, 1976). The contemporary significance of organizational development is indicated by the recent proliferation of terminology such as, organizational renewal, organizational effectiveness, organizational improvement, organizational transitioning, organizational transformation, and so on.

For all the attention given and the significance attributed to organizational change through organizational development, however, few would propose that change phenomena are fully understood. It seems safe to say that neither the essence of nor the nuances of change have as yet been fully explicated. While a plethora of cases, definitions, and normative and descriptive models of change and development exist, each seems to capture only a part of the reality of change. It is this very piece-meal quality of the field that stimulated the work reported in this chapter which seeks to both see more and see differently via a frame-changing exercise.

Specifically, this chapter offers two contributions to seeing more and seeing differently by changing conceptual lenses. One is to bring attention to and systematically begin to elaborate an as-yet little appreciated emerging theoretical perspective on social behavior—social rules theory. The other contribution is to use social rules theory to reconceptualize organizational change and development in a way which both reinterprets and extends existing organizational change thinking and practice—organizational development as facilitating the surfacing, assessing, use, and modification of social rules in social systems.

Our argument in favor of a social rules conception of organizational change and development will proceed in three sections. First, we will briefly review organizational development theory and its criticisms, ending by noting what a more adequate, more appropriate theory might look like. Second, we will introduce social rules theory, its origins and major tenets, hopefully finding some reasonable balance between presenting its essential ideas and being smothered with technical detail. Third, we will examine organizational change and organizational development through the lens of social rules, outlining its diagnostic utility as well as its interventional implications.

ON ORGANIZATIONAL CHANGE THEORY

Since its inception organizational development and change has been conceived in general and multiple ways; organizational development (OD) is commonly synonymous with planned organizational change or combined with it. Even the early labels invented to circumvent this vagueness or to emphasize some feature such as renewal (e.g., Lippitt, 1969) or more recent labels such as organizational effectiveness, organizational improvement (e.g., Golembiewski, 1979) or organizational transformation (Bartunek & Louis, 1988) are understood to refer to OD. Early on, a major criticism was that OD was conceptually undifferentiated (Bennis, 1969). To be sure, there have been efforts to conceptually differentiate OD in several ways, for example, in terms of the fundamental tasks of organizations such as internal adjustment, external adaptation, and future anticipation (Lundberg, 1989); in terms of a continuum of change agent intentionality such as planned change, guided change, and enhancing evolutionary change (Lundberg, 1994); and in terms of environmental differentiations such as relative stability, gradual change, rapid change, and discontinuous change (Lundberg, 1994). For most of its existence, however, OD has emphasized a collection of fix-it (present time), how-to (atheoretical), subsystem (internal) practices (Alderfer, 1977; Beer & Walton, 1987; Friedlander & Brown, 1974; Golembiewski, 1979; Sashkin & Burke, 1990; Strauss, 1976). The archtypical image of OD provided by Kurt Lewin, that is, his unfreezing–change–refreezing model, is essentially one of planned, internal adjustment in a relatively stable environment—an image that OD has had trouble overthrowing.

For over two decades, numerous OD scholars have commented unfavorably or critically on the quality and state of OD theory (e.g., Bowers, Franklin, & Pecorella, 1975; Burke, 1982; Margulies & Raia, 1978; Porras & Paterson, 1979; Porras & Robertson, 1987; Sashkin & Burke, 1990; Woodman, 1989). Theoretically, OD may be faulted in many ways.

- Paradigmatically, OD appears to be dominated by a unidirectional, causal paradigm, one which promotes a preoccupation with determinism. OD thus is causally unsophisticated in that it presumes an overdetermined, one-way flow of influence from cause to effect, from intervention to change. The paradigmatic exemplars of OD, such as survey-feedback and team building, may also be questioned as inadequate for replacing the explicit rules for solving the normal puzzles of normal science (Kuhn, 1970) simply because they refer to a relatively narrow band of phenomena.
- The primitive statements/assumptions behind most OD theorizing, for example, conditions of economic and technological growth, the centrality of work in people's lives, or that emotions are secondary to cognitive processes, appear to be unduly restrictive. Such assumptions blind conceptualizers to the real pluralism of contexts, structural and system differences, inherent politics and conflict, and much else. One could argue that the development of OD theory will in part depend on the identification and questioning of such assumptions.
- OD theorizing also appears to be caught in a complex ideological tension. On the one hand, OD theory often reflects a managerial ideology supporting the interests of and survival of those in power. On another hand, OD theory sometimes reflects a supposedly value-neutral, scientistic ideology. On still another hand, OD theory sometimes reflects a humanistic ideology centered on developing human potentials. Some OD theorists seem to promote a

conflict ideology, which sides with organizational and societal underdogs (Stablin, 1988). These alternative ideologies, singularly and in combination, obviously have considerable impact on the descriptive adequacy of a theory as well as on its normative application.

- Values also subtly condition theorizing because the explication of value alternatives is basic to any theory of social structures or social processes. Among the core values of OD are humanism, rationality, and idealism (Connor, 1977; Walter, 1984) although Cobb and Margulies (1981) would add democracy. Whyte and Wooten (1987) would further add productivity and efficiency. The compatibility among these values is of course problematic (Stephens, D'Intino, & Victor, 1995). While OD values are central foci for conceptualization, they are lamentably more discussed in terms of OD practice than incorporated in its theory.
- OD theorizing has, from its inception, been captive of a rational-goal "problematic" (i.e., a set of related ideas organized around a core idea; Althusser, 1970). Organizations, their subsystems, and their members are almost always viewed as goal-seeking entities. OD is thus "understood" within this problematic and has engendered an "improvement" image, which is at least partial, and thus misleading. It is quite easy to envision organizations and other social agencies as sometimes acting randomly or as ambiguity-laced (e.g., Martin, 1992; Pondy, Boland, & Thomas, 1988), more under the control of external constraints (e.g. Hannen & Freeman, 1977), or experiencing entropy, long-term deterioration, or appreciable organizational dry rot (e.g., Argyris, 1973), and other sorts of changes such as meandering (Weick, 1974), episodic (Gersick, 1991) or continuous (Vaill, 1989). None of the alternatives to improvement, however, has achieved more than a passing glance theoretically.
- OD theory all too often confuses the sort of explanation utilized (Evered, 1976), blurring or combing starkly contrasting teleological explanations (events explained in terms of some subsequent situation) with non-teleological, scientific ones (events explained in terms of earlier events in a predictable way). OD has perhaps too long settled for goal-based "in order to" models instead of empirically based "because of" models that would more accurately and validly underpin change practice.

From the outset of OD, its theoretic aspiration has been to develop a widely accepted, general theory of organizational change through OD. This aspiration, arguably, has not yet been met as noted by Sashkin and Burke (1990, p. 324), "We find no real coherence among the various theoretical contributions... that would lead us to think that the field of OD is approaching a theoretical synthesis." If OD theory development has not succeeded thus far by tactics of extending, elaborating or refining extant conceptualizations, perhaps another developmental stratagem is needed—to step back from the relatively constrained (paradigmatically, phenomenologically, and ontologically) conceptualizations of OD and reframe social phenomena in some other way that explicitly embraces a variety of social settings, social relationships, and social processes, while allowing for social actors individually and/or collectively to maintain and/or change these settings, relationships, and processes. Needed is a theoretical perspective that embraces most conventional social levels of analysis, provides linkage across these levels, focuses on regularities and patterns but allows change in a nondeterministic way, and, probably, keeps power and social ranking central. If such a perspective were available, organizational change and development would necessarily be understood not as phenomenologically unique, not as one or more activities or events, not as a bundle of techniques or competencies, and not as essentially different across cultures. The emerging perspective termed social rules theory seems to meet these theoretic criteria.

THE PERSPECTIVE OF SOCIAL RULES

The perspective outlined by social rules theory has roots in several places. It draws on the rich theoretical traditions of Marx and Weber but is clearly distinct from them and goes beyond them in making use of contributions from organization theory, comparative institutional analysis, theories of social structuration, microsociology, linguistics, and ethnomethodology as well as the theoretical insights provided by a number of contemporary scholars who have developed rule and rule system ideas, for example, Chomsky (1965), Cicourel (1974), Goffman (1974), Harre (1979), Lindblom (1977), and Twining and Miers (1982). The perspective breaks from most prior social theorizing in two important ways. On the one hand, it is explicitly meso, that is, it fosters a cross-level actor-system synthesis (Rousseau, 1985). Reference here is to the two fundamentally different conceptions of man and society that underlie the vast majority of theories of social behavior. One stresses the human agent as the source of social regularities and the forces that structure social systems as well as the conditions of human activity. The other stresses structure or system where either humans are not found or where humans enact roles and function in social structures or systems they cannot basically change. On the other hand, social rules theory embraces both static and dynamic dimensions. As such, social rules theory resembles the attempts at new syntheses by such theorists as Archer (1986), Bourdieu (1977), Crozier and Friedberg (1980), Giddens (1979), and March (1994) who have eschewed the conceptual restrictions that follow from assumptions of social systems as only purposive and functional. Burns and Flan (1987) have provided the most focused and detailed conception of social rules theory at the institutional level. The version of social rules theory presented below is philosophically consistent with the epistemology of Winch (1958) and builds upon the foundations provided by Burns and Flan. It extends the recent micro-applications of Lundberg (1998), and is consciously influenced by the theoretical work of March (1994) and Weick (1988).

All conceptual edifices have one or more primitive statements or assumptions. Social rules theory is no exception. It requires that we accept the following: that human beings often act purposefully; that social activities take place in concrete transaction situations in which the actors involved have unequal resources and opportunities to realize their purposes and interests; that through their actions and interactions, social actors regulate and change their material, institutional, and cultural world; and that behavior is organized in large part on the basis of explicit and implicit, multiple systems of rules. The basic elements and tenets of social rules theory are as follows.[1]

1. There are *social systems*. A social system is composed of a set of persons in a setting involved with one another in action (with shared knowledge and shared ways of knowing) governed by a rule system. A social system, therefore, is different than membership in a social category (Lave & Wenger, 1991). Social systems are constrained by both endogenous and exogenous factors.
2. *Agents* are the members of a social system. Every person has membership in several/many social systems but with varying degrees of centrality, where *centrality* is a function of time spent, role or position occupied, and expertise in the action of the social system. At a point in time, because of their centrality, agents will vary in their status and power, primarily reflecting their sophistication (knowledge and/or capability) about and their commitment to the system of rules that govern the social system.
3. *Rule systems* are made up of a set of related rules, which govern transactions among agents in a social system. *Rules* may be explicit and/or implicit and they specify to a greater or lesser extent who participates (and who is excluded), who does what, where, when, and how in relation to other agents. Rule systems do not account for the psychology of agents; they do regulate behavior but do not determine it. Rule systems are constrained by the biological and physical as well as by social-technical tools and structures and the encompassing sociocultural context.
4. Rule systems exist at four general levels. At the institutional and societal level there are *rule regimes* composed of meta-rules, which may be real or idealized (e.g., moral codes, constitutional laws, or organizational principles, etc.). Rule regimes are essentially rules about rules and social system type, configuration and function. At the level of social systems of varying size and complexity, there are *system rules* for long-term, ongoing social systems, and *project rules* for shorter, time-bounded social systems. At the level of agents there are *personal rules* (individual rule structures commonly known as personality) which are individually inspired or motivated. Rule systems at all levels are never in equilibrium because of changing agent knowledge and capability, agent motivation, and changes in encompassing rule regimes as well as endogenous constraints. In general, higher level rule systems have

Social Rules

generative priority and tend to be more immutable than lower level rule systems.

5. The *functions of rule systems* are multiple although not equal or balanced at any point in time: they differentiate intra- and intersystem activities, they clarify communications among agents, they symbolically provide agent and system identity, they reduce uncertainty, they provide access to resources, and they enable agents to make and justify claims.
6. Compliance to rule systems occurs because of: perceived payoffs to agents; the internalization of rules, habit, and custom; sanctions in a social system; public opinion; and societal meta-rules that instruct agents to suspend serious questioning of rules and rule systems.
7. Rules and rule systems are (a) *learned* by agents through socialization and diffusion processes—though never perfectly, (b) *maintained* in social systems via sanction and coercion processes—again rarely perfectly, and (c) *modified* through persuasion, negotiation, and conflict resolution processes. The fact of multiple agents with varying status and power and multiple social system memberships allows for such changes. As human constructions, rules and rule systems structure the subjective experience of persons, thereby providing meaning to experiences and identity to agents.
8. Agents, individually and collectively, acquire and have identities. An *identity* is a conception of self-defined rules matching actions to situations. Identities are both created by persons and imposed upon them. Identities define the essential nature of an agent in a social system, serve as prepackaged contracts, and frequently come to be assertions of what is good, moral, and true. Social systems tend to select agents with compatible preexisting identities, and impose agent identities specific to themselves. Social systems also possess a collective identity.
9. Rule systems are composed of four types of rules:

    - *Descriptive rules*—non-time bound rules that specify nature and reality. Descriptive rules entail factual or descriptive formulations and refer to objects, states of the world, agents, events and developments. They categorize the world making socially important distinctions as well as make statements about patterns. Descriptive rules constitute knowledge. Their generic form is: under circumstances (x, y, z...), then it is so that (proposition). A "proposition" is a probability statement of the relationship, associative or causative, between two or more ideas (i.e., constructs, concepts, or variables) that has more or less empirical confirmation.
    - *Situational rules*—rules that categorize situations by time, place, and and/or social meaningfulness. While situational rules vary in their scope, they all provide presumably useful distinctions whereby the similarity of apparently dissimilar phenomena can be recognized and the differences among similar phenomena specified. Their generic form is: when propo-

sitions (r, s, t...) are salient, then specify (categories). "Categories" refer to configurations of objects, agents, activities, events, and states of the world, which are socially distinguishable cognitively and linguistically.
- *Valuation rules*—rules concerned with what is or is not desirable, that is, those things pursued or avoided by persons in a sphere of activity. Their generic form is: under conditions (a, b, c...), assign positive/negative value to (categories).
- *Action rules*—rules which specify instructions about how to do something or how to respond to certain problems, circumstances, or situations. Action rules say how social decisions and activities should be organized or coordinated, that is, what to do, how to do it, when to do it, and with whom. Their generic form is: under situations (k, l, m...), then do/do not do (action or procedure).

10. For every social system there will be action rules that specify four *processes* (informed, of course, by the other types of rules): (a) rule creation, (b) rule interpretation, (c) rule implementation, and (d) rule modification.

The theory of social rules is grounded in a logic of appropriateness. Agents are presumed to ask (explicitly or implicitly) more or less constantly three questions: What sort of situation is this? What is my identity in this situation? What rules apply? This questioning process is neither random, arbitrary, nor trivial; rather, it is systematic and often complicated. Interestingly, this search for appropriate rules to follow is conditioned by the rule systems currently salient to agents. The reasoning process of the logic of appropriateness is one of using a rule system (of a social or project system) to clarify identities and situations and matching rules to them. We emphasize that this process is in contrast to the common perspective on organizational behavior where action is based on an assessment of alternatives in terms of their consequences for preferences.

To say that agents and social systems follow rules and have identities, however, is not to say their behavior is always easily predicted. Rule-based behavior is freighted with uncertainty. Situations, social systems, identities and rules can all be ambiguous. Agents use processes of recognition to classify situations; they use processes of self-awareness to clarify identities; they use processes of search and recall to match rules appropriately to identities and situations. While these processes are standard human ones, they do require thought, judgment, imagination, and care (quite different, however, from rational analysis) and they are emotionally imbued.

As portrayed, social rules theory accounts for both stability and change in social and project systems—through the influence of rule regimes and personal rule structures, through the degree of shared knowledge by members of the system, through the degree of familiarity of situations encountered, through the competencies of agents with more or less centrality, through the degree to which members have learned the relevant rule system, through the existence of rules for maintaining and modifying the rule system, through resources and motivation for rule compliance, and, importantly, when there are agents with identities that specify combinations of action rule processes. Social and project systems, seen through the lens of social rules theory, are characterized as being in a state of quasi-stable equilibrium—never really stable, always changing in some aspect to some degree. If rules and rule systems are as central for understanding social behavior as social rules theory suggests, then it follows that they might be a useful perspective for understanding and implementing changes, to which we now turn.

OD THROUGH THE LENS OF SOCIAL RULES

If social rules theory is as phenomenologically encompassing of social behavior as asserted, then we ought to be able to use its terms and tenets to both translate conventional OD's ideas, models, and practices into the theory, and, more significantly, reconceptualize organizations and change as well as OD. These are the aims of this section.

In the conventional conception of what constitutes organizational change, some environmental changes impact the tasks, technologies, structures, operational systems, and personnel of the organization and/or one or more of its subunits, which in turn alters intra- and interorganizational relationships and member roles. OD in this conception promotes the asking of two diagnostic questions—Where are we now? Where would we like to be?—and offers a variety of ways in which members can move their organizations toward the desired state, for example, of improved efficiency, effectiveness, problem-solving, health, and so on. As such, OD is intentional, utilizes change knowledge and competencies, and is value-based.

Organizations Reconceived

The foci and processes of organizations is both widened and sharpened when described using the language and ideas of social rules theory. An organization's environment is reconceived as a set of encompassing social systems: on the one hand, those organizationally relevant institutions (each with a rule regime) that constitute society, and, on the other hand, the set of other social systems (each with a rule system), large and small, mostly other organizations, that transact or compete with a focal organization or subpart. Organizations are now understood to be one or more formal and informal social systems and/or project systems (each with its own rule system) linked in terms of varying kinds and amounts of rule-guided dependency and mutual coordination. Each social/project system has an identity, more or less clear, and a roster of agent members. Agents may, and often do, have membership in more than one social/project system. Agents more or less share an understanding of a social/project system's rule system, have a degree of centrality that reflects this understanding, and have an identity based on their membership. The degree of agent centrality in and identity with a social/project system means that they will more or less follow the system's action rules (always conditioned by relevant descriptive and situational and valuation rules) with regards to the enactment of four system rule processes (creating, interpreting, implementing and modifying rules). Variation in rule following and other rule processes by agents occurs because of multiple system membership, their personal rule structures, how well the rule system has been learned, and the perceived payoffs for compliance, habit, sanctions, and so on. It should be noted that, while variation is common, rule systems are naturally inertial, that is, they tend toward stability without ever quite achieving it.

Organization Change Reconceived

This social rules theory reconceptualization of organizations holds several significant points about change. One is that organizational change is ongoing, always occurring. While the scope, locus and focus of changes will vary a great deal, change is continuously pulsating within and between the social/project systems that constitute formal organizations. Organizational change is thus the product of external, environmental changes, the shifting in the linkages among an organization's social/project systems, the evolving drift of rule systems within these systems and changes in the personal rule systems of members. When intentional change endeavors are initiated, they can now be seen as being either consistent or inconsistent with the natural, ongoing changes occurring at that point in time. The larger the scope of an intentional change, the more social/project systems, and agents, will be affected. The more the identities and rule systems of these organizational components are differentiated, the more intentional change efforts will have to factor in the functionality of maintaining, increasing, or decreasing variety. In general, then, organizational change means change in one or more rules or rule systems. Which rules or rule systems, located where, enacted by which agents, and with what identities, become the central questions for managers and change diagnosticians alike.

If the rule systems of the social/project systems comprising organizations are at the heart of organizational behavior as social rules theory suggests, then the diagnosis leading to organizational improvement and renewal must focus on them. In what ways can rule systems go awry? Consider the following.

- *Ambiguity.* Whether because they are inherently unclear or because agents have incompletely learned them, ambiguous rules and rule systems do not provide adequate behavioral guidance.
- *Conflict.* Two or more existing rules or rule systems many offer conflicting guidance to an agent—either rules in the same rule system or rules from different rule systems (social/project/personal).
- *Too many/few rules.* Too many rules may reduce needed agent flexibility or creativity. Too few rules may leave some needed patterned behavior unguided.

Social Rules   51

- *Superstitious rules.* Rules that gloss cause and effect relations may provide inappropriate behavioral guidance, that is, following the rule will lead to unanticipated consequences.
- *Insufficient agent identity.* Unless committed to the social/project system, its agents will not be motivated to follow its rule system.
- *No agent with centrality.* Because centrality means system expertise and power, action rule interpretation, implementation, and other processes may not occur as needed.
- *Inadequate linking rules.* In part, social/project system effectiveness and survival will depend on the adequacy of its rules by which it relates to other systems.

These then begin to constitute the what or content focus of organizational diagnosis, the set of possible "problems" that may be discovered and corrected.

Change Processes

Organizational change in essence means that one or more of the formal and/or informal component social/project system's rules or rule systems is modified. While the focus of organizational change is constant, that is, rules and rule systems, the scope and locus of rule/rule system modification can vary from one rule in one social/project system to all of the whole rule systems of all organizational component social/project systems. Regardless of the scope or the locus of a change endeavor, however, the general process is comprised of three ordered stages—the surfacing, assessing and modifying, and implementing and institutionalizing of rules.

The first change-oriented activity, *surfacing*, identifies the rules and agents of a social/project system, going on to note the appropriateness and congruency among the rules of the rule system, the degrees to which agents are sophisticated about and committed to the rule system (their centrality), and whether there are action rules for the four action rule processes. As such, surfacing is simply focused information gathering, which may occur naturally in the normal course of events or more consciously utilizing standard techniques (e.g., interviewing, observation, the analysis of documents, surveys, etc.).

Once rule and agent information has been acquired, it is next *assessed*, and, as necessitated, *modified*. Assessment simply means looking for the sorts of "problems" listed above, or, in other words, judging the appropriateness and adequacy of the rule system–agent configuration (and, of course, the relationships between agent requirements/responsibilities and their personal rule systems, and, intersocial/project system rules). When intra- and inter-rule system inadequacies are discovered, attention turns to their modification by either reinterpreting or appropriately redesigning them.

The third change focused activity involves the *implementation* and *institutionalization* of any reinterpreted redesigned rules. Implementation may be as simple as acting in conformance with the new rule or rules (and thus modeling the change). More usual, however, is also informing other social/project system members of the change, especially those agents who interpret and implement those rules. Once known and first utilized, modified rules need to be repeatedly used (i.e., stabilized in practice). The institutionalization of modified rules is accomplished via inducements for rule conformity (i.e., how compliance pays off to agents) and influence by agents with centrality (e.g., appeals to reason, loyalty, fear, etc.).

The first two stages of the organizational change process correspond to what is usually termed diagnosis and the first part of the third stage corresponds to what is usually termed intervention. The three-stage, rule-system focused change process outlined above has thus far sidestepped two important questions—When is it actuated? and Who performs it? In response to the former question, the answer is either episodically, regularly, or continuously. Episodic change typically occurs in response to perceived symptoms of organizational disturbance, in efficiency and/or in effectiveness, that is, in response to those perceived symptoms beyond some learned threshold of magnitude which energizes change actions. Episodic change, therefore, will tend to be irregular over time and related to unique internal and environmental stimuli. While regular change usually is triggered by unique stimuli like its episodic cousin, here a governing social system of the organization has a rule that says that all social/project system components will examine their rule systems on some periodic basis regardless of symptom appearance. Continuous, that is ongoing, change efforts are the mark of what has of late been termed the learning organization (e.g., Senge, 1990). These organizations have either a component social system with the task of continuously surfacing and assessing the rule systems of its sister systems or there are rules built into the rule systems of all component social/project systems that activate regular rule examination (Lundberg, 1996). In each of these change modalities, rule surfacing, rule assessment and modification, and rule implementation and institutionalization may be performed by organization members with or without outsiders. Facilitating the performance of the three organizational change activities constitutes the role and purpose of OD to which we now turn.

The OD Project System

In other than the rare fully operative learning organization, OD can now be understood to be a special type of project system in an organization which is charged with assisting with the examination of the rule systems within or linking one, several, or all organizational components.[2] As a project system, they are by definition time-limited although an OD project system's life may vary from being quite short to relatively long term. Unique to OD project systems is that they may be contained within existing organization social/project systems or overlap with them (these existing social/project systems hosting OD work are nominally referred to as the change target [Lippitt, Watson, & Westley, 1958] or the primary client [Schein, 1998]). OD project systems may vary in size, that is, from one to many agents. Regardless of the size of the OD project system, however, two kinds of sophistication are required. An OD project's agent(s) have to be sophisticated about processes for rule examination and sophisticated about the operations of the host social/project systems. In addition, one or more OD agents must have or potentially must be able to acquire centrality in the host system, that is, the host system's action rules specify that that agent can or may create, interpret, implement, or modify the rule system. While it is conceivable that the two kinds of sophistication just noted and the necessary centrality may be possessed by one social/project system member (e.g., a manager), and thus he or she would constitute the OD project system, the more likely minimal situation is where one member has operational sophistication and centrality and someone else has the rule examination sophistication (a consultant–client dyad). More common in organizations, however, are larger OD project systems composed of multiple agent-members. Regardless of their size, OD project systems may be either formally specified, with an identity known to others and legitimated by the encompassing organization (e.g., as a change task force or in a consulting contract), or informal, unknown to others, and self-authorized (e.g., a cabal).

Given the above general specification of an OD project system, we now turn to their formation and operation. OD projects may be initiated by social/project system agent-members or others. While often framed as an agent's espoused request for help with a perceived problem, sometimes it simply begins with the expressed need to explore or work on something else or when an outsider indicates his or her willingness to assist an agent's consideration of their situation. However begun, such initiations always imply the possible change in an agent's operative rule system. Whether exploratory or more defined, an OD project system has been created which will have its own identity and rule system, roster of agents, and so on.

Once initiated, OD project systems progressively clarify both themselves and the change endeavor. For the OD project system this means identifying its agents, their sophistications and centrality, the scope of the social/project systems worked with, its own rule system and identity, and its authorizations, resources, probable time-span, and so forth, and that their "work" is facilitating the surfacing and assessment of social/project rule systems and their modification and institutionalization as necessitated. Progressive clarification for the change endeavor simply means doing the "work" just alluded to—beginning with either the rules for activity performance (e.g., SOP's routines and those built into technology, structure, etc.) or the rules for managing activity performance (e.g., the rules around goal setting, resource allocation, what are appropriate managerial beliefs, etc.), and then the other, and eventually examining their combination and their relation to the rule systems of encompassing social/project systems and the rule regimes of relevant societal institutions.

CONCLUDING COMMENTS

At the outset, this chapter hoped to bring attention to and begin to elaborate social rules theory, and to begin the reconceptualization of organizational change and development in terms of it. Having done this, it behooves us to step back and comment on the limitations and strengths of these ideas as well as profer suggestions on their implications for theory and practice.

As presented, social rules theory is relatively parsimonious while encompassing a truly large range of phenomena. This is, of course, the consequence of joining two conceptual strategies, that is, the adoption of molar constructs which can be applied at multiple levels of analysis and by using them both statically and dynamically. This theory appears both simple and abstract but is actually neither. For those with a conditioned preference for positivistic-based hierarchical theoretic structures, the concaterrated, mostly subjectivistic form of social rules theory may seem strange, even ungainly. It thus begs both a further conceptual elaboration and justification and further extensive anchoring in rich descriptions of social circumstances. Currently prevalent cross-sectional research designs and survey methodologies are unlikely to be very useful for the confirmation of this line of work. Instead, richly descriptive longitudinal research of change successes and failures should probably be advocated because they would enable the teasing out of both the antecedents to and the resulting systems changes over time.

Social rules theory, when phenomenologically grounded, appears to hold at least three potentials for solving foundational puzzles in the organizational sciences. One is ontological. Current debates about whether reality is more or less objective or subjective will have to be transcended simply because understanding social rules requires it. The four types of rules specified combine objective reality and socially constructed subjective reality to different degrees, for example, descriptive rules are primarily but not exclusively objective and action rules are primarily but not exclusively subjective. A second potential latent in social rules theory is the partial marriage of teleological and non-teleological explanations, thus making scientific knowledge and practitioner knowledge more compatible. Both the "because of" explanations of scientific knowledge and the "in order to" preferred form of practitioner knowledge, when expressed as action rules for a particular context, are essentially identical. The third significant potential of social rules theory lies in the fact that three of the four types of social rules condition action. While we know quite a bit about what organizations and their members actually do, we know considerably less about why they do what they do, what they do not do, why what they do is effective, why the forms of joint action persist, and what the longer term wider effects of more or less coordinated action are. The properties of social rule systems (e.g., directing attention, providing justifications, defining member identity, and generating and guiding action, among others) clearly begin to speak to these action concerns.

The promise of social rules theory for the study of organizational behavior, generally, and for organizational change and development, more specifically, will not be fulfilled without considerable developmental effort on several fronts. One is to examine the linkages of social rules theory to other theories, some seemingly congruent such as appreciative inquiry (Cooperrider & Srivastva, 1987), organizational dialog (Ashkenas & Jick, 1992), and organizations as meaning-action systems (Lundberg, 1996), and others that seem likely to inform social rules theory at different levels of analysis, for example, institutions (Scott, 1995) and organizations (Weick, 1995). Of course, there is the refinement of numerous existing rule-like constructs into generic rule formats, for example, norms, habits, sanctions, customs and conventions. A second front is empirical where the contextual contingencies associated with rule systems and the personal attributes predisposing agent identity, agent rule compliance, and agent centrality beg initial discovery as do the answers to basic descriptive questions such as rule framing (e.g., are OD project systems more likely and/or more successful in social systems which have positive valuation rules regarding change?) and rule distribution (e.g., what is the ratio of proscriptive to prescriptive action rules over the life of a change project?) A third front requiring developmental effort has to do with OD practice. In the short term we will need to do two things—invent a repertoire of techniques for surfacing, assessing, and so on, rules and rule systems, and to translate our library of OD practices into the language of social rule systems.[3]

About a decade ago, Woodman (1989) optimistically predicted that OD, to avoid retrenchment or stagnation, would branch out in new directions. This chapter, in which social rules theory was used to reframe organizational change and development, is one example of the sort of theory-based approach most likely to point to a new direction for OD. The social rules theory perspective advocated in this chapter, while promising in many ways, requires extension and refinement by scholars and practitioners alike, hence we conclude with an invitation to use it and improve it.

NOTES

1. It should be emphasized that the version of social rules theory presented below is offered here for the first time. While believed to be generally consistent with prior conceptualizations by Burns and Flan (1987), Lundberg (1998), and others, it both refines and extends previous work in many ways, for example, actors as agents, social systems as the foundational unit of analysis, the four levels of rule systems, the inclusion of identities, four rather than three types of rules, and the four kinds of processes of action rules.

2. While not explicitly embraced in this definition, OD that serves multiple organizations and their relations will operate in essentially the same way. Also of note is that, as defined, the process and purpose of OD seems identical to most other types of change agents (e.g., therapists and educators).

3. Interestingly, one of the earliest OD manuals (Fordyce & Weil, 1971) took the form of rule-like, how-to-do-it steps, a form frequently followed until today.

REFERENCES

Althusser, L. (1970). *For Marx*. New York: Vantage Books.
Archer, M. (1986). The sociology of education. In U. Himmelstand (Ed.), *Sociology: The aftermath of crises*. London: Sage.
Argyris, C. (1973). *On organizations of the future*. Beverly Hills, CA: Sage.
Ashkenas, R.N., & Jick, T.D. (1992). From dialogue to action in GE work-out: Developmental learning in a change process. In W.A. Pasmore & R. Woodman (Eds.), *Research in organizational change and development* (Vol. 6, pp. 267–287). Greenwich, CT: JAI Press.
Bartunek, J., & Louis, M. (1988). The interplay of organization development and organizational transformation. In W. Pasmore & R. Woodman (Eds.), *Research in organizational change and development* (Vol. 2, pp. 97–134). Greenwich, CT: JAI Press.
Beer, M., & Walton, R. (1987). Organization change and development. *Annual Review of Psychology, 38*, 339–367.
Bennis, W. (1969). *Organization development: Its nature, origins, and prospects*. Reading, MA: Addison-Wesley.
Bourdieu, P. (1977). *Outline of a theory of practice*. Cambridge: Cambridge University Press.
Bowers, G., Franklin, J.L., & Pecorella, P.A. (1975). Matching problems, precursors, and interventions in OD: A systematic approach. *The Journal of Applied Behavioral Science, 12*, 22–43.
Burke, W.W. (1982). *Organization development*. Boston: Little, Brown and Co.
Burns, T.R., & Flan, H. (1987). *The shaping of social organization: Social rule system theory with applications*. Beverly Hills, CA: Sage.
Chomsky, N. (1965). *Aspects of the theory of syntax*. Cambridge, MA: MIT Press.
Cicourel, A.V. (1974). *Cognitive sociology*. New York: The Free Press.
Cobb, A.T., & Margulies, N. (1981). Organizational development: A political perspectives *Academy of Management Review, 6*, 49–59.
Connor, P. (1977). A critical inquiry into some assumptions and values characterizing OD. *Academy of Management Review, 2*, 635–644.
Cooperrider, D.L., & Srivastva, S. (1987). Appreciative inquiry in organizational life. In R.A. Woodman & W.A. Pasmore (Eds.), *Research in organizational change and development* (Vol. 1, pp. 129–170). Greenwich, CT: JAI Press.
Crozier, M., & Friedberg, E. (1980). *Actors and systems: The politics of collective action*. Chicago: University of Chicago Press.
Drucker, P.F. (1989). *The new realities*. New York: Harper & Row.
Evered, R.D. (1976). A typology of explicative models. *Technological Forecasting and Social Change, 15*, 584–602.
Fordyce, J.K., & Weil, R. (1971). *Managing with people: A managers' handbook of organizational development methods*. Reading, MA: Addison-Wesley.
Friedlander, F., & Brown, L.D. (1974). Organization development. *Annual Review of Psychology, 25*, 313–342.
Gersick, C. (1991). Revolutionary change theories: A multilevel exploration of the punctuated equilibrium paradigm. *Academy of Management Review, 16*, 10–36.
Giddens, T. (1979). *Central problems in social theory: Action, structure, and contradiction in social analysis*. Berkeley: University of California Press.
Goffman, E. (1974). *Frame analysis: An essay on the organization of experience*. Cambridge, MA: Harvard University Press.
Golembiewski, R.T. (1979). *Approaches to planned change*. New York: Marcel Dekkar.

Handy, C. (1989). *The age of unreason.* Boston, MA: Harvard Business School Press.
Hannen, M.T., & Freeman, J.H. (1977). The population ecology of organizations. *American Journal of Sociology, 82,* 929–964.
Harre, R. (1979). *Social being.* Oxford: Blackwell.
Kochan, T.A., & Useem, M. (Eds.). (1992). *Transforming organizations.* New York: Oxford University Press.
Kuhn, T. (1970). *The structure of scientific revolutions* (2nd ed.). Chicago: University of Chicago Press.
Lave, J., & Wenger, E. (1991). *Situated learning: Legitimate peripheral participation.* Cambridge, England: Cambridge University Press.
Lindblom, C.E. (1977). *Politics and markets.* New York: Basic Books.
Lippitt, G.L. (1969). *Organization renewal.* Englewood Cliffs, NJ: Prentice-Hall.
Lippitt, R., Watson, J., & Westley, B. (1958). *The dynamics of planned change.* New York: Harcourt, Brace.
Lundberg, C.C. (1989). On organizational learning: Implications and opportunities for expanding organizational development. In R.W. Woodman & W.A. Pasmore (Eds.), *Research in organizational change and development* (Vol. 3, pp. 61–82). Greenwich, CT: JAI Press.
Lundberg, C.C. (1994). Toward managerial artistry: Appreciating and designing organizations of the future. *International Journal of Public Administration, 17,* 659–674.
Lundberg, C.C. (1996). Managing in a culture that values learning. In S. Cavaleri & D. Fearon (Eds.), *Managing in organizations that learn.* Cambridge, MA: Blackwell.
Lundberg, C.C. (1998). Leadership as creating and using social rules in a community of practice. In M.R. Rahim, R.T. Golembiewski, & C.C. Lundberg (Eds.), *Current topics in management* (Vol. 3, pp. 11–30). Greenwich, CT: JAI Press.
March, J.G. (1994). *A primer on decision making.* New York: The Free Press.
Margulies, N., & Raia, A.P. (1978). *Conceptual foundations of organizational development.* New York: McGraw-Hill.
Martin, J. (1992). *Cultures in organizations.* New York: Oxford University Press.
O'Toole, J. (1985). *Vanguard management: Resigning the corporate future.* Garden City, NY: Doubleday.
Peters, T. (1987). *Thriving on chaos.* New York: Knopf.
Pondy, L.R., Boland, R.J., & Thomas, H. (Eds.). (1988). *Managing ambiguity and change.* New York: John Wiley.
Porras, J.I., & Paterson, K. (1979). Assessing planned change. *Group and Organization Studies, 4,* 39–58.
Porras, J.I., & Robertson, P.J. (1987). Organization development theory: A typology and evaluation. In R. Woodman & W. Pasmore (Eds.)m *Research in organizational change and development* (Vol. 1, pp. 1–58). Greenwich, CT: JAI Press.
Rousseau, D.M. (1985). Issues of level in organizational research: Multi-level and cross-level perspectives. In L.L. Cummings & B.M. Staw (Eds.), *Research in organizational behavior* (Vol. 7, pp. 1–37). Greenwich, CT: JAI Press.
Sashkin, M., & Burke, W.W. (1990). Organization development in the 1980's. In F. Massurik (Ed.), *Advances in organization development* (Vol. 1, pp. 315–349). Norwood, NJ: Ablex.
Senge, P.M. (1990). *The fifth discipline.* New York: Doubleday.
Schein, E.H. (1998). *Process consultation revisited.* Reading, MA: Addison Wesley Longman.
Scott, W.R. (1995). *Institutions and organizations.* Thousand Oaks, CA: Sage.
Stablin, R. (1988). *Structure of debate in organizational studies.* Paper presented at the annual meeting of Australia Management Educators, Perth, Australia.
Stephens, C.U., D'Intino, R.S., & Victor, B. (1995). The moral quandary of transformational leadership: Change for whom? In W. Pasmore & R. Woodman (Eds.), *Research in organizational change and development* (Vol. 8, pp. 123–144). Greenwich, CT: JAI Press.

Strauss, G. (1976). Organization development. In R. Dubin (Ed.), *Handbook of work, organizational and society*. New York: Rand McNally.
Trist, E. (1976). Toward a postindustrial culture. In R. Dubin (Ed.), *Handbook on work, organization and society*. New York: Rand McNally.
Twining, W., & Miers, D. (1982). *How to do things with rules* (2nd ed.). London: Weidenfeld and Nicolson.
Vaill, P.B. (1989). *Managing as a performing art*. San Francisco, CA: Jossey-Bass.
Walter, G. (1984). Organizational development and individual rights. *Journal of Applied Behavioral Science, 20*, 423–439.
Weick, K. (1974). Conceptual tradeoffs in studying organization change. In J. McGuire (Ed.), *Contemporary management: Issues and viewpoints*. Englewood Cliffs, NJ: Prentice-Hall.
Weick, K. (1988). Perspectives on action in organizations. In J. Lorsch (Ed.), *Handbook of organizational behavior*. Englewood Cliffs, NJ: Prentice-Hall.
Weick, K. (1995). *Sensemaking in organizations*. Thousand Oaks, CA: Sage.
Whyte, L.P., & Wooten, K.C. (1987). *Professional ethics and practice in organizational development: A systematic analysis of issues, alternatives, and approaches*. New York: Praeger.
Winch, P. (1958). *The idea of a social science and its relation to philosophy*. London: Routledge and Kegan Paul.
Woodman, R.W. (1989). Organizational change and development: New areas for inquiry and action. *Journal of Management, 15*, 205–228.

COLLABORATION AND ALLEGORY
EXTENDING THE METAPHOR OF ORGANIZATIONAL CULTURE IN THE CONTEXT OF INTERORGANIZATIONAL CHANGE

Joseph W. Grubbs and Robert B. Denhardt

ABSTRACT

Organizational change no longer can be thought of as a process limited to individual agencies or firms. Change has become *inter*organizational in nature and carries important cultural implications for participating groups. However, organization theory fails to consider the cultural, mostly symbolic aspects of the shared-change experience. To broaden the understanding of culture and change, we use the literary device, *allegory*, as an approach to interpreting the diverse narratives emanating from change among multiple organizations. We illustrate the allegorical perspective by discussing our experience with a network of public organizations in the State of Delaware. We show how allegory offers a new way of thinking about culture and interorganizational collaboration, as well as how the device informs an intervention strategy that enables us to more effectively inform change across networks.

Organizations face increasing pressure to integrate work processes and systems of leadership across institutional bounds. Consequently, organizational change no longer can be thought of as a process limited to individual agencies or firms. We must think of change as *inter*organizational in nature (Gray, 1990; O'Toole, 1997b; Provan & Milward, 1995; Wood & Gray, 1991) and as having important cultural implications for participating groups (Feldman, 1986; Grubbs, 1998; Lawson & Ventriss, 1992). As it stands, organizational scholars tend to focus on the economic and structural consequences of network relationships (Alter & Hage, 1993; Bluedorn, Johnson, Cartwright, & Barringer, 1994; Elg & Johansson, 1997; Oliver, 1990; Robins, 1987), but neglect the impact of collaboration on the systems of communication, ritual, and belief underlying networked organizations.

To understand the cultural significance of interorganizational change, we propose extending the metaphor of organizational culture (Alvesson, 1993; Frost, Moore, Louis, Lundberg, & Martin, 1985; Schein, 1997; Smircich, 1983) by using the narrative device, *allegory*. In the fictional form of allegory, each character or object brings a unique, symbolic meaning to the story (Madsen, 1994). Applying this premise to networks, cultural attributes from organizations' meaning systems (Ingersoll & Adams, 1992) are represented by "characters" in the change narrative. These characters introduce distinct qualities to the "story," based on attributes from their respective organizational cultures. We will demonstrate this allegorical perspective by discussing our research into a network of public organizations in the State of Delaware. In conclusion, we will describe how the allegorical perspective enabled us to identify tacit barriers to change, allowing us to more effectively facilitate the change process.

ALLEGORY AS A LITERARY FORM

Before discussing our use of allegory in organizational analysis, perhaps we should begin by exploring the narrative form as it has been used in fictional literature. The word *allegory* is derived from the Greek words *allos* or "other" and *agoria* or "speaking," with the concept "other-speaking" referring to a deeper level of meaning within the text (Leeming & Drowne, 1996; Madsen, 1994). As Madsen (1994) explains, two traditions of allegory have emerged in Western literature: the classical tradition, which originated in the interpretations of Homeric myths; and the biblical tradition, which was derived from early hermeneutic studies of the Judaic and Christian biblical testaments. The classical tradition has been, by far, the most popular form of allegory in Western literature, and we relied on this form for the purpose of our research.

In the classical tradition, allegory is an extended metaphor, used most often to convey moral and philosophical themes. The important aspect of the allegory is that the writer employs literary devices, such as symbolism and personification, to shroud his or her themes beneath a veil of abstraction. Whitman (1987, p. 2) wrote,

"It [allegory] seems to refer to something in the fiction, but actually refers to something else in fact. Allegory turns its head in one direction, but its eyes in another. In the traditional [classical] formula, it says one thing, and means another." The meaning itself may not be immediately apparent, at least on the surface of the text, but has been encoded into the narrative.

To illustrate the possible "other meanings" that can be expressed through allegory, let us contrast two well-known examples from Anglo-American literature, Bunyan's *The Pilgrim's Progress* and Hawthorne's *Young Goodman Brown*. Through this discussion, we will see how the authors relied on allegory to communicate vastly different images of early Anglo-American Puritanism. Note that the characters and objects in the narratives represent the various moral or philosophical positions that their names suggest.

In Bunyan's (1987) tale, we have the story of Christian, who personifies the ideal Puritan and whose journey reflects the earthly trials faced by the believers in their quest for heaven. The story begins when Christian, having experienced an alarming dream, meets Evangelist and is sent off on a quest for the Celestial City. After failing to persuade his wife and children to join him, Christian leaves his family and friends in the City of Destruction and embarks on his path to righteousness. As the story unfolds, the path Christian follows becomes fraught with danger, as seen in the metaphor of the mountain, and temptation, as reflected by the Valley of the Shadow of Death. But he remains steadfast on his course.

Along the journey, Christian is enticed to accept a less demanding form of religious faith, one that his tempter, the Worldly-Wiseman, promises will still provide a path to heaven but "without the dangers that thou in this way wilt run thyself into" (Bunyan, 1987, pp. 19–20). He is offered temporal wealth and status, from By-Ends, and experiences those who appear to be among the faithful, such as Talkative. However, as the narrative reveals, these characters represent the worldly vices which prevent the Puritan from reaching heaven.

Bunyan's story concludes with the protagonist and his companion, Hopeful, reaching their resting place, Beulah, then crossing the River of Death into the Celestial City, where they are greeted by the sound of trumpets. It is at once the story of every Puritan, in his pursuit of heavenly rewards, and also the author's own account of the tribulations he faced in his religious life. The characters and symbols in the tale serve as important metaphors which, through their interaction, contribute to the story. While the engagement between metaphors is important, the universal theme emerges only in the deeper meaning of the allegory.

In some respects, Hawthorne's (1967) story of Goodman Brown is similar in that it tells of a Puritan believer who embarks on a quest. However, from the outset, as the young protagonist bids farewell to his wife, Faith, we recognize that his journey will involve a darker purpose. "Methought [says Goodman Brown] as she spoke there was trouble in her face, as if a dream had warned her what work is to be done to-night. But no, no; 'twould kill her to think it" (Hawthorne, 1967, p. 272). Our fears are quickly realized, as Goodman Brown enters the forest, a place symbolizing evil to the early Puritans, and asks that in such a place, "What if the devil himself should be at my very elbow" (p. 272). Just then, Goodman Brown is greeted by his sinister companion.

With the ensuing dialogue, we learn that the travelers have more in common than first imagined. The companion's discourse introduces us to Goodman Brown's family, which in the narrative represents the body of Puritan believers, and we see that his kin for generations have been in league with the evil associate:

> I have been well acquainted with your family [the evil companions says] as with ever a one among the Puritans; and that's no trifle to say. I helped your grandfather, the constable, when he lashed the Quaker woman so smartly through the streets of Salem; and it was I that brought your father a pitch-pine knot, kindled at my own hearth, to set fire to an Indian village ... They were my good friends, both; and many a pleasant walk have we had along this path, and returned merrily after midnight. (Hawthorne, 1967, p. 273)

Here, Hawthorne shows his contempt of what he views as Puritan hypocrisy, that members of the religious group may at one moment preach the importance of faith, and the next subject others to spiritual and physical torture. The author also references the particularly cruel treatment by Puritans of Native Americans, who they viewed as embodiments of treachery.

The journey continues with the companion leading the protagonist deeper into the forest to undertake an evil rite. Along the way, they encounter several members of Goodman Brown's Puritan community, who have ventured out to share in the sinister purpose. The sight begins to take its toll:

> Whither, then, could these holy men be journeying so deep into the heathen wilderness? Young Goodman Brown caught hold of a tree for support, being ready to sink down on the ground, faint and overburdened with the heavy sickness of his heart. He looked up to the sky, doubting whether there really was a heaven above him. (Hawthorne, 1967, p. 278)

As the rite begins, the young protagonist is confronted by his wife, Faith, who has joined the rest of the Puritan community in the dark wood. Goodman Brown becomes consumed by the fury of the ceremony and steps forward as a convert, taking his place among the evil brood. Faith joins her husband at the altar, but as the two prepare to commit their souls the protagonist screams for his wife to resist. Immediately, Goodman Brown finds himself in utter silence. The fury has stopped, and he is left alone in the forest.

The story ends with Goodman Brown, in the light of morning, returning to Salem village. Yet, unlike the ending in Bunyan's tale, Hawthorne's protagonist emerges from the woods a "stern, a sad, a darkly meditative, a distrustful, if not a desperate, man" (1967, p. 284). He sees members of his community and friends, even Faith, as fiendish. As he walks through the streets, he hears not the sound of trumpets, as in Bunyan's tale, but the sounds from the Puritan church—the congregation singing religious hymns and the pastor reading from the bible—which cause him to "turn pale, dreading lest the roof should thunder down upon the gray blasphemer and his hearers" (p. 284). Goodman Brown, who represents every Puritan believer, must live out his days not with joy and heavenly bliss, but filled with evil thoughts and mistrust for all his kind.

Hawthorne's allegory, of course, gives us not a Bunyanesque parable of the spiritual life, but a stern critique of Puritan society. While we must remember that Hawthorne was not overtly anti-Puritan, indeed his family shared Puritan beliefs, the author still penned some of Anglo-American literature's sharpest indictments of this religious sect. A particularly interesting aspect of *Young Goodman Brown* is that it shares a similar plot structure with Bunyan's *The Pilgrim's Progress*. Both allegories place the protagonist on a linear path, with characters and symbols introduced along the way to add different qualities to the experience. That said, we recognize that the underlying meaning of the allegories are starkly opposed. The "other meaning" of the allegories marks a sharp departure between the authors' respective positions on religion and society.

ALLEGORY AND CHANGE

Returning to our purpose, allegory as an approach to research broadens our capacity for organizational analysis, allowing us to interpret the cultural significance of interorganizational change. While we apply allegory mainly as a narrative device (Czarniawska, 1998) to interpret and weave together oral and textual narratives from our organizational inquiry, we actually begin "thinking allegorically" during the early stages of the research. Our purpose in the beginning is to ensure that our conceptual framework for the research program reflects an appreciation of the diverse organizational cultures represented in the network.

In this regard, we attempt to concentrate on those elements of organizational meaning systems that influence human action, what anthropologist Clifford Geertz (1973, p. 44) referred to as "plans, recipes, rules, instructions" or what Ingersoll and Adams (1992) called "meaning maps" for the governing of behavior. Such elements, which we will call cultural attributes, serve as conceptual programs on which members rely to make sense of their organizational surroundings (Weick, 1995). Our research program was designed to study these cultural attributes, not simply for their significance within the organization but for the way they affected each member's interpretation of the change experience.

To use literary terms, the allegorical perspective reveals how different "characters" contribute to the change experience, based on the prevailing cultural attributes of their organizations. As such, allegory offers a way for us to understand relationships among diverse groups and to recognize the distinct "voices" involved in processes of change. Allegory provides a way of "reading" the many diverse interpretations of the narrative introduced by participants, as well as researchers and readers (Clifford, 1986; see also Clifford & Marcus, 1986). The use of allegory, in turn, enables us to identify tacit barriers to change, thus paving the way for an intervention approach that will facilitate more meaningful systems of interorganizational collaboration.

The allegorical perspective we propose stems from a rich narrative tradition in organization theory (Boyce, 1995; Czarniawska, 1997, 1998; Czarniawska-Joerges & Jacobsson, 1995; Morgan, 1997; O'Connor, 1995; Phillips, 1995; Riessman, 1993). Here, we should identify several works that have supported our conceptual and methodological foundation. Czarniawska's (1997, 1998) and Riessman's (1993) narrative approach to organizational analysis proved to be extremely influential, in that it reinforced our belief that inquiry into human action should not end with the action itself but should provide an understanding of the way human beings interpret action—that is, the meaning that we as individuals construct in our organizational lives. As we will see, the allegorical perspective enriches our ability to understand this deeper, constructed meaning in multiple organizational settings.

We also gained important insight from field of organizational ethnography, namely Van Maanen's (1983, 1988) *story-telling* approach and Lincoln and Guba's (1985) *naturalistic inquiry*. Methodologically, we adapted strategies for qualitative inquiry similar to those Van Maanen used in his research on police departments, correctional facilities, and theme parks, including interviewing and participant observation (see also Van Maanen & Kunda, 1989). Our intent was to elicit stories from the participants relating to their personal engagement, and that of their organization, in the change process. Likewise, we borrowed methods from Lincoln and Guba (1985, p. 314), such as "member checks," which provided a way of testing our qualitative data with the participants. We communicated our findings to the network members at various stages in the research, thereby raising member confidence in the credibility of our data and the purpose of our research program.

Finally, our allegorical perspective shares a common belief with Rabinow and Sullivan, who viewed the social research text as "plurivocal, open to several readings and to several constructions" (cited in Riessman, 1993, p. 14). From the framing of the research questions, to the interviewing and observation, and finally to the transcription and composition, the process of social inquiry is a constructed process. The information provided by the participants, and by ourselves, becomes edited, molded at each stage. In turn, our research narrative undergoes a similar process with the audience. Readers serve first as filters, interpreting the story based on their sociocultural context, then as conduits between the text and others around them (Clifford, 1986; Riessman, 1993). The allegorical perspective allows us to more effectively interpret the "voices" that emerge throughout the research process and, more important, channel these diverse interpretations in directions that support our intervention approach.

Despite our indebtedness to these important theories, we must distinguish the allegorical perspective from existing narrative and ethnographic modes of inquiry. First, our use of allegory provides a more appropriate strategy for interpreting the diverse images of change involving multiple organizations. While the works cited above have made an important contribution to theory by exploring the symbolic aspects of organizational life, the inquiry tends to be limited by a concentration on individual organizations—the focus, *one organization at a time*. In contrast, the allegorical perspective broadens our appreciation, allowing us to take account of the symbolic, cultural qualities of interorganizational relationships. Through allegory, we are able to give voice to the vast range of views participants hold of a shared-change experience (see Table 1).

Second, the allegorical perspective contributes to a richer understanding of the way diverse cultural groups come together to form collaborative relationships. We are not only able to conduct inquiry into the cultural significance of interorganizational change, but we are able to apply that knowledge in the study of what has been referred to in cultural archaeology as "cultural contact" (Lightfoot, 1995; Lightfoot, Martinez, & Schiff, 1998)—the intercultural experience between diverse social groups. As we will see, many factors emerge when organizations come together to meet shared objectives. Beneath these factors lie concerns relating to culture that, in much of organization theory, become overshadowed by issues of culture within individual organizations or by structural and economic considerations of the change process. By taking an allegorical perspective, we are able to experience the cultural significance of collaboration among multiple groups.

Third, the allegorical perspective offers a unique approach to organizational intervention. Through the veil of abstraction provided by the narrative form, we are able to communicate to the participants important information relating to the change, without raising individual and group defensive routines (Argyris, 1993; Argyris & Schon, 1978, 1985). Network members experience the otherwise "undiscussable" aspects of the change through a fictional narrative; then, with the presentation of the qualitative data from our research program, they see the connection between the fiction and the constructed meaning of the shared-change process. The participants also may be asked to write their own allegory, as a way of expressing the "plurivocal" nature of the collaboration. As a final step, we use the allegorical narrative to initiate a dialogue among the participants, giving them the opportunity to discuss the many interpretations of the interorganizational relationship.

The allegorical approach, as we will see, broadens our existing narrative and ethnographic modes of inquiry, taking account of the cultural implications of change across multiple groups. However, we should note that interorganizational relationships certainly are not new phenomena. Research on the subject in organizational theory alone spans more than a generation, with topics ranging from organizational set to interorganizational networks (Bluedorn, 1993; Bluedorn et al., 1994) and, more recently, team-based organizations (Katzenbach & Smith, 1993; Mohrman, Cohen, & Mohrman, 1995). Much of this research focuses either on the economic factors surrounding the relationship, such as in *transaction cost economics* (Oliver, 1990; Robins, 1987; Thorelli, 1986), or design and other structural

Table 1. Foundations of Allegory in Organizational Analysis

Current Modes of Organizational Inquiry	Strengths of Mode	Benefit of Allegory
Economic and Structural Modes of Inquiry (e.g., Alter & Hage, 1993; Elg & Johansson, 1997; Mohrman, Cohen, & Mohrman, 1995; Oliver, 1990; Robins, 1987)	• Advances organization theory by exploring chane across networks • Reveals the complexity of design and the transaction costs of networks	• Moves theory away from purely economic and structural considerations • Recognizes the cultural implications of shared-change experiences
Public Policy and Administration Modes of Inquiry (e.g., Milward, 1996; Milward & Provan, 1998; OToole, 1997a, 1997b; Provan & Milward, 1995)	• Measures networks based on their capacity to achieve public outcomescratic governance • Examines the implications of public networks for demo	• Explores cultural, symbolic aspects of networks • Informs intervention approaches for interorganizational change
Narrative Modes of Inquiry (e.g., Boyce, 1995; Czarniawska, 1997, 1998; Czarniawska-Joerges & Jacobsson, 1995; Morgan, 1997; OConnor, 1995; Phillips, 1995; Riessman, 1993)	• Sees human action in organizations as a plot, constructed by the meanings and interpretations of participants • Appreciates the insight generated by fictional and dramatic literature	• Introduces allegory as a narrative device for the study of organizations • Broadens current approach by concentrating on the narratives that emerge in an interorganizational setting
Ethnographic Modes of Inquiry (e.g., Lincoln & Guba, 1985; Van Maanen, 1988, 1983; Van Maanen & Kunda, 1989; also, discussions in Schein, 1997; Smircich, 1983)	• Adapts research strategies from cultural anthropology and sociology • Remains concerned with the constructed meanings in groups	• Studies the long-term impact of intercultural experiences • Applies knowledge to inform processes of interorganizational change
Collaboration and Competing Values Modes of Inquiry (e.g., Gray, 1989, 1990; Denison, 1984; Denison & Spreitzer, 1991; Wood & Gray, 1991)	• Identifies challenges to collaboration from members original organizations • Incorporates issues of individual and group values	• Reveals tacity aspects of the relationships with original organizations • Takes inquiry into values to a deeper, more symbolic level

considerations supporting collaborative teamwork (Katzenbach & Smith, 1993; Mankin, Cohen, & Bikson, 1996; Mohrman, Cohen, & Mohrman, 1995).

Interorganizational relationships also have received attention in public administration theory. Recent examples include a symposium on *the hollow state*, published in the *Journal of Public Administration Research and Theory* (Milward, 1996), and research on policy networks (Alter & Hage, 1993; Milward & Provan, 1998; O'Toole, 1997b; Provan & Milward, 1995). Of primary interest here are the ties between organizations in the public and nongovernmental sectors for meeting public outcomes, with particular concentration on strategies of coordination and joint decision making, as well as overall network effectiveness. A strength of this research is that it moves beyond purely economic questions in its concept of interorganizational relationships, measuring the network not in terms of net gain but on the capacity of the partnership to achieve shared goals (Provan & Milward, 1995) and for its impact on democratic governance (O'Toole, 1997a).

In addition, several scholars have examined the social and normative concerns of network relationships (Denison & Spreitzer, 1991; Gray, 1989, Wood & Gray, 1991). Gray (1989), for instance, reveals the obstacles that arise as network members attempt to pattern constructive ties between their collaborative group and original organizations. She emphasizes the tension between individual interests, the interests of the original organization, and the collective interests of the network. Denison and Spreitzer (1991; Denison, 1984) also make an important contribution by discussing the challenges within and across organizational lines associated with competing values. However, the conceptual base underlying this latter research regards culture as a variable, rather than a root metaphor (Smircich, 1983), with a more instrumental purpose of drawing connections between shared values and organizational performance.

On one hand, organization and public administration theory provides an important foundation for understanding the complexity of network relationships; on the other hand, the existing research fails to address, at a substantive level, cultural concerns relating to interorganizational change. While some research touches on important aspects of organizational life, such as organizational socialization, individual and group interests, and competing values, much of the current research stresses instrumental factors at the expense of substantive inquiry into culture and intercultural contact. Even the works on culture and collaboration fail to delve deeply into the symbolic aspects of interorganizational change. Such issues generally fall beyond the conceptual map of most research programs.

We should mention that a few researchers have used a rationalist approach to study contact between cultural groups. An example is Hofstede's (1980, 1997) attempt to measure intercultural cooperation through a survey research instrument. Yet, as Schein (1997, 185) wrote in his critique, "[Hofstede's] survey labels itself as a culture survey but is, in fact, measuring aspects of the organization's climate or its norms. In that regard the data are perfectly valid artifacts, but they are artifacts that have to be interpreted and deciphered in the same way as other artifacts. One cannot decipher the culture from them alone." Because survey instruments can measure only a limited set of variables, it would be impossible to construct a survey that enabled the researcher to fully experience the impact of engagement over time between multiple groups (see Table 2).

As organizational scholars, we must recognize that change in an interorganizational context becomes interpreted in vastly different ways. Our modes of inquiry must tap the rich source of interpretations of the change, as contributed by all of the actors in the network. Participants serve as reflections of their organizational cultures, and consequently are important at that level, but they also must be seen for the way in which they influence the change process. Organizational analysis must incorporate strategies that enable us to explore these diverse interpretations

Table 2. Researching Culture: Survey Research versus Allegory

Research Strategy	Strengths of Research Strategy	Weaknesses of Research Strategy
Survey Research (Hofstede, 1980, 1997)	• Assesses certain cultural attributes across a broad sample of the organization or network • Generates data which, as artifacts, inform research on issues of organizational norms and behavior	• Remains at the level of artifact rather than exploring the underlying, symbolic aspects of organizational culture • Fails to recognize the meaning human actors construct around the shared-change experience • Neglects the effects over time of the intercultural experience for the participating groups
Allegory (Qualitative Inquiry with Allegory as a Device for Narrative Analysis)	• Recognizes the cultural, symbolic significance of interorganizational collaboration and change • Allows for a rich appreciation of intercultural contact between diverse groups • Informs intervention strategies that enable researchers to more effectively facilitate change across networks	• Potential for researcher and/or reader to focus on the superficial story, rather than on the deeper significance of the intercultural contact • Forces researcher to make decision choices concerning characters, plots, and so on, which may lead to over or limited interpretations • Requires a certain level of creative writing skill on the part of the researcher

and do so across multiple organizations, not simply piecing together findings from individual groups. In this regard, the allegorical perspective offers a way to fully appreciate the complex "webs of significance" (Geertz, 1973, p. 5) that members of networks construct around the shared-change experience (Chapple, 1941; Weick, 1995).

In the following case study, we will show how we use allegory as a narrative device to interpret the cultural significance of change in the Delaware network. We employ allegory as a way of exploring participant narratives—oral and textual—and as a way of understanding how distinct "voices" or "characters" affected the change. We come to recognize how these characters, as metaphors, personify the meaning systems of the organizations in the network. Through allegory, and the symbolic interaction of the characters, we are able to experience how prevailing attributes from organizational cultures influence the nature and form of the change process. In turn, our use of allegory enables us to identify barriers to the change, creating a path for intervention that helps to bring about the interorganizational transformation.

INTERORGANIZATIONAL CHANGE IN THE STATE OF DELAWARE

The State of Delaware's service integration initiative formally began in May 1993, when Governor Tom Carper (1993) challenged the state's human and social service agencies to develop more collaborative systems of service delivery. To guide this effort, Carper created the Family Services Cabinet Council, an executive board consisting of cabinet secretaries from the participating human and social service agencies. The Cabinet Council subsequently placed responsibility of this enormous task to the Service Integration Working Group, which was comprised of executive-level staff from most of the seven Cabinet Council agencies. During the course of three years, the Working Group focused its attention on transforming the state's service delivery systems and ensuring the necessary support mechanisms, including staff training and technological integration.

For the purpose of our research, we concentrated mainly on the Working Group due to its leadership in finding an approach to service integration and as a functioning example of interorganizational collaboration. Our research program featured several diverse strategies. First, we conducted an historical analysis of the Delaware service integration initiative to situate the effort in its broader social and organizational context. The historical analysis was conducted from July 1997 to January 1998 and consisted of a review of all documents in the public record associated with the initiative or other endeavors that related in some way to the current effort. In addition to their substantive relevance, these documents were examined as cultural artifacts, symbols reflecting the underlying significance of the change as perceived by participants.

Second, we completed a series of focus groups with public school personnel to assess views from the service delivery level. The focus groups were conducted from October to December 1996 and yielded important insights into the interpersonal meaning and beliefs relating to service integration, as experienced by "front-line" personnel. This information revealed the current interpretations of the change within schools and the relationship between school personnel and their colleagues in state agencies. Because schools were the primary sites targeted for Delaware's service integration initiative, these focus groups generated valuable information concerning the impact of the change on those who would be charged with implementing the new approach.

Third, semi-structured interviews were conducted from September to December 1997 with the executive staff on the Services Integration Working Group. Once again, the Working Group proved to be an important point of inquiry for two reasons: (1) it was the primary architect of the state's service integration effort and (2) it served as an ideal subject for studying the formation of culture within a collaborative setting. Working Group members for several years had worked together on service integration and other issues. Therefore, they had established a sense of trust and shared communication (Ingersoll & Adams, 1992; Schein, 1997). Exploring the process of change in the context of the Working Group, accordingly, offered a glimpse into what may be on the horizon for participating state agencies.

Fourth, a similar series of semi-structured interviews was conducted from September to December 1997 with Cabinet Secretaries comprising the Family Services Cabinet Council. These individuals represented the primary source of executive leadership for the statewide endeavor. Thus, they offered a unique perspective on leading change and facilitating collaboration for human and social service delivery. In a broader sense, they were crucial sources of information regarding the nature and role of leadership within the new public organizations emerging in the State of Delaware.

THE DELAWARE ALLEGORY

The allegorical perspective presented here explores the State of Delaware's service integration initiative, based on several levels of meaning. First, it begins by introducing the cast of characters engaged in the narrative. The characters serve as metaphors and reflect the cultural attributes of participating organizations. Second, the allegory allows us to examine the deeper significance of interaction between the characters. We see the influence of organizational meaning systems within the broader context of the shared-change experience. Third, we use the allegorical perspective to identify barriers to change in the Delaware network. Our goal here is to develop an approach that would help break down these barriers and facilitate meaningful collaboration among the participants.

Cast of Characters

As we have mentioned, the characters in the Delaware allegory at one level serve as metaphors of the meaning systems from the respective groups. However, we must add that our purpose is not to reduce the otherwise diverse attributes of organizations' cultures nor to construct a set of organizational types based on meaning systems. We want to appreciate this diversity, and recognize the unique contribution made by participants on the change process, not to generalize about culture across organizations. We first attempt to synthesize our findings from the qualitative research program. Based on this synthesis and our interpretations of the narratives, we seek to more effectively understand the relationships between participating groups.

The organizations involved in the Delaware network have distinct sets of values and beliefs. Forms of communication (Adams & Ingersoll, 1990; Ingersoll & Adams, 1992; Schein, 1997), processes of organizational socialization (Louis, 1980; Mignerey, Rubin, & Gorden, 1995), as well as systems of leadership (Schein, 1997), differed substantially from group to group. However, organizations that had undergone a similar formative process, or that shared comparable characteristics in their value systems, exhibited some of the same attributes around the issue of service integration. People from different groups came together with representatives from other groups, based on the shared assumptions and beliefs underlying their otherwise distinct organizational cultures.

Consequently, the characters presented here should not be viewed as representative of a single organization; instead they reflect cultural attributes that prevailed within various organizations on the issue of the change. That is, the characters personify a similar set of cultural attributes shared by more than one state agency. While the belief systems within the different groups varied, important characteristics from the organizational meaning systems were evident between the groups. These attributes had a pervasive impact on the way network members perceived the change initiative and helped to pattern the form of collaboration that started to emerge within the Delaware network.

The characters:

Charity: Personifies the cultural attributes within state agencies that place children, families, and communities above all else. These organizations share, as a prevailing attribute, a humanistic understanding of group and individual action. Primary values are what the group experience means for participants and, ultimately, what the interaction generates for those in the service community. The character, *Charity*, offers a voice to the child and family in the broader service integration effort.

Thrift: Represents the state's most politically oriented organizations, groups which are most accountable for the way in which the state expends public resources. Therefore, the prevailing attributes distinguishing these organizations stems from a rationalist viewpoint (Burrell & Morgan, 1979), in which efficiency is the single most important value. The character, *Thrift*, provides the most rational, cost-conscious voice within the Delaware narrative.

Knowledge: Represents state and other organizations that hold as a common value the practice of learning. Organizations with this at the center of their belief system recognize collaboration as an important step for enhancing services in the state, but their main contribution to the discourse comes in *how* service integration could be done. The character, *Knowledge*, argues that those engaged in collaboration should be prepared with a knowledge base and adequate resources to ensure their success at the delivery level.

Technique: Reflects organizations that have technique—that is, process efficiency facilitated by technological capacity—as a core value. These groups believed that service integration can be done, based on the technique that enables "it" to happen. Collaboration is less of a service-oriented phenomenon and more of an integration of existing information infrastructures. As a result, *Technique*, believes that the main barriers to integration are the "program people," those who concentrate on service-related details.

Suspicion: Personifies state agencies that attempt to retain their respective bases of power. Organizations with such values view collaboration as a direct threat, so the primary course of action is geared toward self-preservation. *Suspicion* introduces elements of "turfism" to the dialogue, elements that translate into either overt barriers to change or more subtle obstructions to the integration process.

Jealousy: Serves as a personification of the ties between individuals working in an network and their original organizations (Gray, 1989). This character has a brief role in the allegory, but one that symbolizes an important concern for intergroup collaboration. In this way, *Jealousy* differs from the other characters in that he does not reflect the prevailing cultural attributes of particular groups. Instead, *Jealousy* stands for the influence of original organizations on interorganizational groups.

Prudence: Represents organizations that serve in a supporting role to the Delaware initiative, but ones that offer a broad, insightful perspective on the dialogue. These organizations share scholarship and public service as common values. *Prudence*, as a result, has the flexibility to provide the perspective of a researcher-interventionist while still enjoying the trust that an "insider" must have to effectively interact.

Presentation of the Allegory

We agree with Morgan (1997), that the use of the metaphor reveals only certain aspects of organizations and that just those attributes associated with the metaphor become highlighted for comprehension. Consequently, it should be noted that the organizations personified in our cast of characters have far greater diversity than reflected in the limited metaphors; they can be appreciated in a much deeper sense than presented here. However, the characters that engaged in the Delaware experience symbolize the prevailing characteristics of their respective organizations, the perceptions and viewpoints that emerged around the issue of service integration. These prevailing viewpoints and how they became manifest within the discourse are our principal concerns.

For our presentation of the allegory, we borrowed a two-column framework similar to the one used by scholars from the Massachusetts Institute of Technology's Center for Organizational Learning, in what they refer to as *learning histories* (Roth, 1996; Roth & Kleiner, 1995). In the learning history, participant narratives are placed side by side with consultant interpretations, as a way of informing the change process. Argyris (1993) also used a two-column approach to contrast between organization's *espoused values* and their *theories-in-use* (see also Argyris & Schon, 1974). Accordingly, we presented the allegory in the left column along side excerpts from participant narratives, as well as our own reflective comments, in the right column.

The reader is encouraged to first read the allegory by itself. By doing this, the reader's first experience with the allegory will be as an abstraction from the constructed social reality in the network. The reader then is asked to reread the allegory with the participant narratives and other comments. These overlays reveal the hidden meaning within the allegory, or the way in which the narrative serves as a reflection of the interorganizational change. Finally, the reader will be offered our detailed analysis, which highlights key themes from the allegory and examines more fully the "other meaning" beneath the story.

The Allegory

Charity awoke from a horrible dream: She had seen a family wandering in the wilderness, weak and without food or shelter. After hours of stumbling across the barren landscape, the family came upon a house.

The current array of services in the State of Delaware, marked by a fragmented and complex system of delivery, as expressed in Governor Carper's executive order (Carper, 1993) and the Lochtenberg Commission (1993) report.

It was not an assuming house, its sides and roof weathered, but it seemed larger than it actually was because of its imposing front door. Cast iron and a full two-feet thick, the door was like an entrance way one might have found on a medieval castle, or a walled city, impenetrable.

The imposing nature of government service agencies, particularly those which remain bureaucratic in form.

The father of the family reached with his tired hand and knocked as hard as he could. But given the weight and breadth of the door, his knock barely rose above the tempest that scoured the wilderness behind them. Again he knocked, but knew that certainly no one would hear. Finally, when the family had turned to leave, a sound of many latches being unfastened came from the other side.

The fact that much of the labor required for accessing services falls upon families, as opposed to agencies offering services where they are needed most.

The door cracked slightly open, and a face peered through. It was not an unfriendly face, but it certainly did not show the warmth the family had hoped for, it was not a face that promised comfort from the storm.

Service delivery systems are not family friendly, instead retaining an agency-based focus.

"May I help you?" asked the face, for the family could see nothing more.

"My family has been ravaged by the storm," the father said, trying to keep his voice from trembling. "We have no food or shelter. Our children are in need. Can you help us?"

"We do not offer food or shelter here," said the face. "And there is no doctor to help your children. We have only water. For the other things, you must continue through the wilderness to the other houses. At each, you will find some of what you need."

The categorical, issue-based form of service delivery, which forces families to search out the support they need by going from agency to agency.

With that, the face moved away from the opening and was replaced a moment later by a hand, which held a small bucket of water. The bucket contained only enough for each family member to get a taste. Placing the bucket on the ground, the hand quickly returned to the darkness. Before the family could move toward the water, the door was closed and the sound of latches closing came from the other side.

Despite their dejection at such treatment, the family one by one went to the water to drink. When they had finished the water, they looked to the father.

"My loved ones," the father said, "I know you are tired and weary. But this face has told us that we must keep on through the wilderness if we want to find the food, shelter and comfort we need. Let us journey together to find the other houses."

So the family trudged off, back into the tempest. And, as the face had promised, with each house they came to, they received portions of the service each of them required. Yet, the storm continued to rage and the wilderness grew dark; the family, though managing to keep moving onward, grew weary. In time, family members needed more and more, forcing the clan to continue its journey through the wilderness.

When, for example, they reached the house of food, they received enough for a single meal, but found themselves thirsty again. So, they had to retrace their steps to the first house in search of water.

On and on the family struggled, at each house gaining only enough to meet their need at that point, never enough to help them overcome their hardship.

As this continued, throughout the long days and nights, the family grew even weaker. The children soon began falling by the wayside, too weary to continue. Then, the father fell ill and died, leaving the mother alone. Burdened by grief and worn out from the storm, she soon succumbed to a lonely death.

A rather bleak picture, but this image suggests that too many families fall "through the cracks" of fragmented delivery systems.

When *Charity* awoke, she immediately recognized the significance of her dream: Families in the Land of Well-being were forced to endure a wilderness of services, which were fragmented along problem lines and, accordingly, remained difficult to structure into a comprehensive system of family support.

Charity concluded that something had to be done, that she would search out a more effective way of supporting families.

Though not sure how this would be achieved, or who could be asked to be a part of this important search, *Charity* knew that she could not allow her dream to be a reality.

As *Charity* left her house, she came across *Thrift*, who had experienced a similar dream. *Thrift*, too, had seen that problems existed in the way people in the Land of Well-being supported families and recognized that something had to be done.

Charity: "*Thrift*, I have had a terrible dream. I saw a family struggling in the wilderness, trying to find the help they needed, until one by one they died in misery. Our many houses did nothing to help them."

Thrift: "I had a similar dream. I saw our houses leave food, water and other things outside for a family to partake on their own. And, some houses even offered some of the same resources. This must be changed."

Charity: "We have to work together, to get our houses to share resources and offer food and shelter in ways that families don't have to wander around through the wilderness. Maybe we can even give them what they need all in one place.

In the State of Delaware initiative, two distinct viewpoints emerged relating to the goal of service integration. These can be seen in the following statements:

Member of the Services Integration Working Group: "We need to do better than we're doing, we need to integrate our services. We need to make it less complex for the client and family to navigate . . . taking the complexity and putting it behind the counter, so that the client can focus on their problems and not focus on mastering the system" (confidential conversation, September 18, 1997).

Participant in the State of Delaware's deliberations:
"It seems to me . . . you're trying to maximize productive uses of your limited resources. And, to the extent that you engage in the duplication of services . . . you're being wasteful and not providing the quantity of services you could optimally provide. That's what's been the problem" (confidential conversation, October 30, 1997).

Working Group member:
"Despite the fact that we have this marvelous system of State service centers, which is providing the government's services for the most

Thrift: "I don't know, *Charity*. I think we need to be more efficient in the way we offer food, shelter and those other things. We can't just keep all these houses around, or let families keep taking what they need. It just costs too much and the houses keep giving the same thing."

Charity: "Oh, *Thrift*. I think we just have to be more supportive. It's not the efficiency we should be concerned about as much as it is how effective we are in supporting families, and how simple we can make it for them."

Thrift: "I still think it's all that duplication, one house doing what some of the others are doing, then the fact that they just leave that precious food and water on the doorstep. I mean, families can just keep taking all they need."

Despite their differences, with *Charity* believing houses needed to be more supportive, and *Thrift* feeling that resources could be saved by cutting availability of support, the two set out together to find a better, more collaborative way of serving families in the Land of Well-being.

Marching down the Path to Integration, *Charity* and *Thrift* were greeted by two others who had houses in the Land of Well-being. They told their new companions, who were named *Knowledge* and *Technique*, about their respective dreams. As they spoke, *Charity* found a small place in the Forest of Deliberation, where the four could talk more comfortably about what was needed in the Land of Well-being.

part . . . [State services] were complex. It was a challenge to manage those . . . it was all out there, but it was very hard for the client to navigate through them" (confidential conversation, September 18, 1997).

Participant from State organization:
"There's a bit of redundancy in government. There are three or four agencies that are touching the same . . . family, and we don't have a sense of whether we're doing the same thing four times or four different things . . . Maybe if we integrate it, then we can do it better, cheaper" (confidential conversation, September 25, 1997)

Unpublished report from the Working Group:
"Families, school and community groups can get easy access to multiple points and interagency personnel work as teams to solve community problems. Because the State is well coordinated, it is better able to form partnerships with communities and stimulate local efforts of self help."

(Memorandum, August 13, 1997, unnumbered):
"Creating a social services delivery system which is easily accessible, rationally organized, and efficient is the core purpose . . . That is what the services integration project is all about."

Charity began the conversation, stating what she believed to be the future for Well-being: "I think families should be able to get food and shelter, or whatever else, at a place that's convenient for them. They should never have to trudge through the wilderness. It's up to us to work together and support them."

Then, *Thrift* cut in and stated his version of the group's mission: "I think we need a system of houses that's more efficient. If all of us keep offering the same things, the Land of Well-being will be the Land of Poverty before too long. We have to be more efficient."

Though puzzled by *Charity* and *Thrift*'s differing viewpoints, *Knowledge* and *Technique* joined the discussion to determine how integration could be achieved. In time, though, they came to view *Charity*'s vision as the best for Well-being and started to work on how this vision could be accomplished.

Sensing that most of the group had become more concerned with providing greater support to families, *Thrift* resolved himself to move outside of the circle there in the Forest of Deliberation. He believed that even as the others chose one direction, he would stay right over their shoulders and stress the point that a more efficient system was needed.

This decision by *Thrift* marked an important turn, since he had positioned himself years ago as the holder of the Land of Well-being's resources. From his new vantage point in the Forest of Deliberation, *Thrift* knew that he held the power. That in order for any of the group's decisions to ever see the light of day, the group would have to gain his approval.

The leader, here, communicates what she believes to be the group's vision and guiding principles.

The participants engaged in the Delaware initiative exhibit a relatively high degree of trust and willingness to work together, despite their differences in viewpoints.

This reflects the beginning of a formation process for a collaborative culture within the Working Group. Note the influence of Charity as an early leader.

Individuals representing organizations that have the value of efficiency as a prevailing cultural attribute have tended to separate themselves from the activities of the Working Group, which has come to adopt a vision more in line with Charity's.

As one Working Group member stated, "Those two visions aren't the same vision at all but they have never been forced into a dialogue that makes the two confront each other" (confidential conversation, September 25, 1997).

Collaboration and Allegory

Thrift, after a while, thought he should further establish this position by sending the group a message.

"I hearby declare," his message stated, "that before those of us in the Land of Well-being move forward, we must decide what types of support we're going to cut back on, which houses are going to provide which services, and in what households we want to remove the redundancy."

Thrift's message concluded by asking the group to respond to his main questions, and he even set out a time frame in which the group must respond.

When they received *Thrift*'s message, the group ceased their discussion and gazed at the parchment in awe. They had just noticed that *Thrift* was no longer among them, that he had distanced himself into a nearby portion within the Forest of Deliberation.

However, the group realized that it must respond to *Thrift*'s questions. The members *Charity*, *Knowledge* and *Technique* respected *Thrift*'s judgement and believed that, in order for anything to get done, they would need his support. So, despite the odd language that appeared in the message, for *Thrift* had begun to use words that the group could barely understand, the group members began thinking about each of his questions.

When they had prepared a series of answers, and sent them along to *Thrift* in language he could understand, the group returned to its discourse. No sooner had they done so, however, than did *Thrift* send down another message from his place in the Forest.

This separation becomes even more critical, however, in that those organizations tend to be in part of the State's central administration and in positions of control over public resources.

(Memorandum, May 28, 1997, 1, 3):
"Before we go much farther, I think we need to decide on some major principles 1) what services we want to integrate; 2) who provides such services; and 3) where we want to provide them."

As the Working Group continues to work together, it forms a unique language that enables group members to share important aspects of the group's own culture. This further distinguishes the group from other State agencies.

"We have reached an important turning point, here in the Land of Well-being," said *Thrift*'s message, "where we've tried a lot of different things, but now it's time to cut all that nonsense out and get something done. Too many of us are trying to hang on to the types of support that we offer in our own houses, but that are also offered elsewhere. Those things must be cut. We must be more efficient."

Charity took special attention to respond to this message, assuring *Thrift* that time and precious resources would not be wasted, but that the group needed to make sure that the change in State services would be effective.

However, *Knowledge* feared that such messages from *Thrift* could begin to stand in the way of the group's efforts. The concern was that the group would spend all its time answering to *Thrift*, and not enough time on helping families. But while the group agreed with *Knowledge*, they recognized that *Thrift* controlled the resources, and so they would have to keep responding.

Once the group had answered *Thrift*'s message, it returned to its conversation. *Charity* explained her dream to the other members, describing what she had seen with respect to the experience of the family.

(Memorandum, August 13, 1997, unnumbered):

"We are at a very important juncture right now where the opportunity to create and test pilots' is over, and where it is time to institutionalize a system of social services delivery which will be in place [into the future] That means learning from the multiple service integration pilots we've started and daring to put in place a statewide service integration system based on those lessons. What it does not mean is starting additional pilots and simultaneously preserving the other ongoing projects. That is services proliferation, not services integration."

Working Group member:

"It [the on-going communication from central administration] becomes distracting . . . because we're moving down the line, we're doing the work, and we're being thoughtful. I mean it's like we're kind of going off and having a beer and saying, Well, that sounds like a great idea. Let's do it.' I mean we're really spending a lot of time and energy, we argue with one another sometimes, and what comes out at the other end is pretty good stuff. And so sometimes those memos are a distraction. Then, we have to spend time with the distraction [which] keeps you from moving steps ahead in the process" (confidential conversation, September 10, 1997).

Collaboration and Allegory

The group gasped in horror at the family's sad plight. Although the members had not witnessed the dream for themselves, nor had they witnessed in any substantial way the way Well-being's houses supported families, they agreed that *Charity*'s vision must reflect the current condition.

Then, *Charity* raised the question of how such a terrible system could be corrected, how could they ensure families the support they need. *Knowledge* answered that the group should send word from the Forest of Deliberation, asking other groups, which had journeyed along the Path to Integration, how they achieved their mission.

Charity and *Technique* agreed this would be a good place to begin. And so *Knowledge* prepared a message and had it delivered to groups far beyond the Forest, for the group agreed that the only useful information would come from far away.

Within days, story upon story was returned, expressing the challenges and successes of similar groups. A treasure of information arrived in a short time, with *Knowledge* ready to share. However, in the days it took for these stories to arrive, the group's conversation had moved on. Group members hardly paused to take into consideration the insights from other groups.

In the months ahead, *Knowledge* would lament this fact, for it represented a missed opportunity to learn. All the insights, all the lessons from far off lands, were soon tucked away, not to be seen again.

The days and nights passed, and group members continued to huddle together in the Forest of Deliberation, talking about how to enhance life in the Land of Well-being. Then, one evening, another of their

The image of service integration within the Working Group, by and large, either comes from the group's leadership, or from members' past experiences. As a result, it does not reflect the input from key stakeholders.

The Working Group engaged in an extensive review of "best practices," which featured the experiences of several states nationwide.

The Working Group has not returned to the best practices review as it proceeded in its planning.

Working Group member: "We've never revisited . . . [the best practices review], we've never looked at the recommendations and tried to figure out . . . are any of these ones we want to try and put forward. So at some point, I think we need to go back and look at what we've done so far" (confidential conversation, October 15, 1997).

colleagues entered the Forest and joined the group. This newest member, whose name was *Suspicion*, had learned about *Charity* and *Thrift*'s dreams and wanted to determine the group's intentions.

As *Charity* explained what they had decided to do, *Suspicion* became alarmed. He began to feel that the group already had decided how the different houses in the Land of Well-being would be brought together, and that the end result would destroy his own house.

"She can't mean what she's saying," *Suspicion* said to himself. "If *Charity* succeeds in her mission, it'll be the end of my house. Everything I've put into place will be lost. That just can't happen."

Aspects of "turfism" and traditional divides between State agencies have contributed to protective behavior among some participants.

Working Group member:
"The . . . [service integration] model equates to me the idea that we are a train on a track with a predetermined destination . . . there's a core direction that . . . [the group] wants to go. That core direction . . . is a direct threat to [my agency] . . . So I sit . . . planning and . . . being an active participant in a process that just puts bullets in a gun aimed at the very existence of . . . [my agency]" (confidential conversation, September 16, 1997).

Suspicion recognized that *Charity* and the other group members expected him to engage there in the Forest of Deliberation. However, *Suspicion* decided to do so in a way that would allow him to preserve his own house.

Organizational defensive routines employed by group members to protect their respective agencies.

As months passed, the group decided it was time to leave the Forest of Deliberation and to prepare a plan for the Land of Well-being that would take them down the Path to Integration. But, each member of the group had a different image of how this would be done.

Working Group members conceptualized internal and external factors relating to service integration, based on the meaning systems of their original organizations. The framework for integration reflects these diverse viewpoints.

Knowledge argued that those leading the journey should be trained, based on the wisdom of the group. That is, the group should share its insights with those that will be breaking the Path to Integration. She proposed training these individuals in ways that the group determined to be the best course.

Technique, on the other hand, stated that the Path to Integration could be achieved by using information to link the different houses. He told the group, "We can work together to get around the different issues in our houses. The houses themselves may be tough to change, but we've got everything we need to make the houses communicate better, to make them actually talk to each other."

Charity, believing that each of these approaches had a role to play, challenged the group members to create a plan that reflected their unique perspectives. *Knowledge* and *Technique* accepted the challenge and quickly started working on their plans.

Upon learning of this new course, *Thrift* sent off a series of messages, which he intended for *Knowledge* and *Technique*. To *Knowledge*, he suggested that learning not be the only goal, that the plan include all individuals who are responsible for opening the door to children and families.

To *Technique*, *Thrift* wrote, "Be sure that when you're putting together this plan to make the houses talk the same language that you recognize everything that's been done on this front. We don't want redundancy. We want efficiency. We want a system that will reduce the time, energy and resources needed to support families."

Working Group member:
"By working together we're able to get around some of the bureaucracy. Bureaucracies are difficult to change . . . but we can do this through our networking of [information systems]" (confidential conversation, October 14, 1997).

The form of leadership taken by Charity shows the type of empowerment necessary to support collaboration (Denhardt, 1993). She builds upon the unique contributions made by each organization.

Central administrative agencies have continued to convey their position, most often in memoranda or other types of formal communication.

(Memorandum, August 13, 1997, 1–2):
"First, a note as to coordination. There are several technology initiatives ongoing at this time. It is critical that these not proceed in isolation . . .

Thus, as members laid out their respective Path to Integration, *Thrift* again established his position of influence.

The group responded to *Thrift*'s messages by structuring the plans around the various provisions, as surely they were expected to do. That is, all except for *Suspicion*, who believed the result of *Thrift*'s message would be the end to his house's base of power.

Indeed, the very notion of sharing information from inside his house drove *Suspicion* into a frenzy! "That information is confidential," he shouted. "Families and children have legal rights. That's it, they have legal rights that we won't share such information."

Suspicion also claimed that people that would be asked to prepare the Path to Integration would not know how to manage the complexity.

Even more important, I want to set forth succinctly two fundamental purposes of . . . the technology initiative . . .

The technology initiative must, to the extent practical, enable (the State of Well-being) to develop a client tracking system which will permit multiple agencies to obtain access to the complete service history of an individual . . .

In addition, the technology initiative must enable agencies to develop and share common financial intake instruments, which will reduce caseworker time and also enable the State to consider more common eligibility criteria. . . ."

The issue of confidentiality has been one of the most sensitive among Cabinet Council agencies. However, these agencies over the past several years have reached a number of agreements that allow information to be shared based on informed consent by the family.

Working Group member: "Customers are very skeptical about their personal information being shared . . . [However,] One of the most important things that was done early . . . on was an effort . . . to develop a confidentiality agreement for the departments to sign and share information" (confidential conversation, October 14, 1997).

Much of the issue of confidentiality, then, is reflective of organizational defensive routines.

"Do you think people who have to open the door to these families will know what to do with this?" *Suspicion* asked. "They'll be afraid. You all know that I don't care about preserving my own house. That's the last thing on my mind. I'm just worried about those who we ask to serve."

"I agree, families and those who open the door will be concerned," said *Technique*. "But remember, the houses have already started talking to each other more often. They even have agreed to share information about families they serve. A lot of those questions already have been answered."

Charity asked what could be done to resolve these questions, to which *Technique* reiterated that such matters had already been resolved through cooperation. The first steps on the Path to Integration had been prepared.

"The houses have started to put these things into place. They're just waiting on those who open the door. And, well us, they're waiting on us. Everything else's in place," *Technique* said.

Gradually, *Suspicion* started to recognize the sharing that was occurring among the other group members. He asked himself, "Should I do as the others have done and open my agency to others, even to families?" No sooner had he asked this than did someone from his house enter the Forest of Deliberation.

These defensive routines are also evident in the types of concerns, raised by executive- level staff, for delivery-level personnel.

Working Group member:
"I don't think the issues are turfism . . . [Instead, it's] Fear, fear, fear . . . It's fear [for the front-line staff]" (confidential conversation, September 16, 1997).

It is interesting that such issues can be raised, when this latter group of individuals has not been heard from in any substantial way on the issue of service integration.

Working Group member:
"Our databases, our systems are based upon the same technology, so whenever we reach that point, and that's what the services integration process is all about, the infrastructure's in place" (confidential conversation, October 14, 1997).

Suspicion: "Why did you search me out?"

Jealousy: "To remind you of the threat those others represent to our house. Don't you remember that *Charity*, there, believes that power should be given to outsiders; and *Technique*, he wants us to give up our information. This would surely be our end. Refuse to give in, *Suspicion*, the house depends upon you."

With that, *Jealousy* left the Forest, and in time *Suspicion* remained steadfast in his belief that integration would spell disaster for his house.

On and on the debate raged in the Forest of Deliberation, until it caught the attention of *Prudence*, who happened to be passing by. *Prudence* did not own a house in the Land of Well-being, but was recognized as being a caring and wise person.

Charity, seeing this opportunity to gain valuable insight, asked *Prudence* to join the group in creating a Path to Integration. As *Prudence* sat down, *Charity* began to explain the efforts taken thus far. She told of the dream she and *Thrift* had experienced, then of *Thrift*'s removing himself to view the group from a distance, of *Knowledge* and the stories from other lands, and of *Technique* and the first steps on the Path to Integration.

Prudence, being a contemplative person, sat and listened to the entire history. Then, looking kindly at the group, she spoke: "I mean this in the best way, but it doesn't seem like you all agree to how you're going to work together. You've been saying some interesting things about your houses, many of which I believe are needed. Yet, you don't seem to agree on how working together can be done, or for that matter what it even means," *Prudence* said.

The exchange here reveals the competing demands faced by members of collaborative teams. On one hand, they retain ties to their original organizations—the systems of resource allocation, accountability, and management. On the other hand, they are asked to be contributing members of the new group (Gray, 1989).

Given the pervasive nature of more traditional systems in government organizations, the original organization often secures the strongest hold.

The Working Group has sought input from a series of consultants. The character, Prudence, personifies the constructive influence of these different consultants within the State initiative.

Consultant's report:

"In this regard, the group has been using the term 'services integration' to describe a change in the delivery of state services... Yet there does not seem to be a clear consensus as to what this means or how it will translate to the administrative, agency or building level...

She continued, stating that the group had made some important decisions about what families need and how they will be better served, what supports they need and how their children can be successful. While recognizing these as noble decisions, she asked why the group had not asked the children and families themselves.

Group members paused, gazing at *Prudence* with puzzled looks. Then, as quickly as it had stopped, the group rang in again with the clatter of voices. And, ever so slightly, they moved to close the circle, leaving *Prudence* just outside of the ring.

Standing to leave, *Prudence* was stopped by *Charity*, who asked her to stay and work with the group. *Charity* particularly asked *Prudence* to play a part in assessing how successful the group is, over time, in reaching the Path to Integration.

Thrift quickly sent word, supporting *Charity*'s request. "We need to keep *Prudence* around," he said. "She can keep track of this integration thing, while we get back down to the actual work that our houses need to do."

And so, as darkness fell, *Prudence* resumed her place just outside of the group. *Charity* returned to the discourse, and *Thrift* watched from afar there in the Forest of Deliberation.

There are also questions that arise in services integration activities concerning the relationship between policy and implementation. The literature on organizational change strongly recommends that both lower-level personnel and external stakeholders be included in the planning and design of change efforts. With the exception . . . [of some efforts, we] have seen very little evidence of such front-line involvement . . ." (Center for Community Development and Family Policy, 1997, 5, 8).

Although the State has implemented many of the consultants' suggestions, recommendations to bring families and front-line staff into the deliberation process have not been followed in any substantive way.

(Electronic mail, August 12, 1997): "Most of all, we want independent credibility and someone who will do this work [of supporting the State] and allow us to govern."

The Working Group continues to struggle with design, while crucial issues lay ahead for implementation and evaluation.

THEMES FROM THE ALLEGORY

To fully explicate the notion of allegory as a way of understanding interorganizational change, it may help to examine some of the themes that emerged from the narrative. This will involve reading the organizational allegory as one would a piece of fictional literature (for discussion, see Eco, 1990). Paying attention to the use of symbolism and metaphor, we see that the allegory presents important themes that may not be immediately apparent in the text. Through such "other meaning," and the varied interpretations of the change, we may benefit from the informative power of the allegorical perspective.

The story begins with *Charity*'s dream, a bleak picture of the existing service environment and the way families must navigate through a complex web of agencies. In the dream, the family is forced to struggle through the wilderness to find the support it needs, and even then the support is fragmented according to what is offered at each house. The experience at the first house exemplifies, albeit in the extreme, the agency-based focus of service institutions: the imposing door symbolizing the enormous cognitive barrier that families must overcome to even set foot inside service agencies; the limited access to staff characterized by the figure only allowing the family to see parts of it at a time; and, the provision of only enough water to cure the family's thirst for that single visit, thus creating an environment of dependency in which the family must continuously return.

From this foundation, the narrative reveals the two distinct perspectives on the concept of service integration as they have emerged in the Delaware network, symbolized in the visions of *Charity* and *Thrift*. *Charity*'s viewpoint on integration is more family-oriented, while *Thrift*'s is dedicated to trimming cost and achieving service efficiency. As the story continues, the other characters organize themselves according to these distinct visions. *Thrift*, who personifies the politically oriented groups, sets himself apart from representatives of the more service-oriented houses. He chooses to influence the process by imposing his power over resource allocation, instead of constructively engaging in the group's formation process.

On the other hand, the characters interacting around *Charity*'s vision create a more cohesive group. This new group develops its own system of communication, one that is different from the language of *Thrift*. The interaction is important in that it represents the type of engagement at the heart of effective collaboration. So, not only does the group in the allegory come together in a shared vision of its intended outcome, it also symbolizes collaboration as it occurs in a shared-power setting. Each character makes some unique contribution to the process, with all members of the group playing a leadership role.

However, as the story unfolds, we experience the many challenges that affect such collaboration. First, the group bases its understanding of the issues on limited sources of information. That is, the group attempts to structure an initiative without gaining substantive input from key stakeholders, relying on its own judgment. Second, information generated through the experiences of other jurisdictions gets set aside. The group has at its fingertips useful input from other areas, but by and large does not take account of this input as its creates a Path to Integration. Third, guidance offered toward the end of the story by *Prudence*, although valued on the surface, is not factored into group deliberations. Group members respect the wisdom of the character; however, they close ranks and continue to shape the initiative without taking into consideration *Prudence*'s important insight.

A significant aspect of the allegory, though, comes in the contribution each character makes to the meaning of the story, in literal and figurative terms. Recognizing that the characters serve as metaphors for their respective organizations, embodiments of prevailing elements within their organizations' cultures, the allegory reveals the characters' influence on the new culture and subsequent discourse of the interorganizational group. For instance, the distinct visions of service integration held by *Charity* and *Thrift* have a profound impact on the group's very structure, which in turn shapes the overall focus of the group. *Charity*, who serves as the group's early leader, communicates her vision and thus provides a critical ingredient for the group's internal integration. Yet, *Thrift* retains his influence by exercising his control over resources and oversight for accountability. The group is forced to respond to his demands as a way of surviving in its external world (Schein, 1997).

Likewise, other members contribute to the group, based on their organizations' cultural attributes. *Knowledge*'s original organizations have learning as a core value, a value that becomes instilled within the interorganizational group due to the character's role in the shared-power relationship (Bryson & Crosby, 1992). In turn, this contribution translates into a tangible output: the group begins to understand collaboration in the context of training and structures this form of learning into its future implementation plan. *Technique* has a similar influence on the group, given the value his organizations place on technological capacity. Members of the group, as a result of *Technique*'s influence, view a key aspect of collaboration as the convergence of information infrastructures. Creating a plan to share information then marks a course taken by the group in its Path to Integration.

Through the allegory, however, we see that not all of the characters support the interorganizational group in its mission. In this respect, *Suspicion* personifies some of the organizational barriers that prevent meaningful collaboration. The character on the surface participates in the group's dialogue, thus suggesting his willingness to interact. Yet his manifest actions run counter to the group's driving purpose. *Suspicion* effectively veils his opposition, using props such as unfounded concerns over confidentiality and for delivery-level staff. Such defensive routines become reinforced as *Suspicion* is confronted by a colleague from within his agency; the influence of *Jealousy* represents the pervasive impact of the original agency within the interorganizational group (Gray, 1989).

As the narrative suggests, tension between members of the interorganizational group and their original organizations represents a key concern (Gray, 1989, 1990). Group members must form a collaborative culture within their new group, one that will enable the group to integrate internally and succeed in its external world, but they also face pressure to retain links to the organizational culture of their original group. Such competing demands pose enormous challenges for the level of organizational change necessary for collaboration. Moreover, overcoming these challenges requires that interorganizational groups take account of the diverse meaning systems reflected in each character's unique contribution. The allegorical perspective becomes more than a new way of appreciating organizational culture; it serves as an approach to intervention into the change process.

AN ALLEGORICAL PERSPECTIVE ON CHANGE

The first step in facilitating change in a network setting involves recognizing the varied, complex challenges to the change. Barriers of this nature stem from organizational defensive routines, most of which are never discussed or associated with manifest behavior (Argyris, 1993). These barriers to change become exacerbated in an interorganizational context. The allegorical perspective shows that, within collaborative groups, it is not simply the organizational defensive routines of the group members that must be overcome but also those from each member's original organization. Consequently, the organizational learning necessary to bring about and sustain meaningful change will need to cut across organizational lines. The change must be effected both in the collaborative group and ultimately within each participating organization.

Here, we turn to the research of Argyris (1993) and Argyris and Schon (1978, 1985) on organizational learning and its role in sustaining planned organizational change. In one respect, the characters in the allegory share an *espoused value* that supports the move in the Land of Well-being toward service integration. They all state, at least on the surface, that collaboration offers a framework that will allow them to more effectively serve families. Yet, as the allegory continues, the characters' *theories-in-use* become apparent. *Suspicion*, for instance, constricts the discourse on integration to those parameters that do not threaten his organizations' survival. Similarly, the group as a whole rejects the insight of *Prudence*, when it appears to challenge the way members had proceeded with the deliberations. The narrative, when supported by an interpretation from the standpoint of organization learning, thus reveals that the characters' manifest behavior contrasts sharply with their espoused values.

We see the "disconnect" between espoused theories supporting collaboration and the everyday practice of human and social service delivery. Characters employ a vast number of organizational defensive mechanisms to protect themselves and their organizations, ranging from the subtle, as in the overall group, to the extreme, as with *Suspicion*. More important, such defensive routines become played out not only in the interaction of the collaborative group but also within the organizational cultures of the original organizations. The actions of *Suspicion*, which some could pass off as "bureaucratic politics," may in fact be symptoms of deeper divisions between participating organizations. When taken in the context of power-sharing and distribution of resources, as in a fully collaborative environment, such divisions may spell disaster.

Other barriers equally may inhibit the interorganizational group. Take for example *Suspicion*'s "identity crisis" (Senge, 1990), which causes the character to remain more concerned with preserving his original organizations than making substantial contributions to the group (Gray, 1989). Identity crises of this nature are especially prevalent when they involve organizations with more traditional systems of accountability. In these institutions, the objective remains on each agency, not with the part each agency plays in the broader collaborative effort, nor on the long-term objectives of the interagency group. Individuals faced with these situations concentrate more on answering to their respective *Jealousy*, or overseers from their original groups, instead of accomplishing public outcomes.

Several action steps are available to help collaborative groups overcome such barriers to interorganizational change. First, the intervention strategy should help the network members establish a shared vision (Schein, 1997). Through the leadership of *Charity*, we gain an image of how this may occur within a interorganizational setting, but the challenge becomes translating the allegorical image into practice. For our work in the Delaware network, we used both formal presentations and more informal work sessions with the Working Group to facilitate this process (for a discussion of alternatives, see Bryson & Crosby, 1992; Gray, 1989).

Regardless of the approach, the key is to initiate a constructive dialogue that enables network participants to identify the underlying barriers to the change. Here, the allegory with its abstraction from the constructed social reality of the group, can be used as a way of introducing the obstacles within the network, without raising individual or group defensive routines. The ultimate goal will be to ensure that the change occurs not merely at the level of individual or group behavior, but in a way that transforms the systems of beliefs interacting within the network.

Another approach based on the allegorical perspective, and one which builds upon the "plurivocal" nature of the change, involves having network members write their own allegory—that is, once they have read the interpretation of the researcher, to prepare an allegorical narrative that reflects their own interpretation of the change. This approach continues the dialogue within the network, but further reveals the diverse "readings" of the shared-change experience. As a result, network members come "face-to-face" with the characters in the story, or introduce new characters and attributes which may have been overlooked by the researcher. The desired outcome of this approach is to further engage the participants by encouraging their interpretation of the change, thereby supporting a deeper understanding of collaboration within the network.

However, the researcher must guard against concentrating so much on the allegory that he or she loses sight of the purpose, or the complex nature of the network relationship becomes glossed over by a superficial story. While allegory offers a unique approach to exploring the attributes of intercultural contact, a principal limitation lies in its simplicity. By making the narrative immediately accessible to the participants, the allegorical perspective opens the door to a deeper level of meaning. Yet, such simplicity requires that the researcher incorporate only a limited number of characters into the narrative. The researcher, as a result, must establish decision rules to help guide the selection process, and then ensure that the character development and selection supports the purpose of the inquiry.

As the collaborative group comes together around a shared vision, the next challenge becomes gaining commitment for this vision within each original organization (Gray, 1989). The original organizations, in turn, must extend to the network a requisite level of power, if the network's vision is ever to become a reality. From the allegory, characters such as *Jealousy* will need to be transformed, to become conduits that support a learning environment as opposed to overseers who ensure adherence on the part of network members to their original groups. Without this support, the collaborative group will be nothing more than a representative body of the original organizations. It is such a power-sharing relationship that demands a change in the very notion of leadership (Bryson & Crosby, 1992).

In this regard, Denhardt's (1993) concept of shared leadership offers a viable beginning point. Power is distributed laterally, as opposed to being concentrated in a centralized, hierarchical structure. The principle of shared leadership, in contrast to more traditional forms of organizing, stems from a recognition that "[i]nvolvement of individuals throughout the organization is now considered essential to organizational survival" (Denhardt, 1993, p. 130). Returning to the allegory, *Charity* to some degree reflects this type of shared leadership. She offers the collaborative group a point of origin by expressing her vision of service integration; and, she then empowers the group to work as one to accomplish its shared outcome. The other characters interact both as contributors to *Charity*'s vision, as well as in leadership roles based on their respective areas of influence.

On the other hand, the allegorical perspective also reveals that the group may have relied too heavily on *Charity*'s vision, without taking into account the viewpoints from other sources. This violates the underlying principle of shared leadership, which states that leadership should be shared throughout the group and members should have the opportunity to contribute based on their respective skills and interests (Denhardt, 1993). Given a learning environment to overcome organizational barriers to change, in addition to adequate base of power, *Knowledge* may have shaped the vision of service integration to reflect a stronger focus on learning, not just service delivery. Or, in a similar environment, perhaps *Suspicion* could have learned to move beyond his agency perspective and contribute positively to the group's outcomes.

The allegory reveals that the notion of leadership also must change with respect to the way collaborative groups adapt in their external environment, namely in their connection with original organizations (Bryson & Crosby, 1992; Gray, 1989). If the original organization is structured along traditional forms of management, the leadership system will need to be transformed to empower the collaborative group and support its commitment to a shared vision. As one might expect, such power-sharing goes against the prevailing view of management, which sees interagency relationships as a direct threat. Managers often respond to pressures to integrate by falling back on protective measures, as seen in the exchange between *Suspicion* and

Jealousy. With collaboration becoming the norm, such management will need to be replaced by integrative forms of shared leadership.

The interorganizational change process will require power to be handed directly to members of the collaborative group, whose vision will become integrated with that of the original organizations. Leadership will be shared horizontally within the group and vertically between original organizations and networks. As a result, participants in the narrative of collaboration will have the capacity to share unique perspectives, based on attributes from their original organizations. They will bring to the relationship systems of communication, of values and beliefs, which reflect their original groups. It is from this cultural background that the participants will enhance the network through constructive engagement. They will, in time, enjoy a foundation for a more meaningful, intercultural experience.

CONCLUSION

Our research into the Delaware network continues to be a work in progress. Currently, the discourse among participants centers around the issues of facilitating more effective communication between state agency and school personnel, creating a system of training for those providing services to children and families, and developing an approach to evaluating the collaborative effort following the implementation period. The leadership in the state certainly has provided its support for the collaboration. Indeed, state leaders recognize that the traditional lines that structured state agencies has limited the effectiveness of Delaware's service delivery system. The challenge becomes effecting the interorganizational change necessary to establish a more integrated system of services.

The allegorical perspective offers valuable insight into the obstacles faced by collaborative groups as they move along their respective Paths to Integration. In contrast to the prevailing viewpoints in organization theory, the use of allegory allows researchers to consider the mostly symbolic, unseen aspects of interorganizational change. It reveals the deeper significance when diverse groups, each with unique systems of communication and values, come together in a shared-change experience. Based on this deeper appreciation, researchers may incorporate the allegorical perspective into intervention strategies geared toward facilitating planned change in network settings. Furthermore, as our social world becomes characterized more by networks than singular organizational domains, such an approach represents a responsible, culturally sensitive course of organizational inquiry.

REFERENCES

Adams, G.B., & Ingersoll, V.H. (1990). Culture, technical rationality, and organizational culture. *American Review of Public Administration, 20*, 285–301.

Alter, C., & Hage, J. (1993). *Organizations working together.* Newbury Park, CA: Sage.

Alvesson, M. (1993). *Cultural perspectives on organizations.* Cambridge, UK: Cambridge University.
Argyris, C. (1993). *Knowledge for action.* San Francisco, CA: Jossey-Bass.
Argyris, C., & Schon, D.A. (1974). *Theory in practice.* San Francisco, CA: Jossey-Bass.
Argyris, C., & Schon, D.A. (1978). *Organizational learning.* Reading, MA: Addison-Wesley.
Argyris, C., & Schon, D.A. (1985). *Strategy, change and defensive routines.* Cambridge, MA: Ballinger.
Bluedorn, A.C. (1993). Pilgrim's progress: Trends and convergence in research on organizational size and environments. *Journal of Management, 19,* 163–191.
Bluedorn, A.C., Johnson, R.A., Cartwright, D.K., & Barringer, B.R. (1994). The interface of convergence of the strategic management and organizational environment domains. *Journal of Management, 20,* 201–262.
Boyce, M.E. (1995). Collective centring and collective sense-making in the stories and storytelling of one organization. *Organization Studies, 16,* 107–137.
Bryson, J.M., & Crosby, B.C. (1992). *Leadership for the common good.* San Francisco, CA: Jossey-Bass.
Bunyan, J. (1987). *The pilgrim's progress.* London: Penguin. (Original work published 1678)
Burrell, G., & Morgan, G. (1979). *Sociological paradigms and organisational analysis.* Portsmouth, NH: Heinemann.
Carper, T.R. (1993). *Executive order number six.* Dover: State of Delaware.
Center for Community Development and Family Policy. (1997). *Interim report.* Newark, DE: Author.
Chapple, E.D. (1941). Organization problems in industry. *Applied Anthropology, 1,* 2–9.
Clifford, J. (1986). On ethnographic allegory. In J. Clifford & G.E. Marcus (Eds.), *Writing culture* (pp. 98–121). Berkeley: University of California Press.
Clifford, J., & Marcus, G.E. (Eds.). (1986). *Writing culture.* Berkeley: University of California Press.
Czarniawska, B. (1997). *Narrating the organization.* Chicago: University of Chicago Press.
Czarniawska, B. (1998). *A narrative approach to organization studies.* Thousand Oaks, CA: Sage.
Czarniawska-Joerges, B., & Jacobsson, B. (1995). Political organizations and *commedia dell'arte. Organization Studies, 16,* 375–394.
Denhardt, R.B. (1993). *The pursuit of significance.* Belmont, CA: Wadsworth.
Denison, D., & Spreitzer, G. (1991). Organizational culture and organizational development: A competing values approach. In R.W. Woodman & W.A. Pasmore (Eds.), *Research in organizational change and development* (Vol. 5, pp. 1–). Greenwich, CT: JAI Press.
Eco, U. (1990). *The limits of interpretation.* Bloomington: Indiana University Press.
Elg, U., & Johansson, U. (1997). Decision making in inter-firm networks as a political process. *Organization Studies, 18,* 361–384.
Feldman, S.P. (1986). Management in context: An essay on the relevance of culture to the understanding of organizational change. *Journal of Management Studies, 23,* 587–608.
Frost, P.J., Moore, L.F., Louis, M.R., Lundberg, C.C., & Martin, J. (Eds.). (1985). *Organizational culture.* Beverly Hills, CA: Sage.
Geertz, C. (1973). *The interpretation of cultures.* New York: Basic Books.
Gray, B. (1989). *Collaborating.* San Francisco, CA: Jossey-Bass.
Gray, B. (1990). Building interorganizational alliances: Planned change in a global environment. In R.W. Woodman & W.A. Pasmore (Eds.), *Research in organizational change and development* (Vol. 4, pp. 101–). Greenwich, CT: JAI Press.
Grubbs, J.W. (1998). *Interorganizational collaboration: An allegorical perspective.* Unpublished doctoral dissertation, University of Delaware, Newark, DE.
Hawthorne, N. (1967). Young goodman brown. In F.C. Crews (Ed.), *Great short works of Hawthorne* (pp. 271–284). New York: Harper and Row. (Original work published 1835)
Hofstede, G. (1980). *Culture's consequences.* Beverly Hills, CA: Sage.
Hofstede, G. (1997). *Cultures and organizations.* New York: McGraw-Hill.
Ingersoll, V.H., & Adams, G.B. (1992). *The tacit organization.* Greenwich, CT: JAI Press.
Katzenbach, J.R., & Smith, D.K. (1993). *The wisdom of teams.* Boston, MA: Harvard Business School.

Lawson, R.B., & Ventriss, C.L. (1992). Organizational change: The role of organizational culture and organizational learning. *Psychological Record, 42*, 205–219.
Leeming, D.A., & Drowne, K.M. (1996). *Encyclopedia of allegorical literature*. Santa Barbara, CA: ABC-CLIO.
Lightfoot, K.G. (1995). Culture contact studies: Redefining the relationship between prehistoric and historical archaeology. *American Antiquity, 60*, 199–218.
Lightfoot, K.G., Martinez, A., & Schiff, A.M. (1998). Daily practice and material culture in pluralistic social settings: An archaeological study of culture change and persistence from Fort Ross, California. *American Antiquity, 63*, 199–222.
Lincoln, Y.S., & Guba, E.G. (1985). *Naturalistic inquiry*. Beverly Hills, CA: Sage.
Lochtenberg Commission. (1993). *Report on early education and social services*. Wilmington, DE: Business/Public Education Council.
Louis, M.R. (1980). Surprise and sense-making: What newcomers experience in entering unfamiliar organizational settings. *Administrative Science Quarterly, 25*, 226–151.
Madsen, D.L. (1994). *Rereading allegory*. New York, NY: St. Martins.
Mankin, D., Cohen, S.G., & Bikson, T.K. (1996). *Teams and technology*. Boston, MA: Harvard Business School.
Mignerey, J.T., Rubin, R.B., & Gorden, W.I. (1995). Organizational entry: An investigation of newcomer communication behavior uncertainty. *Communication Research, 22*, 54–85.
Milward, H.B. (Ed.). (1996). Symposium on the hollow state: Capacity, control, and performance in interorganizational settings [Special issue]. *Journal of Public Administration Research and Theory, 6*(2).
Milward, H.B., & Provan, K.G. (1998). Principles for controlling agents: The political economy of network structure. *Journal of Public Administration Research and Theory, 8*, 203–221.
Mohrman, S.A., Cohen, S.G., & Mohrman, A.M., Jr. (1995). *Designing team-based organizations*. San Francisco, CA: Jossey-Bass.
Morgan, G. (1997). *Images of organization* (2nd ed.). Thousand Oaks, CA: Sage.
O'Connor, E.S. (1995). Paradoxes of participation: Textual analysis and organizational change. *Organization Studies, 16*, 769–803.
O'Toole, L.J., Jr. (1997a). The implications for democracy in a networked bureaucratic world. *Journal of Public Administration Research and Theory, 7*, 443–459.
O'Toole, L.J., Jr. (1997b). Treating networks seriously: Practical and research-based agendas in public administration. *Public Administration Review, 57*, 45–52.
Oliver, C. (1990). Determinants of interorganizational relationships: Integration and future directions. *Academy of Management Review, 15*, 241–265.
Phillips, N. (1995). Telling organizational tales: On the role of narrative fiction in the study of organizations. *Organization Studies, 16*, 625–649.
Provan, K.G., & Milward, H.B. (1995). A preliminary theory of interorganizational network effectiveness: A comparative study of four community mental health systems. *Administrative Science Quarterly, 40*, 1–32.
Riessman, C.K. (1993). *Narrative analysis*. Newbury Park, CA: Sage.
Robins, J.A. (1987). Organizational economics: Notes on the use of transaction cost theory in the study of organizations. *Administrative Science Quarterly, 32*, 68–86.
Roth, G. (1996). *Learning histories*. Working paper, Center for Organizational Learning, Massachusetts Institute of Technology, Cambridge, MA.
Roth, G., & Kleiner, A. (1995). *Learning about organizational learning*. Working paper, Center for Organizational Learning, Massachusetts Institute of Technology, Cambridge, MA.
Schein, E.H. (1997). *Organizational culture and leadership* (2nd ed.). San Francisco, CA: Jossey-Bass.
Senge, P.M. (1990). *The fifth discipline*. New York: Doubleday.
Smircich, L. (1983). Concepts of culture and organizational analysis. *Administrative Science Quarterly, 28*, 339–358.

Thorelli, H.B. (1986). Networks: Between markets and hierarchies. *Strategic Management Journal, 7,* 37–51.
Van Maanen, J. (Ed.). (1983). The fact of fiction in organizational ethnography. In *Qualitative methodology* (pp. 37–55). Beverly Hills, CA: Sage.
Van Maanen, J. (1988). *Tales of the field.* Chicago: University of Chicago Press.
Van Maanen, J., & Kunda, G. (1989). "Real feelings": Emotional expression and organizational culture. In B.M. Staw & L.L. Cummings (Eds.), *Research in organizational behavior* (Vol. 11, pp. 43–104). Greenwich, CT: JAI Press.
Weick, K.E. (1995). *Sensemaking in organizations.* Thousand Oaks, CA: Sage.
Whitman, J. (1987). *Allegory.* Cambridge, MA: Harvard University Press.
Wood, D.J., & Gray, B. (1991). Toward a comprehensive theory of collaboration. *Journal of Applied Behavioral Science, 27,* 139–162.

MAKING CHANGE PERMANENT
A MODEL FOR INSTITUTIONALIZING CHANGE INTERVENTIONS

Achilles A. Armenakis, Stanley G. Harris, and Hubert S. Feild

ABSTRACT

Increasing global competition has accelerated the rate of organizational changes, such as reengineering, restructuring, and downsizing. As a result, organizational leaders find themselves faced with growing cynicism among employees that the current wave of changes is nothing more than the *program of the month* that will pass as those that preceded it. We address the issue of how to make changes permanent by providing a model developed from theory and research on organizational change and from successful practices implemented in numerous organizations worldwide. The model can serve at least three purposes. First, the model can assist change agents in planning for and assessing progress toward institutionalizing organizational change. Second, the model can help focus efforts of organizational scholars to study the change process. Third, the model offers the basis for hypothesis testing regarding the success or failure of change efforts.

Increasing global competition and changing political ideologies worldwide are some of the causes for the accelerated rate of organizational changes. Managers are being required to change virtually every aspect of the way organizations function. Employees' attitudes about work, their jobs, and their psychological contracts with their employer are being changed. As the pace of change increases, employees are continually faced with evidence that some changes are simply passing fads or quick-fix attempts, implemented with little commitment for their long-term success. It is no wonder that announced changes are met with skepticism and a *program-of-the-month* reception. In such circumstances, it makes perfect sense to wait to see if the organization is serious about the change before going through the machinations it requires. Clearly, such a reception to planned changes by those ultimately responsible for implementing them is detrimental to the change effort's timeliness and success. Thus, the change has little or no chance of being *institutionalized*; that is, becoming accepted, permanent, stable and/or normative.

Change efforts fail to become institutionalized for varied reasons. However, much of the problem revolves around the failure to shepherd the change effort through the entire process of change from diagnosis to institutionalization. Moreover, some changes are implemented simply based on the desire to be in *fashion* or to create the impression that the organization is being proactive. The goal is image rather than substantive change and such efforts generally lack the organization's commitment and follow-through to succeed.

But what about change efforts implemented with sincere intentions to improve the organization? We contend that two primary reasons those who are responsible for planning and implementing organizational change fail to follow through on such change efforts are (a) their impatience and assumption that successful change introduction and implementation guarantees institutionalization, and (b) their simple neglect of seeing change through to institutionalization. We suggest that the success rate for planned organizational changes could be improved by giving change agents and students of organizational change a better appreciation of the institutionalizing phase of the change process—by describing the numerous pieces that must be understood, acted upon, and integrated before an organizational change can be successful.

Therefore, the purpose of this chapter is to shed some much-needed light on the institutionalization of planned organizational change. After exploring the meaning of institutionalization, we propose and develop an integrative model of the process of institutionalization. In developing our model, we integrate scientifically rigorous research findings with numerous practical examples taken from our experiences as consultants and from the popular press. These practical examples are excerpted from organizational change experiences of noteworthy companies like Whirlpool, GM, Goldman-Sachs, Control Data, Ford, Merck, Xerox, Allied-Signal, GE, and Chrysler. Thus, our intent is to propose a model a change agent can use to answer the question "What must we do to facilitate the adoption and institutionalization of change?" The model can also serve to guide diagnosis and modification of change efforts. For change scholars, the model provides a framework to stimulate additional research and theory on the dynamics of organizational change and management of the institutionalization process.

WHAT IS INSTITUTIONALIZATION?

The issues of permanence and stability are central to Lewin's (1947) unfreezing–moving–freezing metaphor. From this metaphorical perspective, an institutionalized change is one that is frozen and the process of creating that institutionalization is freezing. Beer (1976) describes freezing as follows: "the stabilization of change at a new equilibrium state through supporting changes in reference group norms, culture, or organizational policy and structure" (p. 939). As organizations are constantly undergoing change and experiencing flux, talking about literal *permanence* to describe institutionalization is unrealistic. However, experience tells us that some changes have a longer life than others. Therefore, it seems reasonable to talk about degrees of institutionalization as reflected in the duration of a state. But how should such degrees be conceptualized?

The Role of Commitment

Institutionalization is reflected in the presence of resistance against deviating from the current state. Resistance to change is the same as commitment to the current state. In his pioneering research on the sources of individuals' normative conformity, Kelman (1958) operationalized commitment into the three dimensions of compliance, identification, and internalization.

Kelman defined *compliance commitment* as that which occurs because an individual expects to receive specific rewards or avoids punishment by conforming. Resistance to change due to organizational structure, resource limitations, fixed investments, interorganizational agreements, threats to power, and economic and interpersonal vested interests all reflect resistance due to external pressures, fears, or constraints. The appropriateness of the change, that is, its *rightness* or *wrongness* for the organization, is not a concern. These sources of resistance represent compliance-based commitment to the system.

Identification commitment occurs because an individual wants to establish or maintain a satisfying self-defining relationship to another person or group. The individual adopts the induced behavior because it is associated with the desired relationship, but the content of the responses may be irrelevant. Culture, group cohesiveness, and other social system vested interest sources of resistance all reflect identification-based commitments.

Internalization commitment occurs because the content of the induced behavior, that is, the ideas and actions of which it is composed, is intrinsically appealing and seen as proper. Thus, the behavior is adopted because it is congruent with the individual's values. Paradigms, fear of the unknown, habit, and organizational culture sources of resistance reflect internalized commitment, that is, people and the system believe the present state to be appropriate. In general, internalized commitment because of its unconscious, preconscious, or automatic nature is viewed as being a more powerful and persistent determinant of behavior.

Since Kelman's research, numerous researchers have made extensive contributions to our understanding and use of organizational commitment. Research indicates foci of commitment can be particular entities (e.g., an organization, people, values) to whom a person is attached (cf. Becker, 1992; Fishbein & Ajzen, 1975; George, 1990; Mathieu & Kohler, 1990; Reichers, 1985). Moreover, Becker's research expanded Kelman's typology to include organizational identification, organizational internalization, supervisor-related identity, supervisor-related internalization, work group identity, work group internalization, and overall compliance. He found supervisor-related and work group-related commitments were distinguishable from the more general organizational commitment. This conceptualization points out the importance of realizing that organizational members can identify with and internalize the values of, not only the organization, but also their supervisors and their work groups. As described below in more detail, we incorporate the importance of change agent (including supervisor) and organizational member (including work group) attributes in the proposed model of the change process. Thus, Becker's (1992) expanded conceptualization of commitment fits neatly into the model for institutionalizing change.

In a recent study of organization development (OD) practitioners, Church and Burke (1995) found that practitioners thought OD should begin to focus more on system-wide organizational issues and less on the traditional OD foci of individual and group processes and interpersonal relationships. In a sense, our focus on institutionalization represents such a shift. However, we prefer to think of our approach to institutionalization as bridging and interweaving the distinction between the system and individual OD concerns. Specifically, we suggest that the process of institutionalization at the system level is the process of building commitment to the changed state (or building resistance to changing from it) at the individual level. To create compliance-based commitment, a change agent must tie the change to organizational structure, interorganizational agreements, sunk costs, and reward systems. In order to create identification-based commitment, a change agent must tie changes to association with their supervisor and membership in their work group.Furthermore, to create internalization-based commitment reflected in individuals' paradigms, a change agent must tie changes to current employee beliefs and values as they relate to the organizational culture.

In most circumstances, a change agent will probably want to create institutionalization based on compliance, identification, and internalization commitments. To the extent a change agent wants a recent change to be more easily changed (unfrozen) in the future, institutionalization based on compliance may be sufficient. In cases where it is clear that many, if not most, employees will resist internalization, it may also be appropriate to focus mainly on creating compliance commitment, particularly in the short term. However, over time, behavior resulting from compliance-based commitment can become normative, that is, *the way we've always done things*. The commitment perspective on institutionalization encourages the change agent to consider and plan for the degree of institutionalization desired.

Factors Affecting Institutionalization

While little work exists on the process of facilitating institutionalization, efforts have been made to determine the types of changes that are most easily institutionalized and the types of organizational factors that are most conducive to institutionalizing change. In general, interventions that are received positively by organizational members are more easily institutionalized than those received negatively. For example, Tornatzky and Klein (1982) found that innovations that (a) produced a relative advantage, (b) were more compatible with an organization, (c) were relatively less complex, (d) were lower cost, and (e) could be implemented on a trial basis were more likely to be adopted and institutionalized.

Regarding organizational factors, Damanpour (1991) analyzed the findings of 23 empirical studies that investigated the role of 13 content, contextual, and process factors on institutionalization. Among the 10 factors found to be associated with innovation were (a) functional differentiation (i.e., content), (b) technical knowledge resources (i.e., context), and (c) communication (i.e., process). The logic to this research focus is that successful change may depend more on the fit between content, context, and process considerations than the nature of the change.

Although research on the attributes and organizational factors provides interesting explanations for the success and failure of change efforts, practical application of the findings is limited. What options are open when relative advantage is not an obvious attribute of the intervention? What options are open to those organizations that do not fit the profiles? Finally, even if an intervention offers competitive advantage and an organization fits a successful profile, the findings do not provide insights into change processes and dynamics. Are there other forces that may contribute to institutionalization? To address such questions, we propose a model to aid in understanding and in ultimately facilitating the institutionalization of change.

THE MODEL

Our process model builds off Lewin's (1947) stages of change and social learning theory (Bandura, 1986). The model, depicted in Figure 1, comprises the following constructs: The three generic stages of change, the change message and its components, commitment (as explained above), the attributes of the change agent and the organizational membership, reinforcing strategies, institutionalization, and assessment.

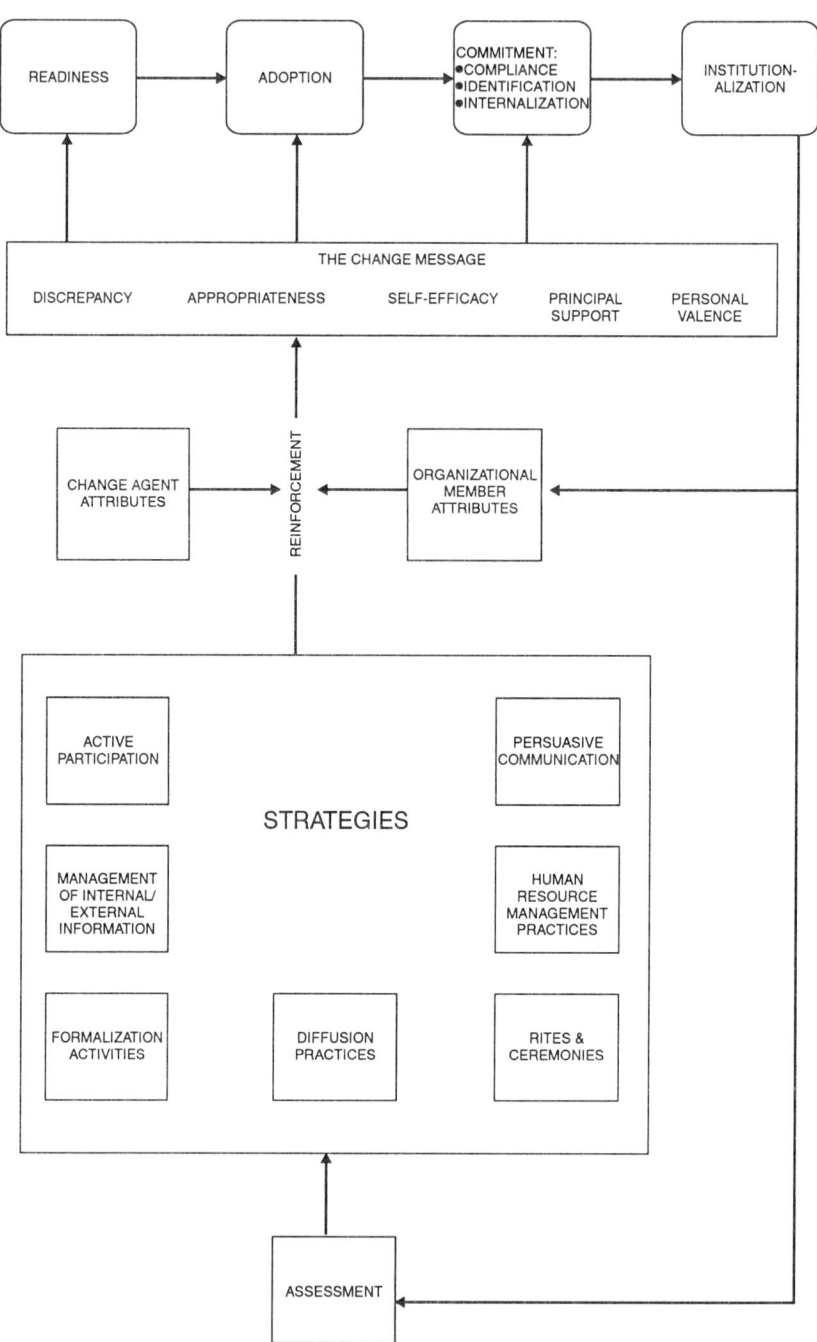

Figure 1. Institutionalizing Change

The Stages of Change

Most models of the change process are built around Lewin's (1947) stages of unfreezing, moving, and freezing. Paralleling Lewin, Bridges (1991), in his work on transitions, frames the process in terms of endings, transitions, and new beginnings. In Figure 1, we have used labels consistent with recent change literature to describe the three stages as readiness, adoption, and institutionalization. *Readiness* is the cognitive state comprising beliefs, attitudes, and intentions toward a change effort. When readiness for change exists (cf. Armenakis, Harris, & Mossholder, 1993), the organization is primed to embrace change and resistance is reduced. Organizational members will embrace the change and the adoption stage begins. If organizational members are not ready, the change may be rejected, and organizational members may initiate negative reactions, such as, sabotage, absenteeism, and output restriction. *Adoption* is the act of behaving in the new way, on a trial basis. That is, the change can still be rejected. As discussed earlier, *institutionalization* is reflected in the degree of commitment to a new way, that is, the post-change state of the system.

The Change Message

At the core of our model is the *message* required to build commitment to a change effort. All efforts to introduce and institutionalize change can be thought of as sending a message to organizational members. The introduction of change creates a great deal of uncertainty and confusion. Essentially, the purpose of the change message is to create certain core sentiments in members of the organization by answering a set of five key questions they have about the change. The first question is "Is change really necessary?" The question is answered by the *discrepancy* component of the message. Discrepancy refers to information regarding the need for change as reflected in the discrepancy between the current and ideal state in the organization. The second key question raised by organizational members is "Is the specific change being introduced an appropriate reaction to the discrepancy?" The *appropriateness* component of the change message provides the response to this question. A third question generated during a change is "Can I/we successfully implement the change?" The *efficacy* component of the change message answers this question by providing information and building confidence regarding the individual and group's ability to successfully implement the change. A fourth question has to do with organizational support for the change and reflects skepticism resulting from previous *program-of-the-month*, half-hearted change interventions. The purpose of the *principal support* message component is to provide information and convince organizational members that the formal and informal leaders are committed to successful implementation and institutionalization of the change. Finally, organizational members will want to know "What is in it (the change) for me?" By clarifying the intrinsic and extrinsic benefits of the change, the *personal valence* component of the change message addresses this question. Imbedded in the concern for personal valence is the intrinsic desire for change fairness and justice. As Cobb, Wooten, and Folger (1995) note, employee perceptions of justice during periods of organizational change encompass assessments of the fair distribution of positive and negative outcomes, of the fairness of change procedures, and of the appropriateness of change agents' treatment of them. Perceptions of justice are particularly important for encouraging the type of extra-role behavior generally required of change efforts (Cobb et al., 1995).

The degree to which organizational members receive adequate answers to their core questions is a prime determinant of the nature of their ultimate commitment to the change. While the emphasis in this chapter is on the institutionalization of change, it is important to note that the change message also has implications for the creation of readiness for change (Armenakis et al., 1993) and its adoption. In fact, to the degree that the core questions are answered adequately in early stages of change, sentiments central to institutionalization may already be established.

The role of these five messages in generating positive change momentum is exemplified in an investigation conducted by Nutt (1986). Nutt studied the change implementation tactics of hospital executives in 91 case studies. The most successful tactics (labeled intervention and participation) described the change agent as demonstrating early support for the change (principal support), communicating the need for change (discrepancy and appropriateness), and involving organizational members throughout the change process (efficacy and personal valence). The least successful tactic (labeled edict) described the change agent as not discussing change plans with organizational members, not justifying the need for change, and using control and personal power to mandate adoption.

Before examining the strategies that can be employed to send the five key message components, it is important that we briefly examine attributes of the change agent that affect the persuasiveness of the change message and characteristics of the organizational members who are the targets of the change that affect their receptivity to the message.

Change Agent Attributes

Anyone involved in initiating, implementing, and supporting a change can be considered a change agent. Initially, this change agent may be the head of the organization (i.e., *global* change agent). People in all leadership positions will ultimately be expected to support and help drive the change throughout the organization. For large-scale change programs involving numerous organizational levels and departments, executives and other managers are extensions of the change agent, thus, serving in a role of *local* change agents. Finally, nonmanagerial organizational members can serve as change agents (i.e., *horizontal* change agents). These horizontal change agents can be those persons who interact socially (on the job, as well as, off the job) with colleagues as opinion leaders and can reinforce the favorable interpretation of the message. Naturally, organizational members' perceptions of the change agents' attributes will influence the persuasiveness of any change message and ultimately commitment, hence, institutionalization of organizational change.

The single most important attribute that a change agent should possess is credibility. Kouzes and Posner (1993) identified the primary components of credibility to be honesty, competence, vision, and inspiration. The importance of credibility in changing cognitions and behaviors has been researched for several decades and continues to be of interest in current research investigations. According to research cited by Slater and Rouner (1992), changes in cognitions of organizational members have been linked empirically to the credibility of the change agent. Nystrom (1990) found that the quality of the relationship between a change agent and the organizational members was a significant factor in determining commitment. Similarly, Eisenberger, Fasolo, and Davis-LaMastro (1990) concluded that organizational member perceptions of being valued and cared about were instrumental in influencing innovativeness. In the Buller and McEvoy (1989) investigation, trust in the change agent was significant in institutionalizing a new performance appraisal system. Communicating a shared vision through speeches, memos, and newsletters along with executive visibility were significant in developing organizational commitment (Niehoff, Enz, & Grover, 1990). Larkin and Larkin (1996) refer to numerous surveys of employees from large organizations that revealed the preferred information source was frontline supervisors, implying, frontline employees had little confidence in upper-level executives. Likewise, Cobb et al. (1995) summarize the results of several studies that clearly demonstrate the importance of the perception among employees that the change agent can be trusted and is viewed as being, or attempting to be, fair and just in the way in which the change is being managed.

The obvious conclusion from these findings is that a required, although not sufficient, condition for institutionalization is that organizational members should perceive the change agent and his/her representatives as credible. Change agents develop credibility through their behaviors.

Attributes of the Organizational Membership

The organizational membership is the collection of individuals who must modify their cognitions and behavior to achieve the objectives of the change effort. Ultimately, it is the commitment of these individuals that determines the institutionalization of a change. Confronted with the same information and intervention, individuals can still be expected to react differently. *Listening* to organizational members, even those who may not be sold on the change, can result in a better mutual understanding of the concerns (Weisbord, 1988). Then, this understanding can be used to anticipate differential reactions of organizational members and can be helpful in orchestrating strategies intended to build their commitment for change initiatives. Research on two factors, namely, individual differences and organizational differentiation, provides some guidance in the process of institutionalization.

In terms of individual differences, two scales are valuable for understanding change dynamics. Kirton's (1984) Adaption-Innovation Inventory, a paper-and-pencil instrument, has been used to categorize individuals as either adaptors or innovators. Innovators are more likely to embrace fundamental change while adaptors are less likely to embrace fundamental change (Kirton, 1984). A second scale, the Self-Monitoring Scale (Snyder, 1974) was used by Burkhardt (1991) who found that high self-monitors' (i.e., those who were more attentive to social comparison information) attitudes toward change were more influenced by opinion leaders and other individuals in their work groups. In contrast, low self-monitors were more influenced by those individuals who were performing jobs considered to be hierarchically similar.

The second factor to be considered in change institutionalization is the organizational differentiation that exists throughout the organization. Organizational change that originates outside of a group may be perceived as a threat. Defensive mechanisms are mobilized to ward off the intervention that will undoubtedly upset the norms. Plans to address the needs of these diverse groups (e.g., union members, professional classifications) and enlist the support of some as change agents can enhance the success of institutionalization.

Organizational differentiation has been researched in the change literature as subcultures (cf. Van Maanen & Barley, 1984) and cultural ecology (Baba, 1995). Baba's research explained how 15 work groups reacted to an organizational transformation aimed at commonizing the tools and methods used in the product development process of a large manufacturer. Baba's data revealed that the differential reaction of the work groups was related to the cultural ecology (i.e., the type of work they do, relationships with other internal and external organizational groups, and availability of resources) of each group. Thus, the cultural ecology of the work group determines how individuals within a work group will react to an organizational change.

Researchers investigating diffusion of agricultural innovations (Ryan & Gross, 1943) found that individuals who were highly respected influenced the willingness of others to institutionalize change. Therefore, it is critical that opinion leaders within the cohesive groups support the change and, consequently, influence others to embrace it, thus building momentum. Organizational commitment and commitment to change initiatives can be generated throughout the organization by peers serving as role models and providing social support and becoming horizontal change agents (e.g., giving encouragement and positive feedback).

Implications of the findings regarding individual differences and organizational differentiation are that a change agent needs to be familiar with organizational members and should attempt to enlist support for change. For example, organizational differentiation, including unions, engineers, and technical specialists, should be considered when identifying opinion leaders. Although it may be impractical to assess organizational members by administering paper-and-pencil tests to identify innovators/adaptors or high/low self-monitors, these findings are important in understanding change dynamics.

Institutionalizing Strategies

Naturally, during organization change programs, change agents initiate actions that are intended to redefine norms and establish new ways of thinking and behaving. Their actions (or lack of action) can have both real and symbolic consequences with regard to sending and reinforcing the five core message components. In short, everything a change agent says (or does not say) and does (or does not do) can reinforce (or contradict) the change message. Executing the seven influence strategies shown in Figure 1 is intended to transmit and reinforce the five core message components.

Active Participation

Participation strategies can enhance the relationship between a change agent and organizational members, build the credibility of the change agent, and establish ownership in and reinforce commitment for organizational change (cf. Nutt, 1986). Furthermore, active participation's effectiveness as a change strategy is based on the concept of self-discovery, that is, the learning that results through personal experiences. Three active participation tactics are particularly useful in building commitment to change: (a) enactive mastery, (b) vicarious learning, and (c) participative decision making.

Enactive mastery. Enactive mastery is accomplished by gradually building competence and skills through successively accumulating activity blocks; that is, chunks of the total responsibilities. In enactive mastery, more complicated behaviors are not introduced until the previous behaviors have been mastered. In this way, individuals experience *small wins* (Weick, 1984) and are not confronted with the magnitude of the overall behavioral change requirement all at one time. In some situations, enactive mastery may mean initiating a new practice, such as quality meetings to discuss very simple issues on an infrequent basis. Over time, however, the frequency and intensiveness of the meetings can be increased as people become more accustomed to them. Through enactive mastery, organizational members can experience and realize any advantages of the new procedures and can gradually develop a sense of efficacy with regard to those procedures. Through prolonged practice and exposure, it is also possible that enactive mastery can be a source of demonstrating the appropriateness of a change.

Vicarious learning. Vicarious experiences are those during which organizational members observe others, most preferably respected colleagues, in the performance of the new behaviors. Benchmarking, the process of identifying and emulating the excellent practices of other companies, is another type of vicarious learning experience and results in enhanced efficacy (e.g., "if they can do it, so can I") and appropriateness, because individuals will be able to observe any advantages of the new method. Furthermore, principal support for the intervention is observed as respected colleagues initiate and continue the adoption of the organizational change. Desire to adopt the new behavior results from wanting to be identified with the adopters. Consequently, the vicarious observer is encouraged to continue adoption and ultimately institutionalizes the new behavior.

Participative decision making. Participative decision making is an effective method to change cognitions and behaviors (cf. Beer, Eisenstat, & Spector, 1990; Dalton, 1970; Neumann, 1989; Pasmore & Fagans, 1992). To the extent that participative decision making enhances members' psychological ownership of change in the organization, individuals' dispositions toward promoting such change will likely be enhanced (cf. Dirks, Cummings, & Pierce, 1996). Individuals can be involved in decision making for all phases of the change process. Naturally, there are limits to participation, and there are situations in which participation is simply not possible. Some aspects of a change program may involve participation, such as strategic planning, but other aspects, such as downsizing, may involve only selected decision makers. Through their involvement in making change-related decisions, organizational members self-discover a clearer understanding of the problems and issues confronting their organization (discrepancy) and the improvements that can be realized from implementing change (appropriateness). Likewise, having a say in the changes that are ultimately introduced is likely to make individuals feel efficacy for those changes. As shown in the goal-setting literature, people have more confidence in their ability to meet a goal that they participated in setting (cf. Locke & Latham, 1990). Through participation, individuals observe the support of other organizational members. Participation also increases the likelihood that changes consistent with intrinsic/extrinsic motives are selected and that the change procedure is viewed as being more fair and just because participants are given a *voice* (Cobb et al., 1995; personal valence).

Persuasive Communication

Persuasive communication strategies are useful in efficiently communicating information relevant to all five core message components. Oral transmissions can be formal, such as speeches, or informal, such as chance face-to-face encounters. In addition, oral transmissions can be transmitted live as well as through audio- and videotape technology. Written messages can take the form of memos, e-mail messages, annual reports, letters/memos, and newsletters.

Cobb et al. (1995) make it clear that persuasive communication is a particularly important vehicle by which change agents provide *social accounts* which shape perceptions of the change and its fairness and justice and can be used to build support for the change. Drawing off the work of Bies (1987), Cobb et al. (1995) describe four such social accounts. In causal accounts, change agents articulate the reasons and rationale behind the change and engage paradigms that provide a more *acceptable* interpretation of the change. In ideological accounts, change agents justify the change within the context of superordinate goals or visions and clarify the values which will guide change decision making and set the standard for definitions of what is just. Referential accounts clarify benchmarks for the change and clarify how things will be worse if change is not made and how things will be better once the change is made. Finally, penitential accounts are used to honestly address the difficulties of the change, make apologies, and demonstrate empathy with the employees.

In addition to the verbal content of the messages sent, the nature by which those messages is sent can have symbolic consequences as well. For example, the time, energy, and resources utilized in giving verbal communication provide symbolic evidence of the change agent's support for the change effort. In addition, change agent attempts to communicate powerful metaphors to summarize the change process often helps clarify events and processes that might otherwise be confusing for organizational members.

Because of people's inherent stance of "I'll believe it when I see it," persuasive communication cannot be expected to be as powerful as a strategy based in self-discovery, like enactive mastery or vicarious learning. However, persuasive communication is a basic and required strategy of all change agents because the act relates to one or all of the attributes of credibility. Furthermore, the symbolism emanating from the persuasive communication incidents can bolster the message transmitted by other strategies. To illustrate the integrated use of various persuasive communication strategies, examples from Whirlpool and General Motors are briefly outlined below.

Whirlpool example. During 1988, Whirlpool Corporation leaders changed the corporate strategy and reorganized from a functional to a strategic business unit (SBU) organizational structure, followed by a 10 percent reduction in managers. A significant attempt at persuasive communication occurred in June 1988, when the Whirlpool chairman and CEO, Dave Whitwam, addressed the 250 officers, directors, and managers (and spouses) in an 80-minute, after-dinner speech. The purpose of the speech was to demonstrate symbolically his support for the changes, as well as to provide a degree of confirmation for them. Regarding discrepancy, the CEO reiterated that the changes were needed because of increased global competition and the accelerated rate of industry consolidation. The old functional structure inhibited the company's ability to capitalize on opportunities that existed in the market. As for appropriateness, Whitwam pronounced that with the changes, the company could now more effectively compete. To communicate his support for the changes that were implemented, he assured them "there is no turning back. Whirlpool of the past will be no more" (*Confirmation Speech*, 1988).

This persuasive communication incident is significant because it required time and effort in planning, reinforced the change, and was symbolic. For example, making arrangements for all directors, officers, managers, and their spouses for a formal dinner and to prepare a videotape of Whirlpool's *Confirmation Speech* for broader dissemination required more than a casual effort. Furthermore, the event was intended to do more than summarize the process. By incorporating the spouses at the dinner, the intent was apparently to acknowledge the impact of the changes on the whole family and to enlist familial support.

A symbolic innovative attempt, referred to as dramaturgy (Ritti & Silver, 1986), to institutionalize the changes at Whirlpool was the preparation (in 1990) of a videotape of three top executives (the CEO, the CFO, and the Executive Vice President) being interviewed by a French newscaster (*Business in Profile*, 1990). The setting for the staged interview was five years into the future, and the three executives were answering questions related to the reasons for Whirlpool's success during the last decade of the twentieth century. In introducing the executives, the newscaster referred to Whirlpool as the *renaissance company of the 1990s*. In the staged discussion, the executives attributed the company's outstanding performance (in the future) to the organizational changes which were currently being implemented in real life. Thus, this dramaturgy was reinforcing (a) their support for the continuing adoption and institutionalization of the new/improved ways of operating, (b) the need for and appropriateness of change (i.e., the new way proved better than the old ways), (c) efficacy (i.e., Whirlpool was capable of responding more quickly to external environmental changes) and (d) valence (i.e., the evidence indicates that the organizational changes were indeed something to be proud of and had been rewarded in the marketplace).

It is noteworthy to point out that this Whirlpool example is referred to as *video feed-forward* (Dowrick, 1991). Dowrick and associates describe some interesting research findings (with various groups of people, including alcoholics, children with reading disabilities, and athletes) that demonstrate its effectiveness in mastering new skills and enhancing one's ability to achieve higher levels of performance. The idea is to permit subjects to view themselves "not as they performed in the past (including errors), but as they will perform (correctly) in the future" (pp. 240–241). Using video feed-forward with a nationally ranked power-lifting athlete, Franks and Maile (1991) recorded a 26 percent improvement in performance in a 25-week period. A 10 percent increase in a one-year period is considered *remarkable*, especially when one considers the elite performance level of such an athlete. Thus, video feed-forward shows promise for implementing organizational change.

GM example. In 1992, General Motors Corporation announced numerous executive successions in the wake of staggering operating losses. Subsequently, the new leaders initiated a series of organizational changes intended to improve the corporation's performance. In its 1992 annual report, the chairman, John Smale, referring to its austere appearance, emphasized "that this annual report is a dramatic departure from past practices. I believe it demonstrates the Corporation's commitment to meaningful change" (*General Motors Annual Report*, 1992, p. 3). In addition, the CEO, Jack Smith communicated the appropriateness of the changes by stating "I firmly believe your Corporation is, right now, on the way to achieving the strength it once had. We know what we have to do, and we're going to do it. Watch and see" (*General Motors Annual Report*, 1992, p. 3). Therefore, this report reinforced discrepancy (i.e., company performance was unacceptable) and efficacy (i.e., we are capable of returning to the strong company we once were). This annual report was symbolic because of its austere appearance. The statements made by both John Smale and Jack Smith communicated their support. Furthermore, their statements and the strategy were intended to demonstrate their vision and to inspire support from GM personnel.

Management of Internal/External Information

Information from internal and external sources is a powerful lever for reinforcing the message needed to institutionalize change. Examples of internal data are employee attitudes, productivity, costs, and other performance indicators (e.g., scrap) usually considered to be minimally influenced by external environmental fluctuations. Examples of external data include direct contact with customers via telephone surveys conducted by employees or articles in the business press. Clearly, much of the content of the information will be incorporated into messages conveyed using the various forms of persuasive communication. To the extent that individuals affected by the change can actively participate in gathering the information, believability is enhanced. Some information allows for the tracking of change success, so important to celebrating small wins (Weick, 1984) and keeping individuals focused on the desired changes.

Collecting survey data has a long history as a tool for identifying the need for change within the action-research community (cf. Nadler, 1977). After a change is initiated, organizational members can be reinforced to continue adoption by tracking the progress of change with data collected from internal and external stakeholders.

Leveraging and managing information from external media sources, such as the popular business press, can be disseminated to further enhance adoption and institutionalization (cf. Macdonald, 1995). For example, compiling and sharing press releases regarding the activities of competitors can be used to justify change efforts. Other tactics, like contracting with consultants and high visibility speakers can be very effective in reinforcing the ideas among organizational members that the firm is employing leading edge technology and is considered among the elite in their business area.

Human Resource Management Practices

Human resource management (HRM) practices can be used to complement other strategies in the institutionalization process. HRM practices include selection, performance appraisal, compensation, and training and development. While each practice can contribute to reinforcing all five core message components, human resource practices are generally a primary source of extrinsic reinforcement for desired behavior and symbolic evidence of organizational support for the change. It is important to realize that much of the reinforcement of change is accomplished through HRM practices (cf. Tichy, 1982). As evidence of the power of HRM practices, Yeung, Brockbank, and Ulrich (1991) report the results of a study of 1,200 businesses showing that HRM practices support change initiatives. However, care must be taken to consider employee attitudes related to the perceived fairness of change when using these HRM practices to reinforce change implementation. Kilbourne, O'Leary-Kelly, and Williams (1996) studied employee reactions to a shift from seniority-based to performance-based layoffs in a large electronics firm. They theorized that in addition to procedural and distributive justice concerns, transformational justice issues impacted employee attitudes toward the perceived fairness of change. Moreover, they concluded that justice concerns are present during change implementation, and the interrelatedness among HRM practices was critical to employee justice perceptions of change.

Selection. Selection decisions include firing, hiring, transferring, and promoting or demoting employees. At the most basic level, selection systems can be used to reinforce change adoption and discourage resistance. Selection activities can be used to institutionalize change by hiring and promoting those individuals whose values match those represented in the changed state of the organization and removing those individuals whose values do not (Pascale, 1985).

During periods of retrenchment, when organizations are downsizing, the commitment of the *survivors* may be influenced by the treatment of the selected *victims*. Brockner, Grover, Reed, DeWitt, and O'Malley (1987) found that publicized benefits (e.g., severance pay, job placement, and continuance of health insurance) provided to layoff victims significantly affected the commitment of the survivors. Therefore, the quality of the relationship between the agent and organizational members prior to any changes and the actions taken regarding those adversely affected during crisis will influence the survivors' response.

In addition to the usual knowledge, skills, and abilities (KSAs), selection specifications may include the likelihood of being committed to changes introduced. Therefore, an understanding of salient job requirements, plus the commitment of the time necessary to identify appropriate personnel, will contribute to the institutionalization of change programs. Furthermore, this knowledge can be applied in promoting and transferring from within (thus capitalizing on symbolism) those individuals who will initiate and sustain the social dynamics necessary to accelerate the change process. Farkas and Wetlaufer (1996) described how Goldman-Sachs encouraged managers to volunteer for foreign assignments. A managing partner of the company explained "It was just not valued as an attractive career opportunity by most of our US people, and their spouses didn't necessarily want to go, and their dogs couldn't possibly endure living in Tokyo...So we took an exceptionally talented young banker and promoted him to partner two years ahead of his class because he went to Asia at great personal sacrifice" (Farkas & Wetlaufer, 1996, p.121).

Selection can be a powerful practice to reinforce discrepancy, efficacy, support, and valence. The survivors of an effectively executed downsizing should sense the change was needed and that the organization feels they are capable of performing the new behaviors. Job candidates who satisfy the technical and interpersonal job criteria, after extensive screening, may appreciate the thoroughness of the selection procedures and be more likely to contribute to a synergistic relationship with peers, subordinates, and superiors. Furthermore, rewarding individuals for volunteering for unpopular assignments and performing admirably can transmit new meaning to unpopular duty. The rewarding of such volunteering will demonstrate the organization's support and likely create support among organizational members.

Performance appraisal. The importance of feedback in changing one's behavior has been documented in field and laboratory studies (Bretz, Milkovich, & Read, 1992). Ilgen and Moore (1987), for example, found that feedback about quantity led to higher quantity, feedback about quality led to higher quality, and feedback about both led to higher quantity and quality. Including change criteria in appraisal systems serves to constantly reinforce the desired behaviors. Performance appraisal systems can contribute to institutionalization of change by providing appropriate feedback in a timely manner to organizational members. Performance appraisal systems at Control Data (Gomez-Mejia, Page, & Tornow, 1985), Ford Motor Company (Scherkenbach, 1985), Merck (Wagel, 1987), and Xerox (Deets & Tyler, 1986) have been designed to promote behaviors necessary to accomplish changes in corporate strategy.

Of late, a particular form of appraisal system, the 360-degree or multisource feedback (MSF) program, has become popular. MSF is feedback to leaders and managers from superiors, peers, subordinates, vendors, and customers aimed at improving organizational effectiveness (cf. London & Beatty, 1993). Its focus on change renders it appropriate for the adoption and institutionalization process because (a) new visions of the types of leaders an organization will begin to reward and advance can be incorporated and (b) knowledgeable individuals can be involved in the design of the MSF program (London & Beatty, 1993).

Lepsinger and Lucia (1997) describe an MSF process introduced at a bank. The leaders responded to the numerous external environmental changes affecting the banking industry by developing a business strategy that addressed service excellence and customer focus. In order to translate this to the branch-based employees, an MSF program was designed to ensure that individual employees' skills and orientation were realigned to support the new strategy. An MSF instrument was completed by self, superior, direct reports, and peers. Personalized feedback reports were presented to each employee. Skill gaps for each individual were identified and during one-on-one development meetings, plans were formulated that specified how skill gaps were to be eliminated within an 18- to 24-month time period.

Performance appraisal and MSF provide an opportunity for an employee to receive feedback regarding the adoption and institutionalization of an organizational change. A difficulty in adoption is related to the delay of the new behavior resulting in tangible personal benefits. However, periodically giving feedback about employee performance in recently changed jobs can influence valence and efficacy sentiments. This feedback can serve to extend the adoption (trial) period, thus enhancing the possibility of institutionalization.

A great deal of research has been conducted on performance appraisal focusing on understanding and improving ability, through changing formats and training. The motivation for appraising the performance of others has not been well researched (cf. Harris, 1994). However, rater motivation in performance appraisal is an important variable that must be recognized. If performance appraisal is to serve as a change lever, change agents must recognize the importance of this tool. It appears that the popularity of MSF will stimulate research, hopefully on improving change agent ability and motivation to appraise performance.

Compensation. The acceptable performance of a job is rewarded through some type of compensation program. Compensation typically consists of fixed interval plans (e.g., hourly, bi-weekly, monthly,) and variable plans (i.e., incentives), plus employee benefits (e.g., pensions, medical allowances, annual/sick leave, and severance pay). These tangible rewards can be very effective in reinforcing adoption and ultimately institutionalization of organizational change.

Bandura (1986) argued that extrinsic incentives (e.g., financial compensation, special priviledges, social recognition, advancement in rank, and other status-conferring rewards) should be provided to sustain adoptive behavior until the intrinsic value becomes apparent. When an organization establishes a new compensation plan to encourage organizational members to adopt an organizational change, change agent support is apparent. Furthermore, when a manager ceremonially recognizes a team or an individual, it is obvious that management supports the accomplishment. The likelihood of receiving compensation, in general, can be a significant factor in influencing the valence for an organizational change. The actual receipt of compensation will reinforce efficacy. When this compensation is tied to the goals of the organization change effort, organizational members infer that the change is appropriate and was indeed needed.

Traditionally, organizations have developed vertical hierarchical structures such that advancement within the organization has been perceived as upward movement. Typically, pay structures have matched hierarchical levels. However, the recent trend toward flattening organizational structures is incongruent with a rigid multi-grade pay scale because the pay grades no longer match the reduced number of organizational levels. One way to resolve this incongruency is to introduce broad-banding, the act of collapsing pay grades into a few wide bands, thus supporting today's flatter organization (Abosch & Hand, 1994). By having fewer grades, restructuring, with the concomitant changes in roles and responsibilities, is less difficult. Among the benefits to broadbanding are that it: (a) provides greater opportunity for cross-functional moves; (b) fosters team building and functioning because team members will likely be from the same band; and (c) emphasizes skills/competencies (Abosch, 1995).

Broadbanding is consistent with the concept of teamwork and with various pay delivery strategies. Skill-based pay programs (i.e., rewarding individuals for acquisition and demonstration of work-related skills and competencies), career-development pay programs (i.e., rewarding lateral job involvement to encourage breadth of experience, knowledge, skills, and abilities), and merit cash/gain-sharing programs (i.e., re-earnable lump-sum awards for individual/team performance) are all consistent with the broadbanding concept.

Research on these pay delivery strategies has shown them to be effective in reinforcing organizational change (cf. Bullock & Tubbs, 1990; Murray & Gerhart, 1998). Each provides various degrees of extrinsic rewards (e.g., periodic distributions of monetary payments, minimally influenced by external environmental disturbances) and intrinsic rewards (e.g., autonomy and skill development).

Some factors that are possible with these pay delivery strategies may not be as practical with managerial bonus programs (e.g., eliminating the influence of external environmental fluctuations). Still, however, the logic of reinforcing new cognitions and behaviors with extrinsic rewards applies to managerial personnel. For example, before Whirlpool's SBU reorganization, the bonus system rewarded managers for the performance of the entire corporation. With the SBU reorganization, bonuses were based on the performance of the SBU. Thus, the receipt of the bonus was contingent on the performance of a manager's SBU and all managers were treated equitably, based on the performance of their SBU.

Training and development. To assist in the acquisition of new knowledge, skills, and abilities (KSAs) for recently changed jobs, an organization can initiate focused training programs. Parsons, Liden, O'Connor, and Nagao (1991) found that training in the use of personal computers, when perceived as *instrumental*, contributed to institutionalizing the new behavior. El Sherif (1990) reported that training was a dynamic component in accelerating the institutionalization of decision support systems. Furthermore, Buller and McEvoy (1989) found that training contributed to the institutionalization of a new performance appraisal system. Also, Goodman and Dean (1982) found that training was a significant factor in institutionalizing changes brought about through Quality of Work Life programs.

Not only is training considered to be significant in institutionalizing change but the *trainers* can have an added influence on the outcome. Bandura (1986) proposed the use of respected peers as trainers or facilitators because of the symbolic social dynamics. A recent example illustrates how Coopers & Lybrand trained a cadre of Allied-Signal employees to conduct total quality management (TQM) workshops for all other employees (Stewart, 1992). This tactic is symbolically different from bringing in external trainers; this is more akin to respected peers spreading their influence through the organization, encouraging peers to adopt and to institutionalize the TQM philosophy.

Training and development, if linked to an organizational change, can reinforce all message components. First, the superiority of the new way should be obvious. Thus, need for change and the appropriateness of the change can be sensed. Second, the trainee should experience efficacy in performing the new job, because the KSAs will be related to the tasks. Organizational support should be apparent through the expenditure of funds and by permitting the training to be done on company time. Peer support can be transmitted by using colleagues of the participants as trainers. Finally, valence can be created by associating tangible and intangible rewards to the successful performance of the new job.

While skills training is focused more on the individual KSAs, team building capitalizes on the social dynamics of groups. Through team building, organizational member commitment to the change effort is influenced. Respected peers influence team members to increase their commitment by identifying with the team and internalizing the values of the change effort.

Diffusion Practices

Diffusion is spreading change adoption within one organizational group as well as spreading the adoption to other organizational groups. It is a common practice to establish a pilot program for experimental purposes to test an innovation. If test results are positive, then the innovation may be shared with potential adopters. Although some customization of the innovation may be necessary, the originating unit serves as a valuable resource in demonstrating the benefits, forewarning of adoption barriers, and advising about methodologies to facilitate implementation. Thus, the diffusion practice is an attempt to establish some kind of organizational dialogue whereby organizational members *learn to talk, walk the talk,* and *sustain the talk and walk* (Ashkenas & Jick, 1992).

Research by Johnston and Leenders (1990) on the diffusion of minor technical improvements in operations revealed that diffusion was accomplished through an interpersonal exchange between the potential user and the originating unit. Blakely, Emshoff, and Roitman (1984) identified site visits and diffusion facilitators as significant factors in institutionalization. Stewart (1991) reported that GE's successful Exchange of Best Practices program consists of presentations made by respected peers.

Other means of organizational sponsorship, such as providing research funding for the diffusion of an innovation, have also been shown to be effective. Gannis (1987) reported that financial support was a major factor for the success of Merck & Company, the pharmaceutical firm, in changing its research methodology for the invention of new drugs. The awarding of such funds, especially if handled in a ceremonial manner, may enhance the desire to achieve the high status of the awardees. Thus, the new practice can be diffused throughout the organization.

Bushe and Shani (1991) have described the use of a parallel learning structure (PLS; also referred to as a transition team, collateral organization, and parallel organization), as a vehicle for diffusing change. A PLS can connect and balance all the pieces in a change effort (cf. Duck, 1993). In addition to being a diffusion strategy, a PLS is a form of active participation which develops commitment among organizational members and enhances institutionalization. A PLS typically consists of a steering team (for overall direction) and smaller teams representing the various differentiated groups resembling the overall organizational membership. The operating procedures of a PLS are intended to promote change by representing the various interests and by assisting in customizing the change to that organizational group. Duck (1993) emphasized that someone on the PLS should pay particular attention to the emotional and behavioral issues that can and should surface in conjunction with the organizational change. Consequently, an organizational change can be tailored to be more compatible with the organizational networks, thus facilitating diffusion.

Ness and Cucuzza (1995) provided an example of how Chrysler used a PLS in diffusing activity-based costing (ABC) throughout its organization. A PLS was assembled consisting of 20 employees from the finance, manufacturing, engineering, and information systems. Three critical steps were executed in the diffusion process. First, the PLS persuaded critical opinion leaders to give ABC a fair shake and ultimately embrace the program. Second, the PLS coordinated training programs at all levels dealing with principles and mechanics of ABC. Finally, the PLS set up a pilot program at one plant and then rolled out the program throughout the company making sure local managers were involved and that there were visible successes (Ness & Cucuzza, 1995).

Use of diffusion practices is valuable because it effectively transmits the five message components. The benefits revealed from pilot programs or vicarious experiences from site visits can produce sufficient evidence that a change is needed (i.e., there is a better way). The actions by the change agent to sponsor the site visits or coordinate an exchange of best practices program symbolizes support for the organizational change. Customizing the innovation, practice, or organizational change to the uniquenesses in the adopting organization addresses the appropriateness issue. Knowing that the organizational change is being used by others and that it may be customized for local use can create efficacy. Finally, valence is created through some extrinsic reward (e.g., monetary compensation or social recognition) for adopting the organization change; furthermore, the realization that others are adopting the change may extend the adoption and ultimately result in institutionalization.

Rites and Ceremonies

As noted above, symbolism is an important outcome of workplace activities and is the *expressive* meaning associated with organizational structures, behaviors, and processes (for a thorough treatment see Trice & Beyer, 1993). Rites and ceremonies are considered symbolic practices evident in all organizations. Rites are public performances that signify the consequences of actions and the extent to which these actions are consistent with an organization's values; ceremonies connect two or more rites into a single occasion (Trice & Beyer, 1993).

Rites and ceremonies are powerful shapers of underlying cultural values (Harris & Sutton, 1986; Trice & Beyer, 1993) and powerful sources of the key change message. Several change researchers have emphasized the role of *letting go of the past* in helping organizational members adopt and institutionalize change (cf. Bridges, 1991; *Confirmation Speech*, 1988). The transition to a new state is facilitated by accepting the fact that the old is gone forever. Retirement ceremonies are intended to provide an opportunity to summarize one's accomplishments and contributions to the organization, as a way of expressing gratitude to the retiree and officially ending one's formal association with the organization. However, these ceremonies also focus on the life thereafter, the new challenges, the application of accumulated experience and skills to the many new endeavors that continue to make life fulfilling. Consistent with the notion of letting go of the past, Harris and Sutton (1986) described parting ceremonies in public- and private-sector organizations that were *dying*. These were formally planned functions where employees mourned the passing of the old; a kind of cathartic experience which was intended to confirm an end to a rewarding worthwhile experience. During these ceremonies, participants shared in a final feast and exchanged stories of the good times and perhaps expressing sadness and anger at the organization's death. However, the one indisputable fact was that it was over and it was time to move on. Thus, parting ceremonies can be useful in adopting and institutionalizing the *new* organizational change.

Rites of enhancement are intended to reinforce individuals and groups for adopting new ideas, values, and behaviors. Tunstall (1985) described AT&T's efforts (during the divestiture transition) to recognize and reward individuals for exemplary performance by establishing the Eagle award for new marketing ideas and the Golden Boy award for outstanding customer service.

Another rite of enhancement was Whirlpool's announcement of the joint venture with N.V. Philips (of the Netherlands). In August 1988, Whirlpool leaders quickly scheduled a meeting of the officers, directors, and managers to formally announce the joint venture. This joint venture made Whirlpool the world's largest home appliance manufacturer. Badges with "We're #1" were distributed to all attendees. The atmosphere of the meeting was one of excitement and celebration.

This meeting was effective in transmitting the five message components; these inexpensive badges served as a symbol representing the outcome of the negotiations with Philips and in reinforcing the appropriateness of organizational changes recently experienced by these employees. The audience was then briefed (for approximately 60 minutes) on the strategic logic employed in seeking the joint venture with Philips. This accomplishment reinforced that the recent strategic and structural changes were necessary. The pride of being employed by the world's largest appliance manufacturer was an intrinsically valent sentiment that also contributed to the employees' efficacy. Support from the leaders was apparent from scheduling the meeting and the strategic briefing that followed. The jubilation and excitement of the officers, directors, and managers during and after the meeting were evidence of the support.

While symbols like the Whirlpool badges may seem insignificant and by themselves may not have the effect of more costly awards, they demonstrate the extent to which some organizations will go to influence emotions regarding organizational change. Furthermore, while the cost of these badges was nominal, over time, numerous low-cost symbols can have a cumulative effect.

Because symbolism is in virtually all actions (as well as in *nonactions*), several strategies/tactics explained above qualify as rites. Notable are, for example, Merck and Company's awards for adopting a new pharmaceutical research methodology (i.e., rite of creation), the dramaturgy video prepared by Whirlpool's leaders to celebrate the positive consequences of the reorganization (i.e., rite of transition), and Whirlpool's CEO delivering his *Confirmation Speech* that included the phrase "there is no turning back" (i.e., rite of parting).

Formalization Activities

In practically all organizational change efforts, there are concomitant changes in formal activities necessary to demonstrate emphatic support for the changes being implemented which may play a major role in the overall success of the change. Generally, such formal changes in structure and procedures are substantial and difficult. They are unlikely to be undertaken unless the organization wholeheartedly supports the organizational change. For example, in 1988, Whirlpool initiated a fundamental change in business strategy by changing from a low-cost leader to a product differentiation strategy. Concomitant with this change in strategy was a change from a functional to a divisional (i.e., strategic business unit) organizational structure. To support the new structure, revised job descriptions, policies, procedures and other formal job requirements were necessary.

Other initiatives like ISO 9000 (Uzumeri, 1997) and reengineering (Hammer & Champy, 1993; Vansina & Taillieu, 1996) are implemented concomitantly to change the formal activities, and thus change the way an organization conducts its business.

These formal activities naturally require new behaviors and are intended to support the organizational change. By applying the tactics described within the other strategies, change agents can facilitate the adoption and institutionalization of these new behaviors and send a powerful message regarding management's support of the change.

Use of the Model for Planning Organizational Change

The process of planning and implementing change obviously requires a great deal of time, effort, money, and patience. Summarized in Table 1 are the major questions that should be answered regarding the institutionalization of a change effort. We realize it is common practice for an organization to have multiple changes

Table 1. Checklist for Institutionalizing Change

- What level of commitment is appropriate for this organizational change? Is this commitment level consistent with the time dimension necessary for this change?
- Do we have realistic answers to the five questions used to focus the change message?
- Are the change agents and their representatives credible?
- Can we systematically categorize organizational members into groups? Do we know the opinion leaders within each group? How will they assess the change effort? Can we anticipate their reaction?
- Does this organizational change lend itself to a participative approach? Who do we involve? Can the overall change effort be divided into components, thus permitting sequencing the components, and capitalizing on the concept of enactive mastery? How can we incorporate vicarious experiences to facilitate and extend the adoption of the change effort?
- How should we incorporate the various tactics (e.g., live presentations, written communications, electronic transmissions) of persuasive communication? How frequently should the tactics be used? Are some of the tactics more appropriate for some audiences?
- What internal information can be used to transmit the message? What external information can be used to transmit the message?
- Regarding HRM practices:
  - How can we execute the selection tactic to communicate the message?
  - Is our performance appraisal program in synch with the organizational change?
  - Are we assessing the appraisees in terms of their commitment to the change program?
  - Does our compensation program support the organizational change?
  - What educational programs are needed and who can conduct them most effectively?
- Can we benefit from establishing a parallel learning structure? Who should the representatives be?
- What symbolic actions/events can we execute to transmit the message?
- What formal activities are necessary to support the organizational change?
- How are we going to assess the level of commitment among our employees?
- Does this organizational change resemble a *program-of-the-month*?

being implemented simultaneously. By planning each change, using the model as a guide, a change agent can appreciate the magnitude of the effort required for implementation. Consequently, it should be apparent that *change overload* can be experienced quite quickly. Thus, a change agent can more realistically anticipate the challenges faced in implementing changes successfully. In situations requiring implementation of multiple changes at least two alternatives are apparent. One alternative may be to coordinate the various projects and use the strategies and tactics common to more than one project to facilitate institutionalization. Another alternative may be to prioritize the change projects and initiate a plan to implement those that are most critical early on and those that are less critical subsequently. In either case, by conscientiously answering these questions, change agents can appreciate the magnitude of implementing organizational change. Furthermore, the symbolic effect of using multiple strategies should be that organizational members will realize that the change program is not simply another program-of-the-month.

Commitment and Assessment

The assessment of any organizational change effort is intended to determine the extent to which the objectives have been accomplished. Understandably, a change implemented in response to unacceptable levels of productivity or profitability would be ultimately evaluated in those terms. However, early researchers (cf. Georgopoulos & Tannenbaum, 1957; Likert, 1967) established the link between managerial actions and behavior (i.e., causal variables); perceptions, attitudes, and satisfaction (i.e., intervening variables); and productivity, scrap, and employee absenteeism and turnover (i.e., end-result variables; Likert, 1967). Relying solely on assessment criteria, such as productivity and profitability, may not accurately portray the existing commitment level. The assessment of commitment can be used to predict future trends in end-result criteria. In other words, commitment may be a leading organizational indicator of productivity. That is, internalized commitment is more informative about institutionalization than simply reporting productivity has increased. An increase in productivity could be associated with a higher level of compliance commitment. Thus, assessing the effectiveness with which change strategies are executed and received should be based on level of commitment and selected end-result variables that are relevant to the specific organization change effort.

The research on commitment that we used in developing the model provides useful detail is assessing progress toward institutionalization. This conceptualization of commitment is relevant to the theory, research, and practice of organizational change because it permits a change agent to establish objectives related to each level. In addition, it serves as a barometer of the extent to which institutionalization has been achieved. For example, a change agent may elect to establish time horizons for achieving each level of commitment. In some situations, particularly those involving business turnarounds, a change agent may attempt to accomplish only compliance-based commitment. However, for the long-term the goal might be internalization commitment. Moreover, some of the strategies and tactics are useful in short-term situations while others may be expected to produce changes over the longer term. For example, formalization activities may influence one's behavior in the short term because organizational members will be required to comply with new job duties. However, over time, other tactics, such as vicarious learning, may require the longer term perspective.

The research provided by Becker (1992) has demonstrated that we can refine the assessment of commitment into overall compliance commitment and identification and internalization commitment as each applies to the organization, an organizational member's supervisor, and work group. Furthermore, the other research (cf. George, 1990; Reichers, 1985) implies that we could extend the measurement of commitment to the change agent and the change initiative. Thus, it may be possible to assess one's commitment (i.e., identification and internalization) to each entity, namely, the organization, the change agent, the change initiative, the work group, and the supervisor. A comparison of the various levels of commitment to each entity (e.g., the change initiative with the supervisor) can indicate whether organizational members perceive their supervisor as supporting the change initiative. This added refinement in measurement can assist in pinpointing where additional resources might be needed in institutionalizing change.

In addition to assessing commitment, the extent to which organizational members have *heard* and *believe* the five core message components should be assessed. If an organizational change effort is not having the intended effects on commitment, then it might be valuable to know whether one or more message components has been ignored or lacks believability. Through this part of assessment, a diagnosis can be initiated to identify why the message lacks believability and what actions need to be taken to remedy the situation. Each element in the model must be analyzed. For example, (a) is there something unacceptable about the message components, (b) do some groups have different beliefs about the message, (c) do the change agents lack credibility, and (d) are the strategies being effectively executed?

The methodological issues related to assessment are beyond the scope of this paper. However, one must recognize that assessment of change offers some interesting methodological challenges (cf. Armenakis, 1988; Svyantek, O'Connell, & Baumgardner, 1992; Woodman, 1989).

CONCLUSIONS AND IMPLICATIONS

Organizational change efforts comprise *content* and *process* issues. Content issues are related to *what* is to be changed. Changes in business strategy and organizational restructuring and other formalization activities are examples of content issues. These may be the required changes that must be embraced and effectively implemented by organizational members in order for the organization to improve its performance. On the other hand, process issues deal with *how* the content issues are implemented. We proposed and explained a comprehensive process model useful in institutionalizing organizational changes. Although the model is primarily focused on process issues, we do not want to underemphasize the required changes. Our intent in developing the model was to emphasize that a change agent must be knowledgeable of *both* the content and process of change.

Each construct in the model (see Figure 1) should become a conscious and explicit step in planning, implementing, and assessing organizational change. We selected commitment because we feel it is a natural outcome measure for institutionalization. Most people seek to be committed, in terms of psychological attachment, to some entity in order to satisfy a set of needs. Some interesting findings regarding loyalty and commitment have revealed that because of the organizational responses (e.g., mergers and acquisitions, reengineering, downsizing) to external environmental changes, employees are finding less reason to be committed to their employer than to some professional or trade organization (cf. Rousseau, 1997; and this trend is worldwide; see Kanter, 1991). We think this is unfortunate and unnecessary and thus made commitment a centerpiece in the model.

In realizing this trend in commitment, change agents should build and periodically assess commitment. Not only should change agents be concerned about losing good employees who have significant investment value and who can contribute to an organization's success, they should also be concerned about the commitment level of all employees.

The recent research on commitment has initiated some interesting debates. Several researchers (cf. Becker, 1992; Becker, Billings, Eveleth, & Gilbert, 1996; Harris, Hirschfeld, Feild, & Mossholder, 1993; Meyer & Allen, 1997; O'Reilly, Chatman, & Caldwell, 1991) have refined and extended Kelman's (1958) original conceptualization. While differences in the typologies exist, they are similar and compatible (Meyer & Allen, 1997). Although we used Becker's extension of Kelman, other conceptualizations can certainly be used with the model.

Change agent attributes were included in the model to emphasize not only the importance of the driving force behind the change but the importance of those representatives (e.g., managers and respected peers) who may act as change agent extensions in diffusing change throughout the organization. It was argued that change agents must be perceived as credible if institutionalization is to occur. The believability of the message is related to the credibility of the change agent.

The value in including attributes of organizational members in the model is to create awareness that it is unlikely that change will be institutionalized homogeneously throughout the organization. Understanding the role of individual differences and organizational ecology is important in appreciating the challenges change agents face in institutionalizing change. Change agents should be familiar enough with organizational members to systematically categorize individuals into one or more groups. Furthermore, it is useful to be able to identify the opinion leaders and to anticipate their assessment of and reaction to the change. Recognizing these attributes and planning the implementation accordingly will enhance the probability of success. In order to establish positive momentum for the change and minimize the likelihood of a change initiative being cynically labeled a program-of-the-month, organizational members need to be converted to change agents.

The social distance that exists between the change agent's actions and organizational members is a major issue that must be dealt with in any organizational change. A CEO can initiate a change program, but there should be an explicit endorsement at all levels of the organization; otherwise, the change effort will not succeed. Those in leadership positions should be considered credible, should endorse the change, and should mimic the behavior of the change agents. At the local level these leaders should be expected to use the strategies in the model and apply them in ways that are appropriate for the level. Thus, all strategies in the model should be used at the macro- and micro-levels.

All strategies and tactics should be explicitly linked to the organizational change effort and should be considered to be complementary. In terms of linking tactics to the change effort, the human resource management practices should demonstrate support. For example, criteria related to the change effort may be included in the existing performance appraisal program; a multisource feedback program may be initiated and aimed at supporting the change effort. Similarly, the compensation package must reinforce the change effort.

The factors in the model can be expected to have interactive relationships. Thus, a change agent who was not considered to be credible could negate the potential positive effects of the strategies and tactics. We cannot identify which factors in the model are the most significant in the institutionalization process. Intuitively, we believe that those factors based on the concept of self-discovery have more impact than those that are directly from the change agent. However, active participation cannot be expected to be the sole ingredient in the change process, although it is significant. The other strategies should be considered as supportive and used as appropriate to institutionalize change. Until research can identify which are most effective, we argue that all should be coordinated and applied in all organizational change efforts. We prefer to think about the model in terms of an athletic metaphor and argue that omitting any element in the model is tantamount to fielding a team without all the players. Thus, we advocate use of the *complete* model in the adoption and institutionalization of change.

REFERENCES

Abosch, K. (1995). The promise of broadbanding. *Compensation & Benefits Review*, January–February, 54–58.

Abosch, K., & Hand, J. (1994). *Broadbanding design, approaches, and practices*. Scottsdale, AZ: American Compensation Association.

Armenakis, A. (1988). A review of research on the change typology. In W. Pasmore & R. Woodman (Eds.), *Research in organizational change and development* (Vol. 2, pp. 163–194). Greenwich, CT: JAI Press.

Armenakis, A., Harris, S., & Mossholder, K. (1993). Creating readiness for organizational change. *Human Relations, 46*(3), 1–23.
Ashkenas, R., & Jick, T. (1992). From dialogue to action in GE workout: Developmental learning in a change process. In W. Pasmore & R. Woodman (Eds.), *Research in organizational change and development* (Vol. 6, pp. 267–287). Greenwich, CT: JAI Press.
Baba, M. (1995). The cultural ecology of the corporation: Explaining diversity in work group responses to organizational transformation. *Journal of Applied Behavioral Science, 31*(2), 202–233.
Bandura, A. (1986). *Social foundations of thought and action: A social cognitive theory*. Englewood Cliffs, NJ: Prentice-Hall.
Becker, T. (1992). Foci and bases of commitment: Are they distinction worth making? *Academy of Management Journal, 35*, 232–244.
Becker, T., Billings, R., Eveleth, D., & Gilbert, N. (1996). Foci and bases of employee commitment: Implications for job performance. *Academy of Management Journal, 39*, 464–482.
Beer, M. (1976). The technology of organization development. In M. Dunnette (Ed.), *Handbook of industrial and organizational psychology* (pp. 937–993). Chicago, IL: Rand McNally.
Beer, M., Eisenstat, R., & Spector, B. (1990, November-December). Why change programs don't produce change. *Harvard Business Review, 68*, 158–166.
Bies, R.J. (1987). The predicament of injustice: The management of moral outrage. In L. Cummings & B. Staw (Eds.), *Research in organizational behavior* (Vol. 9, pp. 289–319). Greenwich, CT: JAI Press.
Blakely, C., Emshoff, J., & Roitman, D. (1984). Implementing innovative programs in public sector organizations. In Oskamp, S. (Ed.), *Applied social psychology annual: Applications in organizational settings* (Vol. 5, pp. 87–109). Beverly Hills, CA: Sage.
Bretz, R., Milkovich, G., & Read, W. (1992). The current state of performance appraisal research and practice: Concerns, directions, and implications. *Yearly Review of Management, 18*(2), 321–352.
Bridges, W. (1991). *Managing transitions: Making the most of change*. Reading, MA: Addison-Wesley.
Brockner, J., Grover, S., Reed, T., DeWitt, R., & O'Malley, M. (1987). Survivors' reactions to layoffs: We get by with a little help for our friends. *Administrative Science Quarterly, 32*(4), 526–541.
Buller, P., & McEvoy, G. (1989). Determinants of the institutionalization of planned organizational change. *Group & Organization Studies, 14*(1), 33–50.
Bullock, R., & Tubbs, M. (1990). A case meta-analysis of gainsharing plans as organization development interventions. *Journal of Applied Behavioral Science, 26*, 383–404.
Burkhardt, M. (1991, August). *Institutionalization following a technological change*. Paper presented at the Fifty-first Annual Academy of Management Meeting, Miami, FL.
Bushe, G., & Shani, A. (1991). *Parallel learning structures: Increasing innovation in bureaucracies*. Reading, MA: Addison-Wesley.
Business in Profile. (1990). [Videotape]. Benton Harbor, MI: Whirlpool Corporation.
Church, A., & Burke, W. (1995). Practitioner attitudes about the field of organization development. In W. Pasmore & R. Woodman (Eds.), *Research in organizational change and development* (Vol. 8, pp. 1–46). Greenwich, CT: JAI Press.
Cobb, A., Wooten, K., & Folger, R. (1995). Justice in the making: Toward understanding the theory and practice of justice in organizational change and development. In W. Pasmore & R. Woodman (Eds.), *Research in organizational change and development* (Vol. 8, pp. 243–295). Greenwich, CT: JAI Press.
Confirmation Speech. (1988). [Videotape]. Benton Harbor, MI: Whirlpool Corporation.
Dalton, G. (1970). Influence and organizational change. In G. Dalton, P. Lawrence, & L. Greiner (Eds.), *Organizational change and development* (pp. 230–258). Homewood, IL: Richard D. Irwin and The Dorsey Press.
Damanpour, F. (1991). Organizational innovation: A meta-analysis of effects of determinants and moderators. *Academy of Management Journal, 34*(3), 555–590.

Deets, N., & Tyler, D. (1986). How Xerox improved it performance appraisals. *Personnel Journal, 65*(4), 50–52.
Dirks, K., Cummings, L., & Pierce, J. (1996). Psychological ownership in organizations: Conditions under which individuals promote and resist change. In R. Woodman & W. Pasmore (Eds.), *Research in organizational change and development* (Vol. 9, pp. 1–24). Greenwich, CT: JAI Press, Inc.
Dowrick, P. (1991). *Practical guide to using video in the behavioral sciences*. New York: John Wiley & Sons.
Duck, J. (1993). Managing change: The art of balancing. *Harvard Business Review, 71*(6), 109–118.
Eisenberger, R., Fasolo, P., & Davis-LaMastro, V. (1990). Perceived organizational support and employee diligence, commitment, and innovation. *Journal of Applied Psychology, 75*(1), 51–59.
El Sherif, H. (1990). Managing institutionalization of strategic decision support for the Egyptian cabinet. *Interfaces, 20*(1), 97–114.
Farkas, C., & Wetlaufer, S. (1996). The ways chief executive officers lead. *Harvard Business Review, 74*(3), 110–122.
Fishbein, M., & Ajzen, I. (1975). *Belief, attitude, intention, and behavior: An introduction to theory and research*. Reading, MA: Addison-Wesley.
Franks, I., & Maile, L. (1991). The use of video in sport skill acquisition. In P.W. Dowrick & Associates, *Practical Guide to Using Video in the behavioral Sciences* (pp. 231–243). New York: John Wiley & Sons.
Gannis, S. (1987, January 19). Merck has made biotech work. *Fortune, 123*(2), 58–64.
General Motors Annual Report. (1992). Detroit, MI: General Motors Corporation.
George, J. (1990). Personality, affect, and behavior in groups. *Journal of Applied Psychology, 75*, 107–116.
Georgopoulos, B., & Tannenbaum, A. (1957). A study of organizational effectiveness. *American Sociological Review, 22*, 534–540.
Gomez-Mejia, L., Page, R., & Tornow, W. (1985). Improving the effectiveness of performance appraisal. *Personnel Administrator, 30*(1), 74–81.
Goodman, P., & Dean, J. (1982). Creating long-term organizational change. In P. Goodman (Ed.), *Change in organizations: New perspectives on theory, research, and practice* (pp. 226–279). San Francisco, CA: Jossey-Bass.
Hammer, M., & Champy, J. (1993). *Reengineering the corporation*. New York: Harper Business.
Harris, M. (1994). Rater motivation in the performance appraisal context: A theoretical framework. *Journal of Management, 20*(4), 737–756.
Harris, S., Hirschfeld, T., Feild, H., & Mossholder, K. (1993). Psychological attachment: Relationships with job characteristics, attitudes, and preferences for newcomer development. *Group and Organization Management, 18*, 459–481.
Harris, S., & Sutton, R. (1986). Function of parting ceremonies in dying organizations. *Academy of Management Journal, 29*, 5–30.
Ilgen, D., & Moore, C. (1987). Types and choices of performance feedback. *Journal of Applied Psychology, 72*, 401–406.
Johnston, D., & Leenders, M. (1990). The diffusion of innovation within multi-unit firms. *International Journal of Operations and Production Management, 10*(5), 15–24.
Kanter, R. (1991). Transcending business boundaries: 12,000 world managers view change. *Harvard Business Review, 69*(3), 151–164.
Kelman, H. (1958). Compliance, identification, and internalization. *Journal of Conflict Resolution, 2*(1), 51–60.
Kilbourne, L., O'Leary-Kelley, A., & Williams, S. (1996). Employee perceptions of fairness when human resource systems change: The case of employee layoffs. In R. Woodman & W. Pasmore (Eds.), Research in organizational change and development (Vol. 9, pp. 25–48). Greenwich, CT: JAI Press, Inc.

Kirton, M. (1984). Adaptors and innovators: Why new initiatives get blocked. *Long Range Planning, 17*(2), 137–143.
Kouzes, J., & Posner, B. (1993). *Credibility: How leaders gain and lose it, why people demand it.* San Francisco, CA: Jossey-Bass.
Larkin, T., & Larkin, S. (1996). Reaching and changing frontline employees. *Harvard Business Review, 74*(3), 95–104.
Lepsinger, R., & Lucia, A. (1997). *The art and science of 360 feedback.* San Francisco, CA: Pfeiffer.
Lewin, K. (1947). Frontiers in group dynamics. *Human Relations, 1,* 5–41.
Likert, R. (1967). *The human organization.* New York: McGraw-Hill.
Locke, E., & Latham, G. (1990). *A theory of goal setting and task performance.* Englewood Cliffs, NJ: Prentice-Hall.
London, M., & Beatty, R. (1993). 360-degree feedback as a competitive advantage. *Human Resource Management, 32*(2 & 3), 353–372.
Macdonald, S. (1995). Learning to change: An information perspective on learning in the organization. *Organization Science, 6*(5), 557–568.
Mathieu, J., & Kohler, S. (1990). A cross-level examination of group absence influences on individual absence. *Journal of Applied Psychology, 75,* 217–220.
Meyer, J., & Allen, N. (1997). *Commitment in the workplace: Theory, research, and application.* Thousand Oaks, CA: Sage.
Murray, B., & Gerhart, B. (1998). An empirical analysis of a skill-based pay program and plant performance outcomes. *Academy of Management Journal, 41*(1), 68–78.
Nadler, D. (1977). *Feedback and organization development: Using data-based methods.* Reading, MA: Addison-Wesley.
Ness, J., & Cucuzza, T. (1995). Tapping the full potential of ABC. *Harvard Business Review, 73*(4), 130–138.
Neumann, J. (1989). Why people don't participate in organizational change. In R. Woodman & W. Pasmore (Eds.), *Research in organizational change and development* (Vol. 3, pp. 181–212). Greenwich, CT: JAI.
Niehoff, B., Enz, C., & Grover, R. (1990). The impact of top management actions on employee attitudes and perceptions. *Group & Organization Studies, 15*(3), 337–352.
Nutt, P. (1986). Tactics of implementation. *Academy of Management Journal, 29*(2), 230–261.
Nystrom, P. (1990). Vertical exchanges and organizational commitments of American business managers. *Group & Organization Studies, 15*(3), 296–312.
O'Reilly, C., Chatman, J., & Caldwell, D. (1991). People and organizational culture: A profile comparison approach to assessing person-organization fit. *Academy of Management Journal, 34*(3), 487–516.
Parsons, C., Liden, R., O'Connor, E., & Nagao, D. (1991). Employee responses to technologically-driven change: The implementation of office automation in a service organization. *Human Relations, 44*(12), 1331–1356.
Pascale, R. (1985). The paradox of "corporate culture": Reconciling ourselves to socialization. *California Management Review, 27*(2), 26–42.
Pasmore, W., & Fagans, M. (1992). Participation, individual development, and organizational development: A review and synthesis. *Journal of Management, 18*(2), 375–397.
Reichers, A. (1985). A review and reconceptualization of organizational commitment. *Academy of Management Review, 10,* 465–476.
Ritti, R., & Silver, J. (1986). Early processes of institutionalization: The dramaturgy of exchange in interorganizational relations. *Administrative Science Quarterly, 31,* 25–42.
Rousseau, D. (1997). Organizational behavior in the new organizational era. *Annual Review of Psychology, 48,* 515–546.
Ryan, B., & Gross, N. (1943). The diffusion of hybrid seed corn in two Iowa communities. *Rural Sociology, 8,* 15–24.

Scherkenbach, W. (1985). Performance appraisal and quality: Ford's new philosophy. *Quality Progress, 18*(4), 40–46.
Slater, M., & Rouner, D. (1992). Confidence in beliefs about social groups as an outcome of message exposure and its role in belief change persistence. *Communication Research, 19*(5), 597–617.
Snyder, M. (1974). The self-monitoring of expressive behavior. *Journal of Personality and Social Psychology, 30,* 526–537.
Stewart, T. (1991, August 12). GE keeps those ideas coming. *Fortune, 125*(15), 41–49.
Stewart, T. (1992, November 30). Allied-Signal's turnaround blitz. *Fortune, 126*(12), 72–76.
Svyantek, D., O'Connell, M., & Baumgardner, T. (1992). A Bayesian approach to the evaluation of organizational development efforts. In W. Pasmore & R. Woodman (Eds.), *Research in organizational change and development* (Vol. 6, pp. 235–266). Greenwich, CT: JAI Press.
Tichy, N.M. (1982, Autumn). Managing change strategically: The technical, political and cultural keys. *Organizational Dynamics,* pp. 59–80.
Tornatzsky, L., & Klein, K. (1982). Innovation characteristics and innovation adoption-implementation: A meta-analysis of findings. *IEEE Transactions Engineering Management, EM-29*(1), 28–45.
Trice, H., & Beyer, J. (1993). *The cultures of work organizations.* Englewood Cliffs, NJ: Prentice-Hall.
Tunstall, W. (1985). *Disconnecting parties—managing the Bell System breakup: An inside view.* New York: McGraw-Hill.
Uzumeri, M. (1997). ISO 9000 and other metastandards: Principles for management practice? *Academy of Management Executive, 12*(1), 21–36.
Van Maanen, J., & Barley, S. (1984). Occupational communities: Culture and control in organizations. In B. Staw & L. Cummings (Eds.), *Research in organizational behavior* (pp. 287–365). Greenwich, CT: JAI Press.
Vansina, L., & Taillieu, T. (1996). Business process re-engineering or socio-technical system design in new clothes. In R. Woodman & W. Pasmore (Eds.), *Research in organizational change and development* (Vol. 9, pp. 81–100). Greenwich, CT: JAI Press.
Wagel, W. (1987). Performance appraisal with a difference. *Personnel, 64*(2), 4–6.
Weick, K. (1984). Small wins: Redefining the scale of social problems. *American Psychologist, 39,* 40–49.
Weisbord, M. (1988). Towards a new practice theory of OD: Notes on snapshooting and moviemaking. In W. Pasmore & R. Woodman (Eds.), *Research in organizational change and development* (Vol. 2, pp. 59–96). Greenwich, CT: JAI Press.
Woodman, R. (1989). Evaluation research on organizational change: Arguments for a "combined paradigm" approach. In R. Woodman & W. Pasmore (Eds.), *Research in organizational change and development* (Vol. 3, pp. 161–180). Greenwich, CT: JAI Press.
Yeung, A.K.O., Brockbank, J.W., & Ulrich, D.O. (1991). Organizational culture and human resource practices: An empirical assessment. In R. Woodman & W. Pasmore (Eds.), *Research in organizational change and development* (Vol. 5, pp. 59–81). Greenwich, CT: JAI Press.

TQM AND ORGANIZATIONAL CHANGE
A LONGITUDINAL STUDY OF THE IMPACT OF A TQM INTERVENTION ON WORK ATTITUDES

Jacqueline A-M. Coyle-Shapiro

ABSTRACT

The embryonic stage of total quality management (TQM) theory and the significant lag of academic investigations leave many gaps in our understanding of this current organizational practice. Longitudinal studies examining the effects of TQM are conspicuous by their absence. This case study examines the process, content, and consequences of a TQM change intervention using quantitative and qualitative data. The data suggest that the intervention had a limited effect on employee work attitudes. The reasons for this effect are explored. Among the contributing factors identified was the absence of supporting changes introduced to reinforce a TQM philosophy. A key challenge facing TQM is the incorporation of a systemic view of change into a TQM process thereby increasing the likelihood of generating effective organizational change. In the process, this may dilute the core ideas of the movement's founders.

INTRODUCTION

Total quality management (TQM) has been described as the Holy Grail for organizational survival (Steel & Jennings, 1992). Holding out such promise, it is not surprising that approximately 75 percent of U.S. and U.K. organizations are implementing some form of quality initiative (Mohrman, Tenkasi, Lawler, & Ledford, 1995; Wilkinson, Snape, & Allen, 1993). Yet, despite its pervasiveness, major theoretical and empirical gaps exist. As Reed and Lemak (1998) argue, the current state of the TQM literature reflects a practitioner orientation, with three strands visible. These include: reviews of the works of the movement's founders, anecdotal evidence of organization's experience with TQM, and the "how to recipe" focused literature. In providing an antidote, the critical literature presents the contention that TQM is a modern repackaging of scientific management utilizing sophisticated techniques purporting to reveal the hidden true character of TQM masquerading under the guise of empowerment and customer sovereignty (Boje & Winsor, 1993).

The clear divide between the prescriptive and critical literature has added fuel to the controversy accompanying TQM. Recent contributors (Hackman & Wageman, 1995; Morrow, 1997) have highlighted the strong emotions that TQM evokes to the detriment of the emergence of scientific studies with an emphasis on neutrality. It is not surprising that the issue of evaluation has been comparatively neglected in spite of recurrent calls for this type of research. Little has been published in terms of "traditional" evaluation studies (Macy & Izumi, 1993), and this is accentuated when a comparison is drawn with quality oriented interventions preceding the emergence of TQM. In particular, considerable empirical work exists that evaluates the effects of quality circles (QCs) and semi-autonomous work groups (SAWGs) on a range of criteria. Nonetheless, there are some empirically grounded studies that examine the effects of TQM from distinct perspectives.

The emerging conclusion from the research is that TQM has failed to deliver its promises. This has contributed to the waning of interest in TQM and to the accusation that it is the latest in a long line of fads destined for the managerial scrap heap. Somewhere along the route, TQM is perceived to have lost its magic. Part of the blame is leveled at the practitioner community for naïvely misunderstanding what TQM is and what it involves. Had the difficulties and complexities of changing organizational culture been highlighted, chances are, TQM would not have been presented in such a manner that it could only fail in fulfilling such a set of unrealistic expectations. Reed and Lemak (1998) level some responsibility to the shortsightedness of managers in the search for quick results. Furthermore, the academic community is not beyond blame in its persistent reluctance to investigate this phenomenon (Reed & Lemak, 1998). Greater academic contribution may have informed managerial judgment regarding the appropriateness of adopting TQM in particular organizational contexts.

The purpose of this chapter is to address the noticeable gap in the evaluation of TQM. The chameleon-like features of the conceptualization of TQM have resulted in divergent practices being implemented as part of an organization's TQM efforts. The implications, therefore, for evaluation studies, highlight the need to describe the organizational context, the TQM intervention being examined, and the process of change adopted. Consequently, a case study approach is employed to evaluate the effects of TQM on work attitudes while providing insight into the intricacies of organizational change. The quantitative data, one pre- and two post-measurement occasions are used to evaluate the impact of TQM on employee work attitudes: teamwork, job satisfaction, and organizational commitment. Qualitative data are used to explore issues surrounding the organizational change. Overall, this provides a fuller picture of the context in which TQM was pursued, the rationale behind the adoption of TQM, the reaction of various interest groups to the change, the effects of the change, and the contributing factors to the effects found.

TQM

The diverging conceptualizations of TQM are succinctly captured by Steel and Jennings (1992, p. 31) who conclude "there are as many approaches to TQM as there are TQM practitioners." However, there is now a theoretical convergence on what constitutes the main principles of TQM: customer focus, continuous improvement, and teamwork (Dean & Bowen, 1994). Yet at the same time, the apparent clarity of TQM is being clouded by the entry of, for example, empowerment into the TQM framework. Furthermore, the secondary principles designed to align organizational practices with a TQM philosophy are analogous to shifting sediment. The consequence of these recent developments is a conceptual paradox whereby consensus exists in defining TQM with disagreement on the practices associated with TQM.

A driving concern behind the interest in TQM is its impact on organizational performance. A prominent and problematic strand of research adopts a macro-perspective on this issue but the evidence is inconclusive, with some studies suggesting that TQM can affect organizational performance and other studies failing to demonstrate any such effect (Fisher, 1992; Gilbert, 1992; Mohrman et al., 1995; Wruck & Jensen, 1994). Clearly, correlational studies of the kind typically conducted cannot provide irrefutable evidence of causation. Hackman and Wageman (1995) note that there are a number of problems confronting any study attempting to examine the link between TQM and organizational performance. These include measurement problems associated with standard indices of organizational performance, exogenous factors that may cloud the link between TQM and outcomes and, finally, the problem of time lapse for the effects of TQM to impact organizational performance. In light of these difficulties, the authors argue that the challenge of statistically detecting the direct effects of TQM on organizational outcomes may be too great. In addition, the chameleon-like qualities of TQM continue to be a thorn in the side of this type of study. Powell (1995) concludes that TQM is not necessary for a competitive advantage on the grounds that once the effects of good management practices are accounted for, the effects of TQM tools on performance is negligible. This highlights the potential influence of the conceptualization and operationalization of TQM on the effects found.

A second strand of research is at a micro-level and primarily adopts a qualitative case study approach using ex post facto analysis seeking to address the effects of TQM on employee work attitudes and behavior. A central aspect of this work attempts to uncover the consequences of TQM along the criteria of empowerment, control, and work intensification. A number of case studies provide support for the contention that TQM is the latest in a line of management techniques to lead to work intensification, greater managerial surveillance, and control (Delbridge, Turnbull, & Wilkinson, 1992; Kerfoot & Knights, 1995; McArdle, Rowlinson, Procter, Hassard, & Forrester, 1995; Sewell & Wilkinson, 1992). Other case study-based research presents TQM as enhancing employee empowerment within narrowly defined limits (Hill, 1995; Wilkinson, Godfrey, & Marchington, 1997). Rees (1998), in examining employee responses to quality initiatives in a variety of settings, concludes that the effects of TQM may seem to be contradictory; greater employee discretion with a simultaneous tightening of managerial control; greater work effort and at the same time enhanced job satisfaction. These studies provide rich insight into the complexities of the TQM process and contribute to an important debate on the implications of TQM for the distribution of power and control between management and employees. The inherent grounding of this type of research in ex post facto analysis raises methodological and measurement concerns regarding the effects of TQM.

Longitudinal studies examining the effects of TQM are conspicuous by their absence. A few longitudinal studies have been conducted to investigate the impact of TQM on a number of attitudinal outcomes: attitudes toward quality (Procopio & Fairfield-Sonn, 1996), quality consciousness (Wood & Peccei, 1995), teamwork and commitment to continuous improvement (Coyle-Shapiro, 1996), and organizational commitment (Guest & Peccei, 1994). Together these studies have sought to examine the effects of TQM initiatives on intermediary outcomes, primarily work attitudes. In doing so, they have attempted to bring TQM into the mainstream evaluation research of organizational change. However, these longitudinal studies are limited in the time span allowed for the effects of TQM to occur. Whether the short-term effects are an accurate reflection of the impact of TQM could be disputed on the grounds that a longer term horizon is needed for the consequences of TQM to materialize. Given the potential methodological obstacles in evaluating TQM over a longer period and the conceptual fuzziness surrounding the concept, it is no surprise that a greater number of longitudinal studies have not been conducted.

Predecessors to TQM

Historically, quality circles (QCs) and semi-autonomous work groups (SAWGS) are predecessors to the emergence of TQM. Steel and Jennings (1992) present the three interventions as falling within the domain of quality improvement technologies. Whether TQM represents an overarching framework encompassing both QCs and SAWGs in the pursuit of quality and continuous improvement may be open to debate. The relationship between QCs and TQM is clear cut in the sense that QCs may play an important role in pursuing continuous improvement at the lower organizational levels as part of a TQM endeavor. At the same time, TQM may overcome the acknowledged problems of relying exclusively on QCs, as a bottom-up strategy for achieving ongoing improvements. Consequently, it is common to see QC-type activity as one element of a TQM process (Hill, 1991; Howes, Citera, & Cropanzano, 1995). This would be consistent with the conceptualization of TQM by its founders and does not threaten the distinctiveness of TQM from other broad multifaceted initiatives (Hackman & Wageman, 1995).

SAWGs have been presented as another intervention falling under the rubric of quality improvement technologies (Steel & Jennings, 1992). However, the relationship between SAWGs and TQM may not be as clear as that for QCs. One perspective views SAWGs as an integral part of TQM and hence subsumed under the TQM banner (Dean & Bowen, 1994; Howes et al., 1995). Hackman and Wageman (1995) argue that many organizations implementing TQM extend employee involvement to the creation of self-managed work groups. However, it is not clear whether SAWGs inform the conceptualization of TQM or muddy the divergent validity of TQM. For those advocating TQM as a vehicle for employee empowerment, SAWGs may fulfill this objective by giving the group responsibility and autonomy over a range of issues concerning quality and improvements. Whether the founding fathers of the quality movement envisaged empowerment beyond a motivational definition or the extension of decision-making power to issues other than quality is doubtful. Therefore, if TQM involves greater empowerment, SAWGs may indeed provide the appropriate mechanism to achieve this in the context of TQM. However, this highlights a more fundamental issue regarding the conceptual boundaries of TQM and its distinctiveness from other organizational approaches. An additional obstacle to the integration of SAWGs into a TQM framework is the universalistic perspective of TQM against the contingency view on the appropriateness of SAWGs. In particular, the degree of task interdependence and the type of technology are two determinants of the suitability of SAWGs (Cummings, 1978; Lawler, 1986). Therefore, whether SAWGs become part of TQM in practice may be dependent on the degree to which the conditions for SAWGs exist.

Change and TQM

The type of change involved in implementing TQM is associated with "radical change" (Munroe-Faure & Munroe-Faure, 1992), metamorphosis, total change (Dobyns & Crawford-Mason, 1991), and reflective of large-scale organizational change (Kaplan, Birmingham, & Ferris, 1998). Some commentators view TQM as a distinct management paradigm and, by implication, associated with transformational change (cf. Grant, Shani, & Krishnan, 1994). However, the paradigmatic shift equated with TQM has been challenged by Kanter (1989, p. 35) who argues that TQM is but "a way station on the road to more complete restructuring of corporate strategy and organizational form." In defining radical change, Nutt and Backoff (1997, p. 230) equate this with an organization "providing greater variety, more skill, and increased ability to serve its customers/clients in new and different ways, which allow the organization to cope more effectively with its changed world." Therefore, whether TQM is categorized as involving transformational change is dependent on the existing vision, strategy, and processes pursued by the organization. For some organizations, TQM as an organizational phenomenon may alter the "rules of the game"; for others, it may involve modifications or fine tuning to organizational systems. A more cautious interpretation of TQM is that it aims to "achieve fundamental change without changing the fundamentals" (Hackman & Wageman, 1995, p. 336) and hence may be best categorized as tectonic change (Reger, Gustafson, DeMarie, & Mullane, 1994).

An underlying assumption of TQM is that quality is ultimately the responsibility of top management and hence top management commitment and support is a precondition for the success of TQM. Therefore, a top-down approach to the implementation of TQM is required whereby "both education about TQM and implementation of TQM practices typically take place in cascading fashion, with each layer carrying the message to the next lower level of the organization" (Hackman & Wageman, 1995, p. 316). Inherent in this approach are implicit assumptions concerning how change occurs in organizations. First, change is expected to occur in a mechanical and rational manner, so that the influence of political behavior on the implementation process is not addressed (Kaplan et al., 1998). Thus, the cascading approach to change is treated as unproblematic; change is assumed to occur in a straightforward manner throughout the managerial hierarchy culminating with first-line supervisors involving employees in TQM. In reality, the dependency of change at one level on the level above creates a fragility in the linking pin approach to change which, if broken, could present an obstacle to a smooth and organization-wide change process.

Second, the primary levers for change include training, education, and recognition. Hackman and Wageman (1995) report that the majority of organizations utilize some form of training as part of their change efforts. The underlying assumption is that change occurs as a consequence of education and training, not only in terms of individual attitudes and behaviors but also acts as a stimulus for changes in organizational practices to support a TQM philosophy. The alignment of organizational practices to TQM may partly explain divergences in the practices associated with TQM across organizations and, also, differing degrees of success with TQM. A case in point is gain sharing or profit sharing programs. Whether TQM needs to utilize extrinsic rewards to become firmly ingrained into an organization's activities is a contentious issue. Arguments from a justice perspective suggest that employees need to share in the financial benefits accrued from TQM (Cobb, Wooten, & Folger, 1995). Despite reasoned arguments outside the mainstream TQM literature, many organizations avoid extrinsic rewards choosing to rely exclusively on the intrinsic motivation of its members (Hackman & Wageman, 1995). The literature on TQM could benefit from the existing knowledge on organizational change and, in particular, from the desirability of adopting a holistic and systemic perspective in the planning and introduction of change. Far too often, attempts at organizational change fail due to the lack of consideration of what else needs to change to achieve congruence among organizational subsystems. Why should TQM be any different?

Criteria for Evaluation

The voluminous literature and varying perspectives of TQM have added confusion to what TQM is and, by implication, increased ambiguity as to how TQM should be evaluated. Morrow (1997) argues that in view of the long-term horizon required to study the effects of TQM, researchers may be better served to study the effects of TQM on work attitudes as an intermediary outcome to performance. One approach to evaluation would be to investigate the link between the implementation of TQM and its key principles: teamwork, continuous improvement, and customer focus, as outlined by Dean and Bowen (1994). Alternatively, there is the approach taken by Morrow (1997) who investigates the link between employee perceptions of the principles of TQM and outcomes such as work environment, communication, and job satisfaction. These two approaches to evaluation of TQM are addressing different issues regarding the efficacy of TQM: the former examines the efficacy of TQM implementation in achieving higher levels of teamwork, continuous improvement, and customer focus; the latter concentrates on the effects of teamwork, continuous improvement, and customer focus on other attitudinal outcomes. The latter perspective, to a greater degree, assumes the existence of a link between the implementation of TQM and the achievement of its principles.

As a quality improvement intervention, TQM could be evaluated using the same criteria adopted in evaluation studies of QCs and SAWGs. Job satisfaction and organizational commitment are two common attitudinal criteria employed to examine the outcomes of QCs and SAWGs. Recently, Kirkman and Rosen (1997) have presented these outcomes as potential consequences of empowered teams. These criteria could also be used to evaluate the effects of TQM and, in the process, provide an initial attempt to integrate more fully these improvement interventions. Furthermore, these criteria have been shown to have positive effects on extra-role and organizational citizenship behavior (Coyle-Shapiro & Kessler, 2000; Organ & Konovsky, 1989; Organ & Ryan, 1995).

In proposing a research agenda for TQM, Steel and Jennings (1992) recommend that the contributions of the individual components of TQM to the overall functioning of the program should be examined. This is of particular importance in view of the amorphous character of TQM and, furthermore, has significant implications for the evaluation of TQM. Specifically, any evaluation needs to outline precisely what is being evaluated as part of a TQM intervention—variations may exist regarding the practices implemented as part of TQM or practices introduced to institutionalize a TQM philosophy. Therefore, a significant aspect of evaluating TQM is the description of the content of a TQM intervention and the process of change.

A common approach adopted in the evaluation of organizational change interventions is to evaluate the effects of participation. Research on QCs and SAWGs primarily focus on the effects of participation on hypothesized outcomes. Consistent with this, an evaluation of TQM could focus on the effects of employee participation on work attitudes. However, research on training activities found that trainee assessment was significant in influencing post-training attitudes rather than participation in training per se (Tannenbaum, Mathieu, Salas, & Cannon-Bowers, 1991). Given the emphasis on training and education as a means to bring about change, it may be important to capture an individual's cognitive assessment or judgment of TQM. In particular, employees' assessment of the benefits of TQM may be important in affecting subsequent attitudinal change (Wood & Peccei, 1995) and participation in TQM (Coyle-Shapiro, 1999b). Finally, the process of implementation may also be an important factor; that is, given the close physical and psychological proximity of the immediate supervisor, the extent to which supervisors reinforce the principles of TQM may be important in affecting the hypothesized outcomes. Given the cascading approach to change and the dependency of change at the employee level on first-line supervisors, employee perceptions of the commitment of their immediate supervisor may have a significant influence on employee attitudes. Rather than limit the evaluation to the effects of employee participation per se, the approach adopted in this study also includes employees' assessment of the benefit of TQM as well as their perception of supervisory reinforcement.

Adopting a multifaceted basis to the evaluation illuminates the mechanisms by which subsequent attitudes are affected. While the evaluation criteria and the components of the TQM intervention are loosely based on previous empirical work on QCs and SAWGs and the less developed TQM framework, the evaluation questions are framed in terms of propositions rather than hypotheses because they relate to a research inquiry that has not been explored previously. Reflecting this, the propositions being examined are broad in nature using an employee's experience rather than specific propositions that relate to an employee's participation, assessment of the benefit of TQM, and supervisory reinforcement of the principles of TQM. Figure 1 depicts the evaluation model that examines the effect of employees' experience of the TQM intervention over time.

Teamwork

The importance of teamwork is clearly recognized in the TQM literature (Bowen & Lawler, 1992; Dean & Bowen, 1994; Deming, 1986; Hill, 1991). What is less clear is the meaning of teamwork. Dean and Bowen (1994) adopt an all-embracing interpretation running from teamwork within a natural work group to collaborative activity between organizations. In between, teamwork involves cooperative efforts between managers and employees in addition to collaboration between different functions. While teamwork may apply to different organizational groupings, the underlying theme is a willingness to cooperate. This focus, visible in the TQM literature, emphasizes holistic behavior oriented toward cooperation with fellow organizational members (Bushe, 1988; Drummond & Chell, 1992; Waldman, 1994). Waldman (1994) argues that in the context of TQM, it may be difficult to disentangle organizational citizenship behavior from in-role work performance. In essence, the implication of TQM is to broaden work responsibilities so that previously considered extra-role behaviors become an inherent component of an individual's job. Therefore, individuals, as part of their job, would be required to share information and help fellow employees; that is, engage in citizenship behaviors oriented toward the team.

The impact of TQM on team cooperation and effort may follow a similar route to that hypothesized for QCs. Steel and Lloyd (1988) present a theoretical map relating to the effect of employee involvement in QCs on intragroup cooperation. The authors theorize that employee participation in QCs will positively affect social contact with fellow members and lead to increased identification with the group. Group meetings, which allow individuals to work on shared problems, are hypothesized to accentuate group cohesion which in turn affects cooperation within the group. Given the importance attached to collaborative efforts in the pursuit of customer satisfaction and continuous improvement, the following proposition is examined.

> **Proposition 1.** The more positive the experience of TQM (greater participation, reinforcement, and benefit), the higher an individual's perception of teamwork.

Job Satisfaction

Although the link between the implementation of TQM and job satisfaction has not been subject to explicit theorizing, it is possible to explicate potential links based on what TQM purports to accomplish. TQM allows for greater individual influence, voice, and responsibility in work. In theory, this would have a positive effect on satisfaction with various intrinsic aspects of the job. Morrow (1997) theorizes that TQM would impact extrinsic job satisfaction in terms of providing a more suppor-

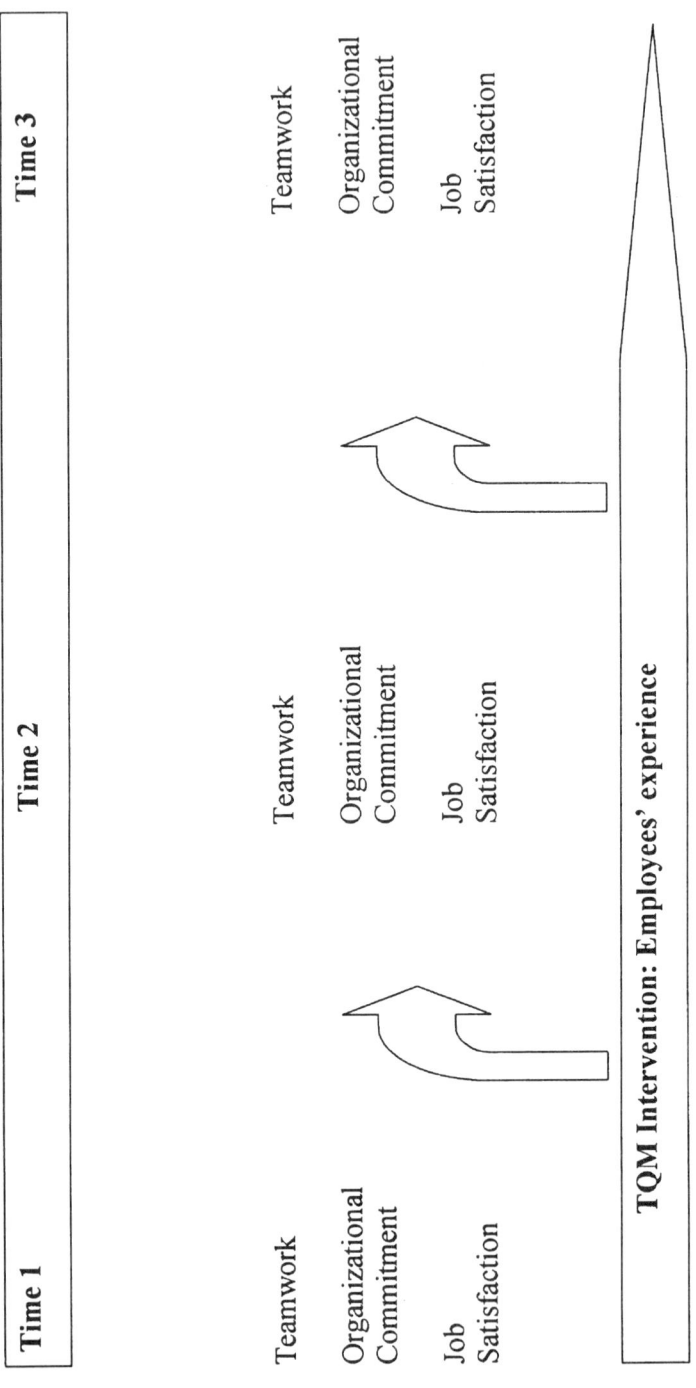

Figure 1. Evaluation Model

tive work environment (satisfaction with coworkers, supervisor, and working conditions). In addition, given that TQM relies heavily on recognition as a mechanism for rewarding contributions to improvement, employees should report greater satisfaction with the recognition they get for their work.

Morrow's (1997) cross-sectional study establishes a positive link between employees' perception of the implementation of the three principles of TQM (customer focus, teamwork, and continuous improvement) and work satisfaction, coworker and supervisor satisfaction (customer focus was not found to be significant in affecting supervisor satisfaction). What these findings support is a link between employees' perception of the operationalization of the principles of TQM and job satisfaction. However, the study did not examine the effect of the introduction of TQM on job satisfaction. Here, it is explored as follows.

Proposition 2. The more positive the experience of TQM, the higher an individual's job satisfaction.

Organizational Commitment

Organizational commitment seems to have achieved the status of a universal criterion in evaluating a range of employee participation programs. In theory, one route to achieving enhanced organizational commitment is through the implementation of employee involvement initiatives (Guest, 1992; Marchington, Wilkinson, Ackers, & Goodman, 1994). Under the umbrella of employee involvement, the nature of activities may vary substantially to include increased communication and consultation, restructuring of work, and changes to the structure of payment systems. Clearly, some of these activities are more likely to affect commitment than others. In the context of TQM, the nature of employee involvement primarily takes the form of participation in an improvement structure as a vehicle for contributing to continuous improvement (Lawler, 1994; Soin, 1992). In addition, under TQM, employees are likely to receive greater information on the overall organizational financial situation; greater information across departments; and upward communication of ideas and suggestions which in turn may enhance employees' commitment to the organization.

Despite the pervasive use of commitment as a criterion variable, the empirical evidence supporting the link between employee involvement and commitment is weak. Guest and Peccei (1994) suggest two possible explanations: weak interventions or a misguided underlying theory as to why these types of interventions could be expected to affect commitment. TQM does not set out to enhance commitment and, therefore, does not directly aim at changing it. However, one component of implementing TQM is the greater involvement of employees in organizational activities. It is this that may affect employees' commitment to the organization. Thus, while commitment is not an explicit outcome of TQM, it may follow from greater employee involvement in the TQM process. With this in mind, the following proposition is explored.

Proposition 3. The more positive the experience with TQM, the higher an individual's commitment to the organization.

METHOD

Organizational Context

The site was one of 34 production units of a U.K. multinational supplier of engineering and electrical components to the automotive and aerospace industry. In a personal message to all employees in 1985, the General Manager of the site stated "despite the strenuous actions on part of all employees over the past few years we are still trading at a loss and, unless this situation is remedied, we must eventually go out of business" (internal documentation). The major reason for the continuing losses of the site was a dramatic decline in the market for heavy-duty electrical components against a background of increased competition. In the previous five years, the U.K. market for heavy-duty electrical equipment fell by 48 percent. This was a direct result of a reduction in the U.K. bus and truck production. While the size of the market shrank, the site was able to increase its market share at the expense of competitors.

To reverse the operating losses, fundamental changes were needed. The objectives were to reduce labor costs, produce the highest quality and most reliable equipment, and reduce the response time to fulfilling customer requirements. The means to achieve this was the introduction of modular manufacturing systems. Module production (originating from the "kanban" system in Japan) is a method akin to "just-in-time" production where stocks and work-in-progress are aligned to fit the production schedule. The system has been described as mini factories within the factory (Turnbull, 1986).

After a pilot scheme, the modular system was introduced throughout the site in 1988. Each module could be described as a mini business producing a single product or a single group of products. All the manufacturing processes associated with that product were placed within the module; also, other activities, which were necessary to the module, were designed to give maximum support. This involved the transfer of activities traditionally carried out by indirect support departments to the module itself. For example, rather than having a centralized buying or purchasing department, this function was devolved and integrated into the module as one of its activities. However, activities such as personnel and management services remained centralized. Depending on the number of products manufactured within the module, the module may have been divided into a number of cells corresponding to each product or component. This cell-based structure, otherwise known as teamworking, became the building block of the new system and represented a radical departure from previous production methods.

A key feature of the new system was that personnel within the cell or module worked as a team. In some cases, this required a wider range of skills from employees. All employees were required to be willing to do whatever they could (taking into account their training and capabilities) to keep their cell or module operating. The aim was to have as much flexibility within the cell or module as possible. In practice (among other things), craftsmen and operators were trained to set up and operate, respectively, as many different machines as the cell or module required to manufacture their component or product. One of the major consequences of this new system was a significant reduction in the number employed. Inherent in the operation of the new system was flexibility not only in technology but also in personnel. What emerged was the disappearance of demarcation, the breeding of "super craftsmen," and the upgrading of operator skills.

In terms of organizational structure, there was a flattening of the hierarchy. The management structure was reduced from a seven-tier to a five-tier structure. Classification of direct production employees was reduced from seventeen to two. Thus, after the manufacturing reorganization, production employees were either craftsmen or operators. Through a combination of redundancies and retraining for new jobs, the number employed at the site was reduced by 50 percent (from 1,100 to approximately 580) by the late 1980s. In parallel to the manufacturing reorganization, changes in terms and conditions of employment were introduced. Specifically, this involved the creation of single status employment with no difference in conditions and benefits between staff and works employees. Methods of payment and pension scheme were harmonized. Pay grades for the shop floor were reduced from 59 to 8 and a simplified payment for skills was introduced.

By 1990, the site had transformed the loss-making situation to one of profitability. The product range was streamlined from 121 product types to 66. Responses to customer needs were reduced from on average 3 months to 1–6 weeks. Overall, the results of the manufacturing redesign meant that the site could match its competitors on quality and delivery. From this point, the main strategic objective was to focus on a narrower product range and to target European vehicle manufacturers for growth. Further changes were introduced to include continuing education and training schemes and an open learning center. In conjunction with health care, these changes signaled a willingness to invest in employees in order that they may reach their potential, which would have beneficial consequences for the site. A comprehensive communications program was launched which served (among other things) to bring what was happening in the marketplace to the door of the manufacturing cells and modules reinforcing the business focus of the new manufacturing system.

Compelled to build on the momentum of previous changes, the site in early 1991 launched continuous improvement groups (similar to the widely known QCs) as a means of continuing the improvements already achieved. From discussions with management and employees, the general view was that this initiative was doomed to failure from the start for a number of reasons. The main managerial objective in introducing continuous improvement groups was to provide a mechanism for employee contribution to efficiency and quality objectives. These groups, however, never gained a strong foothold at the site. From management's viewpoint, participation was voluntary and there was a lack of employee willingness to participate. Some employees felt that they had contributed to the site by accepting the reorganization of production and were unwilling to voluntarily participate in further change. Probably, the primary reason for the failure of these groups was a lack of visible support from managers and supervisors. Aside from the perceived lack of importance attached to these groups, they were interpreted as an add-on activity rather than integrated into the normal activities of the site. Overall, the take up of these groups was sporadic with some groups disbanding while others started.

By the spring of 1992, 20 or so employees were participating in these groups which were slowly fizzling out. In fact, some employees were unsure if their continuous improvement group was still in operation. Overall, the experience and failure of these groups prompted a more serious organization-wide endeavor into TQM. Against the background of the major manufacturing reorganization and the failure of continuous improvement groups, the site embarked upon a TQM intervention under the title of "working together to win" (WTTW).

The TQM Intervention

Senior Management Perspective

Overall, the executive management team at the site saw the intervention as a natural progression of previous changes in work methods, systems, and organizational structure. Furthermore, while the prior changes primarily focused on the hard visible reorganization of manufacturing, it was felt that the culture of the site lagged behind in terms of progress. Consequently, in order to guarantee the survival of the site and to ensure its continued profitability, the key was to change the culture of the site. This type of transition from continuous improvement groups or a similar grassroots improvement structure to TQM is not uncommon. Hill (1991) found that in the initial instance, some organizations experimented with QCs prior to implementing TQM. Even though the foray at the site with continuous improvement groups was far from successful, it signaled to senior management the potential value of employee involvement in continuous improvement and thus, the way forward.

The view of TQM as a philosophy or culture was a key factor in the design of the intervention. The objective was continuous improvement and this was to be achieved by the participative involvement of everyone. Senior management felt that there were lessons to be learned from the previous grassroots approach to change; change needed to occur throughout the site and must begin from the top. By late spring of 1992, the executive team had enlisted the help of a total quality expert from within the overall organization. This individual had considerable knowledge of the site and viewed TQM more in terms of attitudes and values rather than systems and techniques. The role assigned to this individual was to act as a bridge between the executive team and the group of outside consultants that would design and implement a "soft" TQM intervention. Considerable time and energy was given to the design of the intervention, the content of the training program, and its subsequent implementation.

Design of the Program

With the assistance of a total quality proponent and a group of outside consultants, a change program focusing on education and training was designed. The emphasis on training and education as the main driver for change is consistent with the TQM philosophy as espoused by its founders (Deming, 1986; Ishikawa, 1985; Juran, 1989) and consistent with the practice of implementing TQM (Hackman & Wageman, 1995; Hunter & Beaumont, 1993). The philosophy behind the TQM intervention was not too far removed from Xerox's "leadership through quality process" which "aimed at fundamentally changing the way Xerox people work and manage so they can continually improve the way they meet the requirements of their customers" (Ross, 1994, p. 53). The initial target for change were those in the management hierarchy and the prerequisite of change at this level for TQM to succeed is consistent with the TQM literature (Hackman & Wageman, 1995; Hill, 1991). It was assumed that as a consequence of the training and education program, a series of changes would occur throughout the site.

In attempting to create an involvement culture, the training and education covered issues such as theory X and Y, leadership styles, and empowerment. In addition, the tools and techniques of TQM were an integral part of the training. A second focus of the education and training was on leading and managing groups. This entailed team-building exercises, techniques for effective team meetings, and problem-solving techniques. An extension of this was work improvement through leaders facilitating employee involvement in the improvement process. Continuous improvement is the job of all employees who can contribute to small-scale improvements known as "kaizen." Naturally occurring work groups can contribute to "kaizen" but, in addition to cross-functional teams, may make breakthroughs in terms of major improvements. The traditional TQM tools and techniques included the Deming cycle, quality grid, cost of quality, customer-supplier process, brainstorming, fishbone diagrams and force field analysis. These tools were included as guides to achieving work improvements.

The education and training program was perceived to be a sufficient stimulus and inducement for culture change at the site. In theory, supervisors who completed this program would reproduce the training for their subordinates. Following from this, a WTTW team would be set up in that particular work area which would give employees a vehicle for greater say in what happens in their area, provide a mechanism for work improvements stimulated by the internal customer-supplier audit completed periodically with the internal customers of the work area. Furthermore, it would provide a forum whereby problems individuals faced in doing their work could be eliminated and suggestions for improvements could be implemented.

Process of Implementation

The starting point was the training and education program, which the outside consultants ran off site for the general manager and his executive management team. Subsequently, a group of internally selected facilitators were taken off site and undertook a facilitation workshop in addition to attending the training and education program. This group of facilitators, with the assistance of the outside consultants, were given the task and considerable autonomy in selecting the process by which this training and education was cascaded down the organization. Toward the end of 1992, the General Manager, the executive team, and the facilitators had completed the program. A steering committee was set up consisting of an equal number of facilitators and members from the executive team whose task was to launch and oversee the WTTW intervention.

The intervention itself was launched in a blaze of publicity in January 1993. To stimulate interest, a variety of poster and publicity campaigns were launched. The objectives of the intervention were communicated to the union representatives and the entire workforce. The group of facilitators ran the training and education program throughout the managerial/supervisory hierarchy with one-day follow-up workshops. These managers and supervisors were then responsible for training their subordinates. This cascading of the training process is similar to that conducted at Xerox whereby training began at the top of the organization and a manager, once trained, was responsible for training his/her immediate subordinates (Ross, 1994).

By the end of March 1993, all managers and supervisors had completed the program. It was now up to the individual managers to train their direct subordinates and involve them in the intervention. Overall, the feedback given to the internal facilitators on the program was generally positive. The role of the facilitators and steering committee was to guide and oversee the cascading of the training initially and the subsequent formation of teams. Regarding the latter, a manager and his/her subordinates, in theory, would set up a WTTW team in which a facilitator would be present to assist in the process of objective setting, developing action plans, and so forth. The subordinates, if they were supervisors, would subsequently set up their own teams in their work area. Thus, there would be a "linking pin" between different hierarchical teams facilitating the implementation of improvements. For example, a manager would participate in a team (cross-functional or with his/her superiors) which may suggest that certain improvements be made in a manager's work area. These suggestions would be subsequently taken to the manager's own team (in his/her work area) for discussion and subsequent implementation.

It was assumed that after completing the program, managers and supervisors would actively cascade the training to their employees and set up teams. In practice, this led to an uneven cascading process where some managers and supervisors resisted involving their employees in the TQM process. By the time of the second round questionnaire in September/October 1993, in some areas, the training had not only cascaded to the bottom of the organization but also teams had been set up. In contrast, in other areas, employees were still in the dark regarding the intervention apart from the information that had been communicated to them when the intervention was launched nine months previously. The steering committee decided that in dealing with the reluctant mangers, they would launch a "kick start" meeting. This involved each manager presenting to the steering committee the progress they had made in cascading the training and involving subordinates in the intervention. This was viewed as a way to induce change by applying pressure to these managers for action. For some individual managers, this had some effect. While it did not lead to wholehearted enthusiasm and commitment, they did make some progress. For the core resistors, this proved ineffective in stimulating change.

In the autumn of 1993, in view of the pockets of resistance at managerial levels, it was decided that progress in the intervention was to become an integral part of each manager's annual performance objectives and thus part of their performance appraisal. Despite the attempts of one union to bring the intervention into the annual pay negotiations, in keeping with the traditional TQM philosophy, there was no financial incentive offered to employees for their participation. After some debate, the steering committee decided against compulsory participation at employee levels. However, all new employees would be required, as part of their job, to participate in the intervention. During 1994, the site took over the manufacturing of a new product from a different site and consequently hired a new group of employees who were informed of the intervention. Their participation in TQM was made a condition of their employment.

Research Design and Sample

The research methodology employed consisted of a before and after study of the TQM intervention with three measurement occasions; 6 months prior to the commencement of the intervention (time 1); 9 months (time 2), and 32 months (time 3) after the start of the intervention. Prior to the administration of the first round questionnaire, trade union representatives were informed of the research, given the opportunity to raise questions and/or concerns, and were asked to support the research. Subsequently, as part of a quarterly communication's day whereby all employees are given a 40-minute presentation on relevant issues to the site, I introduced myself to the entire workforce and stated my independence from management at the site and the overall organization. In addition, employees were informed that the results of the survey would be communicated to them during a future communication's day. All these steps were taken to facilitate continued cooperation.

From a list of all employees by work area or functional area, a stratified random sample of 40 percent was taken. Having identified the sample, individuals were asked if they would be willing to complete a questionnaire and informed that it was voluntary. Most of the employees completed the questionnaire on a one-to-one basis away from their work area during work time. The first phase of data collection took place in May/June 1992, six months prior to the commencement of the TQM intervention at which stage none of the respondents were aware of the pending initiative. Therefore, the baseline questionnaire was not influenced by individuals' knowledge of the forthcoming intervention. The post-intervention measurements adopted the same administration procedure of the first questionnaire.

At time 1, of the 200 employees asked to complete the questionnaire, 186 did so yielding a response rate of 93 percent. The employee participant sample was reduced to 166 at time 2 and 118 at time 3 due primarily to employees leaving the site in the intervening period. At time 3, the participant group was 95 percent male, with a mean age of 48.0 years, a mean organizational tenure of 18.0 years, and a mean job tenure of 8.85 years. The sample consisted of machine operators (33.3%), craftsmen (26.4%), engineers (14.5%), material/purchase controllers (7.9%), with the remainder of the sample in administrative positions.

Measures

In measuring participation in the intervention, employees were asked to what extent they were participating in the activities of the intervention along a 5-point Likert scale from "not at all" to "a very great extent." Given the nature of the TQM intervention, this method permitted a more accurate representation of the degree of employee participation in the intervention. As a checking measure, when employees responded to this question, they were subsequently asked to elaborate on why they responded in a particular manner. Translating reported extent of participation into more concrete behaviors, the following classification was used. Employees whose response was either "not at all" or "not much" were aware of the intervention and had received communication about the intervention when it was launched. These employees had not received training and effectively were not (as yet) participating in the intervention. Fifty-four out of 117 employees fell into these two categories at time 2 (50 out of 117 at time 3). In contrast, individuals who responded in the "to a very great extent" and "to a great extent" categories had received training by their supervisor and were participating in teams with the aim of making improvements in their work area. Twenty-two out of 117 fell into these two categories at time 2 (31 at time 3). The remaining 41 employees responded in the "to some extent" category at time 2 (36 at time 3). These individuals were not participating in teams. However, they were trained in the principles of TQM by their supervisor and participating at a more informal (unstructured) level such as monitoring internal customer requirements or instigating corrective action based on problems identified by another improvement team.

The second TQM variable assessed respondents' judgment of the extent to which the TQM intervention is beneficial. This four-item scale included, for example, items such as "there is no benefit for me in ___" and "___ is a management initiative to get people to do more work." The final TQM variable measured respondents' assessment of the degree to which their immediate supervisor reinforces the principles of the TQM intervention. This variable taps respondents' assessment of the process of implementation of the TQM intervention. This scale included, for example, "my immediate boss is strongly committed to ___" and "my immediate boss involves me in ___."

The following variables were included in the subsequent analysis as control variables. The measure for supervisory participative style includes seven items assessing respondents' perception of the extent to which their immediate boss is participative and supportive in their behavior. Trust in management and trust in colleagues were each measured with six items taken from Cook and Wall (1980). A four-item quality awareness measure was specifically designed for this study. This measure assesses respondents' awareness of the importance of the quality of their work for others in the organization, the importance of continuous improvement for the success of the organization, and the degree to which respondents have specific ideas to improve the quality of work in their work group.

The measure of teamwork was specifically designed for this study. This three-item measure assesses identification with the work group, willingness to exert effort for the work group, and encouragement of teamwork by other work group members. Organizational commitment was adapted from Cook and Wall's (1980) nine-item scale excluding the three negatively phrased items. This six-item scale assessed an individual's identification, involvement, and loyalty toward the organization. Job satisfaction was measured using a 15-item scale developed by Warr, Cook, and Wall (1979) assessing satisfaction with intrinsic and extrinsic aspects of the job.

Procedure

Factor analysis (principal components with varimax rotation) was conducted on the items measuring perceived benefit of the TQM intervention, supervisory reinforcement of the principles of TQM, and the dependent variables to ascertain the construct independence of the main study variables. Hierarchical regression analyses were used to examine the effect of the TQM intervention on each of the dependent variables. The control variables were entered in step 1. The three elements of the TQM intervention were entered in step 2. This procedure allowed the isolation of the amount of unique variance in teamwork, job satisfaction, and organizational commitment associated with TQM after partialling out other factors.

The control variables included age, gender, organizational and job tenure with dummy variables being created for job title. Respondents were also asked at time 2 and time 3 whether they had experienced a change in supervisor; whether the nature of their job had changed substantially, and whether they had changed jobs in the intervening periods between time 1 and 2 and between time 2 and time 3. The remaining control variables include supervisory participative style, quality awareness, trust in management, and trust in colleagues. These variables need to be controlled for because they potentially could affect the independent and dependent variables. For example, supervisory participative style may affect teamwork, job satisfaction, and organizational commitment and also affect the degree to which employees perceive their supervisor as reinforcing the principles of TQM.

The hierarchical regression analysis was conducted in the following manner. To examine the short-term effects, the time 2–time 1 data was used whereby the control variables and dependent variable at time 1 were entered in step 1 and the TQM variables were entered in step 2. The same procedure was adopted to examine the effects of the TQM intervention in the subsequent period (time 3–time 2) and to examine the overall effects during the period of this study (time 3–time 1). A secondary regression was conducted in which the control variables and dependent variable at time 1 were entered in step 1 and the TQM variables at time 2 entered in step 2 to explain the outcome variables at time 3. This analysis may reflect more accurately the sequence of events involved in the change process: the control and dependent variables at time 1 reflecting the pre-TQM state, the TQM intervention at time 2, and the outcomes measured at time 3.

RESULTS

The descriptive statistics for the main study variables are presented in Table 1. The results of the factor analysis (principal components with varimax rotation) using the time 2 data support the factorial independence of perceived benefit and supervisory reinforcement of the intervention from each of the dependent variables (the results are not reported here). However, the items measuring job satisfaction did not load neatly onto two factors corresponding to intrinsic and extrinsic satisfaction. Consequently, all the items were combined to create an overall measure of job satisfaction. The factor analysis of the time 3 data yielded broadly similar results with one exception. Two of the job satisfaction items were found to have comparable loadings (below .5) on one of the job satisfaction factors and perceived benefit of the intervention. Overall, however, the results lend support to the independence of the main study variables.

Table 2 presents the results of the paired sample t-tests for the sample as a whole. Where there are significant differences over time, these occur between time 1 and time 2 or between time 1 and time 3. There are no significant differences between time 2 and time 3. Second, teamwork has decreased significantly between time 1 and time 2 remaining at that level at time 3. Although it is not possible to ascertain whether the changes over time are due to TQM from these results, they do provide a picture of the extent and direction of change over time. The pattern of change differs among the dependent variables. There was a negative change in teamwork

Table 1. Descriptive Statistics and Correlations Among Main Study Variables[a]

	M	S.D.	1	2	3	4	5	6	7	8	9	10	11	12	13	14	15
1. Teamwork T1	5.74	0.80	(.71)														
2. Job satisfaction T1	4.40	0.77	.34	(.87)													
3. Organizational commitment T1	5.40	0.88	.40	.43	(.78)												
4. Participation in TQM T2	2.54	1.09	.20	.14	.11	—											
5. Perceived benefit of TQM T2	4.20	1.26	.08	.19	.24	.39	(.75)										
6. Supervisory reinforcement of TQM T2	4.58	1.24	.28	.30	.13	.50	.23	(.68)									
7. Teamwork T2	5.59	0.80	.44	.22	.29	.35	.19	.43	(.68)								
8. Job satisfaction T2	4.56	0.78	.28	.59	.29	.29	.30	.44	.43	(.89)							
9. Organizational commitment T2	5.51	0.92	.34	.42	.67	.26	.29	.33	.44	.53	(.84)						
10. Participation in TQM T3	2.73	1.24	.29	.13	.26	.24	.28	.21	.32	.21	.35	—					
11. Perceived benefit of TQM T3	4.15	1.39	.35	.22	.30	.29	.56	.32	.28	.33	.46	.53	(.79)				
12. Supervisory reinforcement of TQM T3	4.50	1.44	.30	.10	.18	.21	.20	.39	.21	.25	.31	.60	.45	(.75)			
13. Teamwork T3	5.52	1.01	.49	.21	.32	.18	.20	.23	.48	.39	.44	.44	.44	.34	(.71)		
14. Job satisfaction T3	4.64	0.79	.36	.47	.27	.12	.31	.22	.33	.59	.47	.31	.50	.35	.55	(.89)	
15. Organizational Commitment T3	5.46	0.97	.28	.33	.57	.15	.18	.21	.28	.38	.72	.35	.45	.26	.49	.53	(.83)

Notes: Correlations ≥ .18 and ≤ .21 are statistically significant at $p \leq .05$.
Correlations ≥ .22 are statistically significant at $p < 0.01$.

Table 2. Paired Sample *t*-tests for the Overall Sample

	Time 1		Time 2		Time 3	
Variables (N = 115)	Mean	(S.D.)	Mean	(S.D.)	Mean	(S.D.)
Teamwork	5.74*+	(0.80)	5.59+	(0.80)	5.50*	(1.02)
Job satisfaction	4.39*+	(0.77)	4.55+	(0.78)	4.64*	(0.79)
Organizational commitment	5.39	(0.88)	5.50	(0.92)	5.46	(0.97)

Notes: *Significant differences at the .01 level between time 1 and time 3.
+Significant differences at the .05 level between time 1 and time 2.

between time 1 and time 2 remaining at that level for time 3, an initial slight positive change in commitment which is not sustained at time 3, and a consistent positive improvement in job satisfaction over time.

Tables 3, 4, and 5 present the results of the hierarchical regressions for the respective time periods: time 2–time 1 (T2–T1), time 3–time 2 (T3–T2), and time 3–time 1 (T3–T1). Turning first to the effects of TQM on teamwork, as Table 3

Table 3. Results of Hierarchical Regression Analyses (Time 2 – Time 1)[a]

	Teamwork T2		Job Satisfaction T2		Organizational Commitment T2	
Variables and Steps	1	2	1	2	1	2
Step 1:						
Control variables Time 1						
Dependent variable Time 1						
ΔR^2, step 1	.34		.45		.58	
F for ΔR^2	4.09**		6.65**		11.16**	
Step 2:						
Participation in TQM Time 2		.18		.02		.04
Perceived benefit of TQM Time 2		.01		.15*		.08
Supervisory reinforcement Time 2		.20*		.23**		.12
ΔR^2, step 2		.08		.07		.03
F for ΔR^2		4.43**		5.22**		2.34
Overall R^2		.33		.46		.55
Overall F		4.49**		7.06**		9.86**

Notes: *Significant at .05 level.
 **Significant at .01 level.
 [a]Beta coefficients are reported in columns.

Table 4. Results of Hierarchical Regression Analyses (Time 3 − Time 2)[a]

Variables and Steps	Teamwork T3		Job Satisfaction T3		Organizational Commitment T3	
	1	2	1	2	1	2
Step 1:						
Control variables Time 2						
Dependent variable Time 2						
ΔR^2, step 1	.36		.46		.61	
F for ΔR^2	4.42**		7.10**		13.42**	
Step 2:						
Participation in TQM Time 3		.15		.00		.07
Perceived benefit of TQM Time 3		.19		.25**		.09
Supervisory reinforcement Time 3		.04		.14		.00
ΔR^2, step 2		.07		.07		.01
F for ΔR^2		3.99**		5.78**		1.14
Overall R^2		.34		.47		.58
Overall F		4.66**		7.63**		11.16**

Notes: *Significant at .05 level.
**Significant at .01 level.
[a]Beta coefficients are reported in columns.

reveals, in the initial, period supervisory reinforcement has a significant positive effect on teamwork ($\beta = .20$, $p < .05$). Together, the TQM variables explain an additional 8 percent of the variance in teamwork at time 2 that was not accounted by the other variables. However, in the subsequent period (T3–T2) and for the overall time period (T3–T1), none of the TQM variables have a significant effect on teamwork. Overall, the results suggest that the impact of TQM is limited to the initial period of implementation but this effect is not sustained in the longer term.

The positive effect of supervisory reinforcement on teamwork sits alongside a significant decrease in teamwork over time. As Morrow (1997) notes, there is little evidence on the potential adverse effects of implementing TQM. The significant decrease in teamwork found here could be thought of as an unintended consequence of the TQM change process. One explanation is that other changes within the site may have counteracted the potential effects of the TQM intervention. As such, changes in the composition of work groups with the movement of individuals across work groups may explain the significant decrease in teamwork. However, this explanation is not supported by the data as none of these respondents reported a

Table 5. Results of Hierarchical Regression Analyses (Time 3 – Time 1)[a]

Variables and Steps	Teamwork T3		Job Satisfaction T3		Organizational Commitment T3	
	1	2	1	2	1	2
Step 1:						
Control variables Time 1						
Dependent variable Time 1						
ΔR^2, step 1	.36		.34		.34	
F for ΔR^2	4.29**		4.38**		4.38**	
Step 2:						
Participation in TQM Time 3		.18		–.02		.08
Perceived benefit of TQM Time 3		.17		.33**		.23**
Supervisory reinforcement Time 3		.04		.20*		.00
ΔR^2, step 2		.07		.13		.05
F for ΔR^2		4.16**		8.65**		3.68
Overall R^2		.34		.40		.42
Overall F		4.59**		5.94**		6.37**

Notes: *Significant at .05 level.
**Significant at .01 level.
[a]Beta coefficients are reported in columns.

change of jobs at either time 2 or time 3. The method in which work was structured in terms of specific work groups organized around specific components remained unchanged during the course of this study. A more plausible explanation directs attention to the TQM intervention. To pursue this line of inquiry further, subgroups of employees are examined using paired sample *t*-tests. For the groups of core participants (employees participating at time 2 and 3), their mean score remained unchanged over time. For the group of core nonparticipants (employees not participating at time 2 and 3), their mean score shifted in a negative direction over time. For employees not participating at time 2 but participating at time 3, their mean score decreased between time 1 and 2 and by time 3 returned to the pre-intervention level. For employees who were participating at time 2 but not at time 3, their mean score remained unchanged between time 1 and time 2 and significantly decreased by time 3.

Therefore, these results suggest that when employees are not participating in the TQM intervention, their perception of teamwork within their work group decreases. The group of core nonparticipants would have been aware of the activities of other groups in terms of meeting regularly to discuss suggestions for improvement and engaging in other cooperative efforts. The awareness of what is happening in other work areas may be providing a benchmark by which these employees are judging teamwork. This may lead them to adopt a more critical view of their own team and how they view their relationship with work group colleagues. For employees who were participating at time 2 but not at time 3, they, in evaluating teamwork, may be using their participation at time 2 as a benchmark and hence adopting a more realistic view of their teamwork at time 3. Employees who did not participate until time 3 followed a similar pattern in terms of their perception of teamwork: an initial decrease between time 1 and 2 that was recouped at time 3.

This raises the issue of why the TQM intervention maintains levels of teamwork for participants and results in a negative shift for nonparticipants. Furthermore, should this be interpreted as a failure of TQM? In distinguishing alpha and beta change, Golembiewski, Billingsley, and Yeager (1976) define the former as variations within a fixed state estimated by intervals that are constant. Beta change involves a recalibration of the intervals used by respondents. The authors further argue that a negative change between pre- and post-measurements of a change intervention may on the surface be taken as an indicator of "failure," if alpha change has occurred. However, it may concurrently be consistent with "success" if beta change has occurred as long as respondents are presenting a more realistic view of reality. The TQM intervention may have provided employees with a standard of what teamwork could be and as a consequence, employees examined their relationship with their work group with a more critical eye.

The effect of the TQM intervention on job satisfaction is strongly supported by the data irrespective of the time period examined. In the initial period, perceived benefit of TQM ($\beta = .15$, $p < .05$) and supervisory reinforcement of TQM ($\beta = .23$, $p < .01$) have a significant positive effect on job satisfaction explaining a further 7 percent of the variance in job satisfaction at time 2. In the subsequent time period, perceived benefit of TQM has a significant effect ($\beta = .25$, $p < .01$) explaining an additional 7 percent of the variance in job satisfaction at time 3. In the overall time period (T3–T1), the TQM variables explain 13 percent of the variance in job satisfaction not accounted for by the other variables. Overall, the results present a consistent picture of the effect of TQM on employees' job satisfaction.

Turning to the effects of TQM on organizational commitment, the results indicate that none of the TQM variables have a significant effect at T2–T1 or T3–T2. In contrast to job satisfaction, the TQM variables explain very little additional variance in organizational commitment (3% and 1% for T2–T1 and T3–2, respectively). However, the results of the T3–T1 analysis suggest that perceived benefit ($\beta = .23$, $p < .01$) has a significant positive effect on organizational commitment. Perhaps, the impact of TQM on organizational commitment takes, in comparison to the other dependent variables, longer to detect. Given that none of the TQM variables were close to attaining significance at T2–T1 or T3–T2, it is necessary to put this finding under closer scrutiny and examine the direction of influence.

The analysis thus far assumes that the dependent variables are endogenous in relation to the TQM variables. It is possible to test this endogeneity assumption through the use of cross-lagged regressions to examine the direction of effect. Cross-lagged regressions were conducted whereby the three TQM variables at time 2 and 3 were regressed independently on all the main variables at time 1 and 2, respectively. The results (not reported here) indicate that job satisfaction and teamwork at time 1 and time 2 have no effect on the TQM variables at time 2 or time 3, respectively. However, organizational commitment at time 1 had a minor effect ($\beta = .20$, $p < .10$) on perceived benefit at time 2. This effect was stronger in the subsequent period where organizational commitment at time 2 had a significant positive effect on perceived benefit at time 3 ($\beta = .26$, $p < .01$). The results support the endogeneity of teamwork and job satisfaction, thus rendering it appropriate to treat these variables as the consequence of TQM. However, this does not hold true for organizational commitment whereby the results suggest that organizational commitment is an antecedent (affecting perceived benefit of the intervention) rather than a consequence of TQM.

Table 6 presents the supplementary hierarchical regression analyses that corresponds to the sequence in the change process: the control variables (pre-TQM) are entered in step 1; TQM variables (time 2) are entered in step 2 to explain the dependent variable at time 3. Arguably, this analysis benefits from utilizing measures over three measurement occasions. The results lend support to the effect of TQM on job satisfaction (perceived benefit $\beta = .26$, $p < .01$). None of the TQM variables have a significant effect on the remaining outcome variables.

Any comparison of the results of this study with the findings of evaluation studies of QCs and SAWGs raises numerous problematic issues. Nonetheless, a few tentative and speculative conclusions may be drawn. First, it seems that TQM, alongside SAWGs, has greater potential to positively affect job satisfaction in contrast to the limited, if any, effect of QCs on job satisfaction (Atwater & Sander, 1984; Bruning & Liverpool, 1993; Head, Molleston, Sorensen, & Gargano, 1986; Rafaeli, 1985; Steel & Lloyd, 1988; Steel, Jennings, & Lindsey, 1990). This may be a result of the nature of changes involved in implementing each of the quality improvement technologies. In other words, the implementation of QCs is an add-on activity to the normal work of employees and the scope to affect job satisfaction may be severely limited. In contrast, SAWGs and TQM involve greater change to the content and context of work, which inherently are more likely to impact job satisfaction.

The limited effect of QCs, and to a lesser extent SAWGs, on organizational commitment is mirrored in this study of TQM (Atwater & Sander, 1984; Bruning & Liverpool, 1993; Cordery, Mueller, & Smith, 1991; Elloy & Randolph, 1997; Steel, Mento, Dilla, Ovalle, & Lloyd, 1985; Steel et al., 1990; Wall, Kemp, Jackson, & Clegg, 1986). Although purely speculative, together these findings suggest that change interventions that do not directly set out to enhance commitment may be unlikely to succeed. None of these interventions have as an explicit goal the

Table 6. Results of Hierarchical Regression Analyses[a]

Variables and Steps	Teamwork T3		Job satisfaction T3		Organizational Commitment T3	
	1	2	1	2	1	2
Step 1:						
Control variables Time 1						
Dependent variable Time 1						
ΔR^2, step 1	.36		.34		.44	
F for ΔR^2	4.29**		4.38**		6.50**	
Step 2:						
Participation in TQM Time 2		−.06		−.14		.03
Perceived benefit of TQM Time 2		.17		.26**		.07
Supervisory reinforcement Time 2		.06		.06		.06
ΔR^2, step 2		.02		.05		.00
F for ΔR^2		1.15		2.91*		0.36
Overall R^2		.28		.30		.36
Overall F		3.71**		4.09**		4.96**

Notes: *Significant at .05 level.
  **Significant at .01 level.
  [a]Beta coefficients are reported in columns.

enhancement of organizational commitment, but in practice entail greater employee involvement, which could be expected to result in enhanced commitment. However, the nature of changes involved in these types of improvement technologies may not be broad or deep enough to enhance commitment levels among employees. Furthermore, the possibility has been raised that organizational commitment among an experienced workforce may be quite a stable construct (Guest & Peccei, 1994). This latter explanation may be subject to one caveat; that is, the purposeful enhancement of organizational commitment through change interventions may be contingent on the existing level of commitment. Beyond a certain level, organizational commitment may become exceedingly difficult if not impossible to enhance. Hence, the issue of stability may be partly dependent on the nature of the "commitment-enhancing" intervention as well as the existing level of commitment among employees.

In summary, the data suggest that the effect of the TQM change intervention was different among the attitudinal outcomes. A potential explanation for the effect of TQM on the proximal outcome of teamwork was the occurrence of beta change whereby respondents' conceptualization of teamwork changed as a result of TQM. On the distal outcomes (not central to the principles of TQM), the effect of TQM on job satisfaction was positive while no effect on organizational commitment was found. Overall, this highlights the importance of adopting multiple attitudinal criteria for evaluation that vary in their proximity to the nature of the change intervention.

DISCUSSION

This case study examined the nature of a TQM intervention, the process of implementation and the consequences on employee work attitudes. The qualitative data provide insight into the nature and process of change while the quantitative data highlight the effect of the intervention on work attitudes. The data suggest that the TQM intervention does have some influence on employee work attitudes, in particular, job satisfaction. Yet, the impact of the intervention may have been severely constrained in its potential to affect work attitudes for a number of reasons. Beginning with the organizational context and the planning of change, there was a thread of factors running through this change attempt that limited its capacity to affect attitudinal change. The change achieved may have been all that could be reasonably expected given the presence of these inhibiting factors.

Effects of TQM

The effect of supervisory reinforcement of the TQM intervention on employees' teamwork is consistent with the human relations argument and the more recent evidence supporting a link between empowering leadership and citizenship behavior oriented toward the team (Cox & Sims, 1996; Salam, Cox, & Sims, 1996). However, the effect of the immediate supervisor on employees' teamwork was one of maintenance rather then development. Essentially, where supervisors were reinforcing the principles of TQM, this had the effect of preventing a negative change in how employees perceived teamwork. Conversely, where supervisors were perceived as not reinforcing the intervention, this had a detrimental effect on employees' perception of teamwork.

The TQM intervention may have influenced employees' interpretation of teamwork by providing a standard of what teamwork could be in practice. In doing so, through expanding employees' conceptualization of teamwork, this may have influenced how employees judged teamwork—given new horizons, employees may have adopted a more realistic and critical view of teamwork. Thus, rather than enhancing teamwork, this intervention may have succeeded in raising the standard of what teamwork could be. If this argument holds true, it suggests that negative change in the context of TQM may need to be interpreted cautiously rather than unduly presented as an indicator of failure. However, this line of argument may be subject to one caveat. It may only hold true when the attitudinal outcomes are proximal to the key principles of TQM that have been integrated into the education and training provided to employees. Thus, the occurrence of beta change may be more likely when change interventions constitute training and education and when the variable of interest is operationalized as part of the training program.

The TQM intervention appears to have a consistent effect on job satisfaction over the different time periods. This finding provides further support to the emerging conclusion that employees gain increased satisfaction from working under TQM despite the potential for TQM to require employees to increase their effort and responsibility over work-related issues. In this study, the results suggest that it is not participation *per se* that is important in affecting job satisfaction but rather how employees assess the intervention in terms of its benefits. From an employee perspective, TQM may provide the mechanism to enhance their responsibility and influence on quality and improvement issues, eliminate obstacles they experience in carrying out their work, and enhance their feelings of job security. Therefore, although the potential gain for employees working under TQM exists, the trade-off is that employees will have to make a greater contribution to quality and improvement issues. For some employees, this trade-off may not necessarily be welcomed.

The TQM intervention was not found to influence organizational commitment. This finding must be interpreted in light of the nature of TQM; its focus is not explicitly designed to enhance employee commitment to the organization. Given the specific limits of employee contributions, this type of narrowly focused participation may limit the extent to which enhanced commitment could reasonably be expected to occur. However, the findings point to a different role played by organizational commitment, not as an outcome of TQM but rather an antecedent of how TQM is judged. As such, prior commitment to the organization creates a positive evaluation bias in judging organizational change efforts and this is consistent with the proposition put forward by Eisenberger, Fasolo, and Davis-LaMastro (1990). Taking this further, the pre-change level of commitment may facilitate the change effort in affecting how individuals assess the change and subsequently influence their disposition toward change.

Limited Effect of TQM?

A potential conclusion to be drawn from this study is the limited effect of TQM on work attitudes. Before pursuing the underlying contributing factors to this conclusion, the role of extraneous influences must be considered. Clearly, it is important to explore the extent to which contextual factors may have influenced the findings. From a detailed knowledge of the site, there were no other planned changes introduced in conjunction with the TQM intervention. Furthermore, unplanned changes such as threatened layoffs, changes in top management, or unanticipated conflict with the trade unions did not occur. Insofar as it is possible to judge, it seems unlikely that exogenous factors could have dampened the impact of the TQM intervention. On the contrary, the site recruited new employees between time 2 and time 3 and during the same period, team leaders were introduced as a way of reinforcing the team concept.

Felt Need for Change

In attempting to unravel the reasons for the limited effect of TQM, a starting point is the organization's readiness for change. As a broad guiding framework, Lewin's (1958) classic three-step process of change can be used to discuss the limiting factors to change. In terms of unfreezing and creating a felt need for change, Cobb and Wooten (1998) present a typology of social accounts that organizational leaders can use in creating the conditions for effective change to occur. Of particular importance in this study is the causal account given by the organizational leader regarding why the site needed to embark upon TQM. In this study, the leader articulated the need for change by relying on the effects of competitors as the primary justification for change. The picture painted by the leader was a future of greater competition among automotive suppliers within an industry that was moving toward developing a long-term relationship with a few quality suppliers. In anticipation, the leader presented the future of the site as contingent upon its ability to continuously improve and change. In addition, internal forces for change came in the form of greater scrutiny from headquarters in terms of the contribution made by each of the production sites to the overall profitability of the organization. Consequently, the forthcoming TQM intervention was presented as a means of addressing and confronting the internal and external pressures facing this production site.

On the face of it, it would seem that compelling reasons were made to organizational members in terms of signaling the necessity of change. However, two issues emerged from discussions with employees that may have significantly reduced the effect of the intended message, at least in their eyes. First, employees relied on the fact that the site was one of the most profitable in the automotive division of the organization and used this as a way of diminishing the need for change. Second, there was a feeling of injustice among employees in terms of what was happening across the different sites. In particular, given that they were (as a site) doing comparatively well, why did they have to change, whereas other sites who were not doing as well financially were not embarking upon a TQM change process. In questioning the need for change, employees cognitively counteracted the message explaining the necessity of change from the organizational leader.

Nature of the Intervention

Although employees were questioning the need for change, this was not the only factor limiting the change effort. Rather, the nature of the intervention may have played a significant role in generating the effects found. In terms of the content of the TQM intervention in this study, the training and education program constituted the foundation of and signified the commencement phase of TQM. While this is consistent with the practice of TQM (Hackman & Wageman, 1995; Hunter & Beaumont, 1993), an exclusive emphasis on training and education may be greatly exposed to the problem of transferability—integrating the principles of TQM training and education into the normal activities of the organization. In the empirical work conducted to date on TQM, there is a general absence of detail outlining what is implemented as part of TQM as well as the process of implementation. Hence, it is difficult to compare the effects of different types of TQM interventions and consequently ascertain the extent to which the effects found are attributable to the type of change implemented as part of a TQM philosophy. Thus, while training and education may be a core element of TQM (Hackman & Wageman, 1995), the difference between how organizations proceed may lie in the subsequent changes adopted to support and reinforce a TQM philosophy and, hence, make the difference between successful and unsuccessful TQM change efforts. For example, Blackburn and Rosen (1993) outline a number of human resource practices designed to support and reinforce a TQM philosophy. Consequently, the outcomes achieved may partly depend on the extent to which and the type of subsequent changes made to other organizational subsystems to achieve congruence and consistency.

Process of Change: Supervisory and Employee Resistance

To compound the potential for limited change, the cascading process of change adopted at this site may have presented a major stumbling block to organization-wide involvement in TQM. The dependency inherent in the process of change and the fragility of the links created an uneven involvement in the change. While this type of approach to change has, in theory, the potential to gather momentum in the managerial ranks prior to reaching employees, in practice, the fragility of the process may lead to considerable variation horizontally and vertically in terms of involvement in TQM.

Supervisory unwillingness to change their behavior has been identified as a primary reason why empowerment initiatives fail (Manz, Keating, & Donnellon, 1990; Stewart & Manz, 1997; Verespej, 1990). Given that TQM requires some degree of employee empowerment over quality and improvement issues, supervisory recalcitrance in involving employees may help explain the limited impact of the TQM intervention investigated here. Supervisory attitudes toward employees and their conceptualization of the nature of supervisory roles are important in understanding resistance from supervisors. In this study, there was a clear divide between the opposers and the promoters of TQM. In the former category, two beliefs were apparent. One negative outcome felt was that the TQM intervention would lead to a loss of control and this was inextricably linked to supervisors' views of their role to direct and control. These supervisors developed a rationalization that there was no point in attempting to involve employees as they would not be interested nor would they be willing to take on board extra responsibility for quality and improvement issues. Other supervisors in this category pointed to a lack of employee ability as a valid reason for not wasting resources and effort in getting them involved.

Thus, the change intervention failed in its attempt to radically change the view of some supervisors on their role and their attitudes toward employees. For those supervisors who saw TQM as encroaching in their arena of control, they perceived little compensation in the form of greater delegation of control from the level above. If one takes this further up the organization, the example setting of devolving control may not have been as forthcoming and visible as it could have been and, arguably, as it should have been if those at the apex of the organization were truly committed. In recognizing this potential obstacle in the context of employee empowerment, Stewart and Manz (1997) argue that if first-line supervisors are to empower employees, they need to be empowered by those higher up the organizational hierarchy. As Nuemann (1989, p. 207) argues, "asking the powerless to empower others does not make sense." Thus, while this was seen as a "supervisory problem," its roots may have run higher.

Reger et al. (1994) argue that a common attribution of TQM failures put forward in the literature is implementation problems. The authors challenge this by offering an alternative explanation based on the assumption that TQM represents transformational change (this is disputed by writers such as Kanter). Thus, change efforts, which are presented as radical departures from the organization's past, are likely to fail due to cognitive opposition by organizational members—failure to understand the change and the perceived undesirability of the change. Based on this cognitive framework, the authors argue for tectonic change that overcomes the inability of incremental change to challenge inertia and also does not cause the cognitive overload of synoptic change. In this study, problems of implementation did surface and present an obstacle to the success (in terms of involvement) of the change effort. However, it is also plausible that the framing of change was a contributing factor. TQM was presented as a natural and evolutionary progression of the previous changes and, thus, closer to the incremental end of the change continuum. Consequently, the potential that the change was not seen as substantial enough to overcome cognitive inertia cannot be ignored. However, the proposed change may be interpreted differently by organizational members contingent on the perceived content and consequences of the proposed change. The recalcitrant supervisors in this study may illustrate the potential resisting force of their beliefs about the organization's identity, which subsequently made elements of TQM difficult to comprehend (in relation to the role of employees) and highlighted the undesirability of change.

A potential reason behind the unwillingness of some employees to embrace the principles of TQM may lie in the previous changes introduced at the site. There were spillover grievances from the introduction of cellular-based manufacturing in the late 1980s that were to come to the fore during the present TQM process. A number of employees felt that they had made a "sacrifice" in accepting the previous changes with little reciprocation from the organization. The TQM intervention, for these employees, presented yet another situation whereby under a different guise, employees felt that they were being asked to accept further change without recompense. Clearly, the TQM change had the unintended effect of jogging employees' memories of previous changes, which became associated with loss or rather imbalance in the exchange relationship. Employees raised the issue of rewards for adopting TQM; the trade union representatives were unsuccessful in bringing this issue into the annual rounds of negotiations. In this case, particularly when employees' experience of previous organizational changes were felt to be unfavorable, these experiences were harbored and surfaced when the next occasion for change was presented. Therefore, how employees react to organizational change needs to be understood in the context of the history of change within an organization and employees' previous experience of change.

It would be naïve to assume that employee resistance to the TQM change process is reducible to their previous experience of change. Rather, how employees respond to organizational change may partly depend on the implications of a particular change intervention for decision making within organizations. Nuemann (1989) draws attention to structural and relational factors as complementary explanations to that of individual personality in determining willingness to participate in organizational change. Essentially, Nuemann (1989) argues that if real decisions follow the chain of command without challenging the existing hierarchy, and if there is ineffective management of the participative effort, the willingness of employees to participate will be greatly reduced. In its original form, TQM does not alter the power structure within organizations and offers employees the opportunity to engage in a managerially defined and limited arena of participation—over quality and improvement issues. It maybe this that thwarts the potential for TQM to achieve desired change in organizations.

Resistance took the form of double jeopardy—resistance from supervisors who were needed to promote the change effort to employees and employees themselves. Regardless of whether the change was perceived to be evolutionary or revolutionary, the decision to embark on TQM was imposed by top management. The consequences of the change were interpreted by those resisting as being subtractive and this is consistent with the propositions put forward by Dirks, Cummings, and Pierce (1996) in understanding the conditions under which individuals will resist change. Change programs, by their very nature, seek to alter the status quo and inevitably some individuals will feel that they lose out. Therefore, for these individuals, what may be of particular saliency is what the change program offers. This resistance due to the perceived inequity of the outcomes of change may have been minimized had a systemic approach to the introduction of TQM been adopted.

Systemic Approach to Change

Successful organizational change requires change in a number of organizational subsystems (Nadler & Tushman, 1983) so that congruent messages are given to those affected by change but also to increase the likelihood that the change intervention generates the desired results (Robertson, Roberts, & Porras, 1993). Similarly, Macy and Izumi (1993) report from their meta-analysis that the strongest financial performance improvement is likely to occur from an integrated and holistic approach to organizational change rather than a "one-discipline" narrow focus change intervention. It is all too clear that the TQM framework is underdeveloped in terms of addressing how change occurs in organizations. Not only is the role of political behavior ignored (Kaplan et al., 1998) but the overly simplistic understanding of the reality of organizational change is not mirrored and supported by the empirical evidence on organizational change. Trade unions may play an important role in conferring legitimacy to a change process and their inclusion as a legitimate interest group may remove strong opposition to a TQM program. Yet, trade unions are rarely mentioned in the TQM literature implying that they fall outside the arena of teamwork, continuous improvement, and customer satisfaction.

In this study, it is the absence of changes introduced to reinforce TQM that may be critical in explaining the limited effects found. This assumes that multifaceted interventions have greater potential to achieve change than single category interventions. Although this proposition receives considerable support from the academic literature, the empirical evidence is mixed (cf. Robertson et al., 1993). However, the equivocal evidence needs to be interpreted with caution as evaluation studies of multifaceted interventions are in an embryonic stage. As recognized by Robertson et al. (1993), the definition of a multifaceted intervention and, hence, what differentiates this type of intervention from a single category intervention, may help explain the inconsistent results found across studies. The present categorization of interventions as either being single or multifaceted may be insensitive to the finer gradations of the possible range of organizational change interventions. Therefore, additional development in the categorization of change interventions may be of benefit in exploring the degree to which multiple change interventions within one category, across categories, or a combination of both generate greater change than a single intervention. Future research needs to explore the conditions, if any, under which different categories of interventions are more or less likely to generate change. These conditions may relate to the type of change and the environment in which the change is introduced. Multifaceted change interventions may elicit greater change than single category interventions when the perceived need for change is moderate. Under this condition, the reinforcing nature of congruent changes may provide the added stimulus needed for individual change to occur.

From a human resource management perspective, Cardy and Stewart (1998) address the issue of systemic change through a theoretical examination of the implications of TQM for selected practices such as job analysis, selection, performance appraisal, and compensation. From an organizational change perspective and recognizing the centrality of culture change to the success of TQM, it may be necessary to use multiple levers for change. The use of training and education *per se* may be effective in generating change for some individuals but not others. Hence, it may be important to supplement this with, for example, the modification of the reward system as a means of eliciting desired change. Recent empirical evidence suggests that profit sharing may be an effective mechanism to enhance employee orientation to continuous improvement, a core element of TQM (Coyle-Shapiro, 1999a). Consequently, some form of team-based reward system may not only facilitate the initial introduction of TQM but also assist in the integration of TQM into the core activities of the organization in the longer term.

A key challenge facing TQM is the incorporation of a systemic view of change thereby increasing its chances of delivering on its promises. In doing so, TQM may have to reconcile a paradox; the introduction of supporting changes to reinforce TQM may elicit greater change but at the same time dilute the core ideas of the movement's founders. Maybe the time has come to recognize that the core ideas and assumptions of the founders of TQM may not succeed in generating effective organizational change. Therefore, one remedy might be to integrate the distinctive features of TQM (customer focus, scientific methods, and process-management heuristics) with more traditional and theoretically grounded change interventions (for example, empowerment through SAWGs and team-based reward systems). Empowered teams may provide the building block within an overall TQM process. If there is an alignment of evaluation and reward systems based on team performance and the hypothesized outcomes associated with empowered teams (proactivity, quality, and customer satisfaction) hold true empirically, empowerment may make a strong contribution to the achievement of TQM outcomes at the organizational level. In developing a greater synthesis between empowered teams, reward systems, and TQM, this may present the means for TQM to recapture some of its lost magic—to achieve change, TQM may have to change.

Although TQM in this study was not introduced with a systemic perspective, the rationale for introducing TQM in this study was to achieve consistency with prior organizational changes. Using the four subsystems of work setting outlined by Robertson et al. (1993), the substantive change preceding the introduction of TQM involved changes to the organizing arrangements, physical setting, and technology leaving social factors largely untouched. Thus, the guiding principle behind the introduction of TQM that emphasized human-social processes was to bring this subsystem into alignment with previous changes. However, the TQM intervention introduced to achieve congruency with previous changes needed to be reinforced by other subsystems that were now out of alignment with the pursuance of a TQM philosophy.

Given the complexity of organizational change, a comprehensive framework may provide a solid basis to the understanding of the consequences, both positive and negative, of change efforts. Such a framework would incorporate factors such as an individual's prior experience of change, the organizational context in which change is introduced, the type of change, and the process of implementation. It is the interplay between an individual's disposition toward change, affected by one's previous experience of change, the type of change intervention, and the process of change that may provide a broader basis to our understanding of the inherent complexities of achieving successful change in organizations. The greater integration between the content and process of change may enhance the development of a change process theory.

CONCLUSIONS AND DIRECTIONS FOR FUTURE RESEARCH

The embryonic stage of TQM theory and the significant lag of academic investigations into this organizational phenomenon leave many gaps in our understanding of this current organizational practice. The central finding of this case study is the limited impact of a TQM intervention on work attitudes. Among the contributing factors presented was the absence of supporting changes introduced to reinforce a TQM philosophy. This, in turn, may reflect an underlying problem with the conceptualization of TQM and in particular, its assumptions about how change occurs in organizations.

As with the majority of studies, the design of the current study is subject to a number of limitations. First is the issue of generalizability of the findings in view of the potentially different guises of TQM in practice. Not to undermine the significance of this limitation, it does indeed reflect an underlying lack of clarity with the conceptualization and boundaries of TQM. Until such stage as TQM becomes more theoretically developed and refined, the issue of generalizability will continue to haunt empirical investigations. More specifically to this study, TQM is a full coverage intervention which renders the establishment of a "no treatment" control group difficult, if not impossible. To insulate a control group from the effects of TQM-related information and knowledge of TQM activities within the same organization is unrealistic. A third limitation is the small sample size due to the mortality effects inherent in conducting longitudinal studies. The low reliabilities of some of the measures used highlights the need for construct development appropriate to TQM research. The limitations of this study need to be offset by its strengths. A notable strength is the time span covered by the study that permitted insight into the planning and process of change as well as capturing the effects of change nearly three years after its introduction. However, those advocating a longer term horizon may dispute whether the time span allowed is adequate for the effects of TQM to materialize. Second, this setting had the advantage of a TQM intervention that was not contaminated with a host of changes designed to reinforce a TQM culture.

The findings point to a number of practical implications for the management of TQM. The need to adopt a systemic perspective in the planning stages of TQM is crucial to the achievement of desired change. It is reasonable to hypothesize that in this study, had the TQM intervention been reinforced with the introduction of some form of team-based rewards, the process of change may have been smoother and consequences of change may have been greater. In view of the pivotal role of first-line supervisors to the TQM process, the transition of supervisory behavior to a more empowering style needs to have visible top management support through example setting. In addition, supervisors need to feel that the change is not eliminating their position but rather changing their role—if TQM is seen as threatening by supervisors, they will be inhibited in promoting the change. In pursuing the objectives of the dominant coalition, management needs to involve trade unions in the introduction of TQM. The risk in bypassing the trade unions may be too great to be ignored if the fruits of a TQM process are to be reaped.

For researchers, if any meaningful conclusions can be drawn from TQM studies, a reconceptualization of TQM is needed with clear boundaries enclosing what TQM is and is not. Greater significance needs to be placed on the context and nature of change in studies investigating the impact of TQM as the question of what is being evaluated as part of TQM becomes even more crucial.

ACKNOWLEDGMENTS

This paper is based on my dissertation, which was supervised by Riccardo Peccei. The author would like to thank Paula Morrow, William Pasmore, Ray Richardson, Richard Woodman, and Stephen Wood for insightful comments on previous drafts. An earlier version of this paper was presented at the annual meeting of the Academy of Management, San Diego, August 1998.

REFERENCES

Atwater, L., & Sander, S. (1984). *Quality circles in navy organizations: An evaluation* (Technical Report No. nprdc tr8453). San Diego, CA: Navy Personnel Research and Development Center.

Blackburn, R., & Rosen, B. (1993). Total quality and human resources management: Lessons learned from Baldrige Award-winning companies. *Academy of Management Executive, 7*, 49-66.

Boje, D.M., & Winsor, R.D. (1993). The resurrection of taylorism: Total quality management's hidden agenda. *Journal of Organizational Change Management, 6*, 57–70.

Bowen, D.E., & Lawler, E.E. (1992). Total quality-oriented human resources management. *Organizational Dynamics, 20*, 29–41.

Bruning, N.S., & Liverpool, P.R. (1993). Membership in quality circles and participation in decision making. *The Journal of Applied Behavioral Science, 29*, 76–95.

Bushe, G.R. (1988). Cultural contradictions of statistical process control in American manufacturing organizations. *Journal of Management, 14*, 19–31.

Cardy, R.L., & Stewart, G.L. (1998). Quality and teams: Implications for HRM theory and research. In D.B. Fedor & S. Ghosh (Eds.), *Advances in the management of organizational quality* (Vol. 3, pp. 89–120). Greenwich, CT: JAI Press.

Cobb, A.T., & Wooten, K.C. (1998). *Social accounts in organizational change: Articulating justice.* Paper presented at the annual meeting of the Academy of Management, San Diego, CA.

Cobb, A.T., Wooten, K.C., & Folger, R. (1995). Justice in the making: Toward understanding the theory and practice of justice in organizational change and development. In W.A. Pasmore & R.W.

Woodman (Eds.), *Research in organizational change and development* (Vol. 8, pp. 243–295). Greenwich, CT: JAI Press.

Cook, J., & Wall, T. (1980). New work attitude measures of trust, organizational commitment and personal need non-fulfillment. *Journal of Occupational Psychology, 53*, 39–52.

Cordery, J. L., Mueller, W.S., & Smith, L.M. (1991). Attitudinal and behavioral effects of autonomous group working: A longitudinal field study. *Academy of Management Journal, 34*, 464–476.

Cox, J.F., & Sims, H.P. (1996). Leadership and team citizenship behavior: A model and measures. In M.M. Beyerlein, D.A. Johnson, & S.T Beyerlein (Eds.), *Advances in interdisciplinary studies of work teams* (Vol. 3, pp. 1–41). Greenwich, CT: JAI Press.

Coyle-Shapiro, J. A-M. (1996). *The impact of a TQM intervention on work attitudes: A longitudinal case study*. Unpublished Ph.D. dissertation, London School of Economics.

Coyle-Shapiro, J. A-M. (1999a). *TQM and profit sharing: A test of two methods of achieving change in employees' orientation to continuous improvement*. Paper presented at the annual meeting of the Academy of Management, Chicago, IL.

Coyle-Shapiro, J. A-M. (1999b). Employee participation and assessment of an organizational change intervention: A three wave study of Total Quality Management (TQM). *The Journal of Applied Behavioral Science, 35*, 439–456.

Coyle-Shapiro, J. A-M., & Kessler, I. (2000). Consequences of the psychological contract for the employment relationship: A large scale survey. *The Journal of Management Studies*, forthcoming.

Cummings, T.G. (1978). Self-regulating work groups: A sociotechnical synthesis. *Academy of Management Review, 3*, 625–634.

Dean, J.W., & Bowen, D.E. (1994). Management theory and total quality: Improving research and practice through theory development. *Academy of Management Review, 19*, 392–418.

Delbridge, R. Turnbull, P., & Wilkinson, B. (1992). Pushing back the frontiers: Management control and work intensification under JIT/TQM factory regimes. *New Technology, Work and Environment, 7*, 97–106.

Deming, W.E. (1986). *Out of the crisis*. Cambridge, MA: MIT Center for Advanced Engineering Study.

Dirks, K.T., Cummings, L.L., & Pierce, J.L. (1996). Psychological ownership in organizations: Conditions under which individuals promote and resist change. In R. Woodman & W. Pasmore (Eds.), *Research in organizational change and development* (Vol. 9, pp. 1–23). Greenwich, CT: JAI Press.

Dobyns, L., & Crawford-Mason, C. (1991). *Quality or else: The revolution in world business*. Boston: Houghton-Mifflin.

Drummond, H., & Chell, E. (1992). Should organizations pay for quality? *Personnel Review, 21*, 3–11.

Eisenberger, R., Fasolo, P., & Davis-LaMastro, V. (1990). Perceived organizational support and employee diligence, commitment and innovation. *Journal of Applied Psychology, 75*, 51–59.

Elloy, D.F., & Randolph, A. (1997). The effect of superleader behavior on autonomous work groups in a government operated railway service. *Public Personnel Management, 26*, 257–272.

Fisher, T.J. (1992). The impact of quality management on productivity. *International Journal of Quality & Reliability Management, 9*, 44–52.

Gilbert, J.D. (1992). TQM flops: A chance to learn from the mistakes of others. *National Productivity Review, 11*, 491–499.

Golembiewski, R.T., Billingsley, K., & Yeager, S. (1976). Measuring change and persistence in human affairs: Types of change generated by OD designs. *Journal of Applied Behavioral Science, 12*, 133–157.

Grant, R.M., Shani, R., & Krishnan, R. (1994, Winter). TQM's Challenge to management theory and practice. *Sloan Management Review*, pp. 25–35.

Guest, D. (1992). Employee commitment and control. In J. Hartley & G. Stephenson (Eds.), *Employment relations* (pp. 111–135). Oxford: Blackwell.

Guest, D., & Peccei, R. (1994). *A test of the feasibility of changing organizational commitment*. Paper presented at the British Occupational Psychology Conference, Brighton.

Hackman, J.R., & Wageman, R. (1995). Total quality management: Empirical, conceptual, and practical issues. *Administrative Science Quarterly, 40*, 309–342.

Head, T.C., Molleston, J.L., Sorensen, P.F., & Gargano, J. (1986). The impact of implementing a quality circle intervention on employee task perceptions. *Group and Organization Studies, 11*, 360–373.

Hill, S. (1991). Why quality circles failed but total quality management might succeed. *British Journal of Industrial Relations, 29*, 541–568.

Hill, S. (1995). From quality circles to Total Quality Management. In A. Wilkinson & H. Willmott (Eds.), *Making quality critical* (pp. 33–53). London: Routledge.

Howes, J.C., Citera, M., & Cropanzano, R.S. (1995). Total quality teams: How organizational politics and support impact the effectiveness of quality improvement teams. In R.S. Cropanzano & K.M. Kacmar (Eds.), *Organizational politics, justice and support: Managing the social climate of the workplace* (pp. 165–184). Westport, CT: Quorum.

Hunter, L., & Beaumont, P.B. (1993). Implementing TQM: Top down or bottom up? *Industrial Relations Journal, 24*, 318–327.

Ishikawa, K. (1985). *What is total quality control? The Japanese way.* Englewood Cliffs, NJ: Prentice-Hall.

Juran, J.M. (1989). *Juran on leadership for quality.* New York: Free Press.

Kanter, R.M. (1989). *When giants learn to dance.* New York: Simon and Schuster.

Kaplan, D., Birmingham, C., & Ferris, G.R. (1998). Influence and politics in organizational quality contexts. In D.B. Fedor & S. Ghosh (Eds.), *Advances in the management of organizational quality* (Vol. 3, pp. 287–320). Greenwich, CT: JAI Press.

Kerfoot, D., & Knights, D. (1995). Empowering the "quality" worker: The seduction and contradiction of the total quality phenomenon. In A. Wilkinson & H. Willmott (Eds.), *Making quality critical* (pp. 219–238). London: Routledge.

Kirkman, B.L., & Rosen, B. (1997). A model of work team empowerment. In W. A. Pasmore & R.W. Woodman (Eds.), *Research in organizational change and development* (Vol. 10, pp. 131–167). Greenwich, CT: JAI Press.

Lawler, E.E. (1986). *High-involvement management.* San Francisco, CA: Jossey-Bass.

Lawler, E.E. (1994). Total quality management and employee involvement: Are they compatible? *Academy of Management Executive, 8*, 68–76.

Lewin, K. (1958). Group decision and social change. In E.E. Maccoby, T.M., Newcomb, & E.L. Hartley (Eds.), *Readings in social psychology* (pp. 163–226). New York: Holt, Rinehart, and Winston.

Macy, B.A., & Izumi, H. (1993) Organizational change, design and work innovation: A meta-analysis of 131 North American field studies—1961-91. In W.A. Pasmore & R.W. Woodman (Eds.), *Research in organizational change and development* (Vol. 7, pp. 235–313). Greenwich, CT: JAI Press.

Manz, C.C., Keating, D.E., & Donnellon, A. (1990). Preparing for organizational change to employee self-management: The managerial transition. *Organizational Dynamics, 19*, 15–26.

Marchington, M., Wilkinson, A., Ackers, P., & Goodman, J. (1994). Understanding the meaning of participation: Views from the workplace. *Human Relations, 47*, 867–894.

McArdle, L., Rowlinson, M., Procter, S., Hassard, J., & Forrester, P. (1995). Total Quality Management and participation: Employee empowerment, or the enhancement of exploitation. In A. Wilkinson & H. Willmott (Eds.), *Making quality critical* (pp. 156–172). London: Routledge.

Mohrman, S., Tenkasi, R.K., Lawler, E., III., & Ledford, G. (1995). Total quality management: Practice and outcomes in the largest U.S. firms. *Employee Relations, 17*, 26–41.

Morrow, P.C. (1997). The measurement of TQM principles and work-related outcomes. *Journal of Organizational Behavior, 18*, 363–376.

Munroe-Faure, L., & Munroe-Faure, M. (1992). *Implementing total quality management.* London: Pitman.

Nadler, D.A., & Tushman, M.L. (1983). A general diagnostic model for organizational behavior: Applying a congruence perspective. In J.R. Hackman., E.E. Lawler, & L.W. Porter (Eds.), *Perspectives on behavior in organizations* (pp. 112–124). New York: McGraw-Hill.

Neumann, J.E. (1989). Why people don't participate in organizational change. In R.W. Woodman & W.A. Pasmore (Eds.), *Research in organizational change and development* (Vol. 3, pp. 181–212). Greenwich, CT: JAI Press.

Nutt, P.C., & Backoff, R.W. (1997) Transforming organizations with second-order change. In W.A. Pasmore & R.W. Woodman (Eds.), *Research in organizational change and development* (Vol. 10, pp. 229–274). Greenwich, CT: JAI Press.

Organ, D.W., & Konovsky, M. (1989). Cognitive versus affective determinants of organizational citizenship behavior. *Journal of Applied Psychology, 74*, 157–164.

Organ, D.W., & Ryan, K. (1995). A meta-analytic review of attitudinal and dispositional predictors of organizational citizenship behavior. *Personnel Psychology, 48*, 775–802.

Powell, T.C. (1995) Total quality management as competitive advantage: A review and empirical study. *Strategic Management Journal, 16*, 15–37.

Procopio, A.J., & Fairfield-Sonn, J.W. (1996). Changing attitudes toward quality: An exploratory study. *Group & Organization Management, 21*, 133–145.

Rafaeli, A. (1985). Quality circles and employee attitudes. *Personnel Psychology, 38*, 603–615.

Reed, R., & Lemak, D.J. (1998). Total quality management and sustainable competitive advantage in service firms. In D.B. Fedor & S. Ghosh (Eds.), *Advances in the management of organizational quality* (Vol. 3, pp. 121–159). Greenwich, CT: JAI Press.

Rees, C. (1998). Empowerment through quality management: Employee accounts from inside a bank, a hotel and two factories. In C. Mabey, D. Skinner, & T. Clark (Eds.), *Experiencing human resource management* (pp. 33–53). London: Sage.

Reger, R.K., Gustafson, L.T., DeMarie, S.M., & Mullane, J.V. (1994). Reframing the organization: Why implementing total quality is easier said than done. *Academy of Management Review, 19*, 565–584.

Robertson, P.J., Roberts, D.R., & Porras, J.I. (1993) An evaluation model of planned organizational change: Evidence from a meta-analysis. In W.A. Pasmore & R.W. Woodman (Eds.), *Research in organizational change and development* (Vol. 7, pp. 1–39). Greenwich, CT: JAI Press.

Ross, J.E. (1994). *Total quality management: Text, cases and readings* (2nd ed.). London: Kogan Page.

Salam, S., Cox, J., & Sims, H.P. (1996). *How to make a team work: mediating effects of job satisfaction between leadership and team citizenship*. Paper presented at the annual meeting of the Academy of Management, Ohio.

Sewell, G., & Wilkinson, B. (1992). Empowerment or emasculation?: Shopfloor surveillance in a total quality organization. In P. Blyton & P. Turnbull (Eds.), *Reassessing human resource management* (pp. 97–115). London: Sage.

Soin, S. (1992). *Total quality control essentials: Key elements, methodologies, and managing for success*. New York: McGraw-Hill.

Steel, R.P., & Jennings, K.R. (1992). Quality improvement technologies for the 90s: New directions for research and theory. In W.A. Pasmore & R.W. Woodman (Eds.), *Research in organizational change and development* (Vol. 6, pp. 1–36). Greenwich, CT: JAI Press.

Steel, R.P., Jennings, K.R., & Lindsey, J.T. (1990). Quality Circle problem solving and common cents: Evaluation study findings from a United States federal mint. *The Journal of Applied Behavioral Science, 26*, 365–381.

Steel, R.P., & Lloyd, R.F. (1988). Cognitive, affective and behavioral outcomes of participation in quality circles: Conceptual and empirical findings. *The Journal of Applied Behavioral Science, 24*, 1–17.

Steel, R.P., Mento, A.J., Dilla, B.L., Ovalle, N.K., & Lloyd, R.F. (1985). Factors influencing the success and failure of two quality circle programs. *The Journal of Management, 11*, 99–119.

Stewart, G.L., & Manz, C.C. (1997). Understanding and overcoming supervisor resistance during the transition to employee empowerment. In R.W. Woodman & W.A. Pasmore (Eds.), *Research in organizational change and development* (Vol. 10, pp. 169–196). Greenwich, CT: JAI Press.

Tannenbaum, S.I., Mathieu, J.E., Salas, E., & Cannon-Bowers, J.A. (1991). Meeting trainees' expectations: The influence of training fulfillment on the development of commitment, self-efficacy, and motivation. *Journal of Applied Psychology, 76*, 759–769.

Turnbull, P.J. (1986). The "Japanisation" of production and industrial relations at Lucas Electrical. *Industrial Relations Journal, 17*, 193–206.

Verespej, M.A. (1990). Worker involvement-Yea, teams? Not always. *Industry Week, 289*, 104–105.

Waldman, D.A. (1994). The contributions of total quality management to a theory of work performance. *Academy of Management Review, 19*, 510–536.

Wall, D.T., Kemp, N.J., Jackson, P.R., & Clegg, C.W. (1986). Outcomes of autonomous workgroups: A long term field experiment. *Academy of Management Journal, 29*, 280–304.

Warr, P.B., Cook, J.D., & Wall, T.D. (1979). Scales for the measurement of some work attitudes and aspects of psychological well-being. *Journal of Occupational Psychology, 52*, 129–148.

Wilkinson, A., Godfrey, G., & Marchington, M. (1997). Bouquets, brickbats and blinkers: Total quality management and employee involvement in practice. *Organisation Studies, 18*, 799–819.

Wilkinson, A., Snape, E., & Allen, P. (1993). *Quality and the manager* (an IM report). Institute of Management, Corby.

Wood, S., & Peccei, R. (1995). Does total quality management make a difference to employee attitudes? *Employee Relations, 17*, 52–62.

Wruck, K.H., & Jensen, M.C. (1994). Science, specific knowledge and total quality management. *Journal of Accounting and Economics, 18*, 247–287.

IMPLEMENTING EFFECTIVE CROSS-FUNCTIONAL TEAMS
A MULTILEVEL FRAMEWORK FOR ANALYSIS

Long W. Lam, Sheri J. Bischoff,
La Verne H. Higgins, and D. Lynne Persing

ABSTRACT

The implementation of cross-functional teams (CFTs) is a significant organizational change. Using a multilevel perspective as a lens, we develop a conceptual framework for CFT implementation that analyzes contextual requirements, the role of interventions, and the effects on organizations. Our analysis suggests that using CFTs should be viewed as a double-edged sword: Certain management interventions are needed to unleash benefits to organizations, groups, and individuals involved in CFTs, but these interventions can produce counterproductive effects that in turn require remedial solutions. Additionally, interventions that focus on one level of CFT effectiveness can produce unintended multilevel outcomes.

Increased industry competitiveness and time pressures led Honeywell Corporation's Building Controls Division to abandon its sequential approach to product development in the mid-1980s. Honeywell shifted to cross-functional teams (CFTs) that used specialists from manufacturing, marketing, sales, and engineering. These CFTs were charged with directing the development of new products from the conceptual to the manufacturing stage, coordinating multiple functional efforts simultaneously. One Honeywell manager commented:

> The team system does not allow people to single-mindedly defend the position of their functional area, of what's easiest, or best, or cheapest for their functional area. It forces people to look at a bigger picture. (Margolis & Donnellon, 1991, p. 4)

Top-level management maintained a high expectation that this new design would result in faster product development and better financial returns. However, using CFTs in the quest to improve competitiveness and return-on-investment led to unintended consequences. Functional barriers were not lowered, and the shift to CFTs exacerbated conflict between departments. One member of a particular Honeywell CFT explained:

> Larry (engineering) and Phil (marketing) are the two most prone to clash. It's because of their roles. The two roles are adversarial. Larry and Phil have to represent different interests. Larry's is what's fastest and least expensive. Phil's is what's best for the customer. (Margolis & Donnellon, 1991, p. 11)

In the ongoing quest to improve effectiveness, organizations continually search for and adopt new and modified managerial tools and techniques. This paper focuses on one of these tools for organizational change—cross-functional teams. We define CFTs as "groups of functionally diverse members both internal and external to the organization who work together as coordinated units to accomplish specific projects that generally require a time-limited effort" (Cohen & Bailey, 1997; Denison, Hart, & Kahn, 1996; Pinto, Pinto, & Prescott, 1993; Uhl-Bien & Graen, 1992). CFTs are usually temporary work teams that pool specialists from different functional areas to work on single projects.

Researchers have regarded CFTs as the solution to many organizational problems, including overly long cycle times in product development, poor coordination among divisional units, and a lack of firm competitiveness in the global arena (Ancona & Caldwell, 1990b; Dean & Susman, 1989; Nonaka, 1991; Parker, 1994; Takeuchi & Nonaka, 1986). Hull and Azumi (1989), in their study of U.S. and Japanese laboratories, concluded that multifunctional teams were key in accelerating the technology life cycle. In addition, CFTs were found in companies engaging in continuous improvement. Almost one-half of the companies studied by Lillrank, Shani, Kolodny, Stymne, Figuera, and Liu (1998) used some form of temporary CFTs to examine how their business processes could be redesigned.

When properly implemented, CFTs can promote greater interfunctional coordination than traditional organizational structures, resulting in decreased cost and time for product development. Because of these perceived benefits, CFTs have been used extensively by organizations including high technology (e.g., Canon, Hewlett-Packard), automobiles (e.g., Honda), health care, and the military (Ancona & Caldwell, 1990a; Denison et al., 1996; Takeuchi &Nonaka, 1986). For example, Boeing relied heavily on 250 CFTs in designing and manufacturing its Boeing 777 aircraft (Galbraith, 1994; Yang, Oneal, & Troy, 1990). However, CFTs are not a panacea for all organizational ills. As the Honeywell example demonstrates, the adoption of CFTs can lead to negative as well as positive results. Consequently, it may be advantageous to view CFTs as a double-edged sword—implementation can lead to both benefits and detriments. While the benefits of CFTs have been studied extensively, it remains unclear how and when the detriments of CFTs are produced. We argue that oftentimes the detriments occurring in some CFTs may be unintended consequences of managerial interventions aimed to unleash the potential benefits. Therefore, we believe that it is necessary to explore the additional interventions needed to curb the detriments of CFTs and optimize this organizational form.

An additional shortcoming in both theoretical and empirical work related to CFT implementation is that it tends to adopt a single level of analysis, ignoring the multilevel effects on individual, group, and organizational outcomes. For example, while adopting CFTs, do organizations improve their financial returns at the expense of their employees' satisfaction? Does the CFT experience more absenteeism and turnover than a conventional group—and is this justified by higher returns in innovation? Is it possible to design CFTs such that positive outcomes can occur simultaneously at the individual, group, and organizational level? Cohen and Bailey (1997) recognize the importance of this multilevel focus in their review of small group research; they indicate that environmental factors affecting CFTs had received little research attention compared to the heavy emphasis on internal group processes.

To address these gaps we will integrate and extend previous research by building a multilevel conceptual model of CFT effectiveness. Our conceptual model focuses on the context, interventions, and outcomes of CFT implementation in organizations. After providing the theoretical background for our analysis, we will address *when* CFTs should be implemented by examining the effect of industry and organizational context. Next we will explore *how* CFTs should be implemented by discussing the role of managerial interventions. Finally we will address how individual, group, and organizational effectiveness will be affected after CFTs are implemented.

THEORETICAL BACKGROUND

Our understanding of the multifaceted nature of CFTs can be broadly divided into four areas. The first is the domain of small-group researchers. By adopting the social psychological perspective, this line of research has traditionally been interested in exploring the dynamics of internal group processes as the precursor to effective CFTs (e.g., Ancona & Caldwell, 1992c; Gladstein, 1984; McGrath, 1984). Recently, this stream has expanded to include the boundary between CFTs and their external constituents (Ancona, 1990; Ancona & Caldwell, 1992a). The second area of CFT research concentrates on new product development teams (e.g., Eisenhardt & Tabrizi, 1995; Hull & Azumi, 1989; Katz & Tushman, 1979; Kazanjian & Drazin, 1986; Takeuchi & Nonaka, 1986). This research has addressed the timeliness and cost of developing new products, promoting innovativeness, and creating new designs (Brown & Eisenhardt, 1995). A third line of research addresses the functional composition and characteristics of top management teams (e.g., Finkelstein & Hambrick, 1990; Glick, Miller, & Huber, 1993; Korn, Milliken, & Lant, 1992; Simons, 1995; Smith, Smith, Olian, Sims, O'Bannon, & Scully, 1994). These researchers concentrate on the upper echelon's perspective and analyze how the functional composition of top management teams affects organizational performance (Hambrick & Mason, 1984). For example, when analyzing 53 high-tech companies, Smith et al. (1994) found that group processes of top management teams in terms of integration and communication accounted for a portion of their financial performance. In addition to direct effects on firm performance, the demography of top management teams also affected firm performance indirectly by influencing how group processes operated among team members. A fourth broad area of CFT research focuses on CFTs as tools of organizational development and change in the contexts of (1) creating team-based organizations (Mohrman & Mohrman, 1997), (2) effecting integration between R&D and marketing (Souder & Sherman, 1993), and (3) contributing to organizational dialogue (Ashkenas & Jick, 1992).

Mohrman and Mohrman (1997) suggest that organizations might benefit from the adoption of team-based structures to deal with the growing complexity and competitiveness of the environment. The team-based structure is an organizational design that is made of "loosely coupled, dynamic teams that form and disband to accomplish the array of tasks that emerge as the organization defines and pursues its dynamic strategy in an unfolding environment" (Mohrman & Mohrman, 1997, pp. 198–199). While Mohrman and Mohrman suggest that this type of organizational form promises great flexibility, successful examples are still relatively scarce because the structure requires large-scale organizational change. An intermediate change that is more widely used by organizations is to shift from traditional and comparatively static functional teams to more dynamic CFTs. The authors thus suggest that the change to team-based organizations can be viewed as an iterative process. That is, individual CFTs can serve as stepping stones toward a network of teams and, ultimately, team-based organizations. Indeed, the importance of making successful intermediate change is supported by investigations of continuous improvement. For example, after comparing continuous improvement efforts in eight companies around the globe, Lillrank et al. (1998) concluded that organizational change need not be large, far-reaching, and radical to succeed. They found that small and incremental adjustments also could lead to positive organizational change in the longer run.

CFTs are also explored in terms of the integration between marketing and R&D (Souder & Sherman, 1993). Most organizations feature two different groups of specialists—marketers and technologists. Each group has its own training, culture, and orientation. Firms that can integrate these two separate groups are therefore able to become successful in product innovation and commercialization simultaneously. To do so requires certain organizational development tools and techniques (Souder & Sherman, 1993). Not surprisingly, one of these tools is the cross-functional integration of marketing and R&D. This structural solution can take the form of temporary cross-functional task forces with members resuming their formal duties after the project is accomplished. Alternatively, there exists the option of creating permanent multifunctional teams or departments in charge of new product development. Souder and Sherman (1993) suggest that these structural arrangements may be costly, but that significant advantages derive from the increased capability to manage the process of innovation.

Finally, CFTs have been discussed in the context of what Ashkenas and Jick (1992) contend is one of the largest and most successful organizational development interventions ever undertaken—General Electric's (GE) "Work-Out" program. With this intervention, employees from different levels of the company participated in a two- or three-day event to develop new ideas to reduce unnecessary work and to make recommendations regarding business processes. In essence, the "Work-Out" program comprised numerous temporary CFTs that were encouraged to use brainstorming to identify new ideas for improvement. Ashkenas and Jick (1992) believe that because at the end of each "Work-Out" session every team's idea was actually considered by top management for final implementation, this form of cross-functional collaboration led to fundamental change in GE's culture—a change toward greater employee empowerment. The authors stress that this type of cross-functional effort can lead to effective "organizational dialogue." That is, the employees break through their functional barriers and ultimately learn to communicate effectively beyond them.

In the prior research, an additional and related issue is whether or not CFTs should be included in a broader definition of groups or teams. Commonalties do exist among CFTs and conventional work teams, such as inclusiveness (members generally know if they belong to a work group or not) and interaction to share information among different types of teams (e.g., expert, self-managing, problem solving, cross-functional). However, notable differences do exist. We believe there are five core characteristics that can differentiate CFTs from conventional work teams that exist in functional and divisional organizations (Hitt, Hoskisson, &

Nixon, 1993; Mohrman, Cohen, & Mohrman, 1995). Core differences between CFTs and conventional work teams are summarized in Table 1.

First, conventional work teams are generally made up of members with the same functional background. Examples are production, marketing, and accounting teams. Compared to these teams, CFTs tend to have higher diversity of expertise in terms of functional background, experiences, knowledge, and skill sets after pooling different specialists to work on a single project. In the Honeywell case, members of CFTs were drawn from the area of marketing, manufacturing, and engineering. Second, work teams in functional organizations tend to process work sequentially (e.g., design teams → manufacturing teams → marketing teams). CFTs have the capability to process their tasks simultaneously instead of sequentially. Third, members of CFTs have a dual focus: individuals maintain memberships in their functional departments and also in their CFT. For example, CFT members in Honeywell "still reported to their functional managers, who continued to supervise and evaluate all employees" (Margolis & Donnellon, 1991, p. 2). These first three characteristics (diversity of expertise, parallel processing, and dual focus) primarily distinguish CFTs from work teams in organizations that group people along functional lines.

In divisional organizations, work teams tend to be permanent arrangements consisting of different functional specialists who share the same objective and task. For example, a firm can group its functional specialists based on its product categories. Other examples are customer and geographical teams. CFTs differ from this type of work team in terms of a fourth characteristic: temporary duration. Unlike teams in a divisional structure in which the members work for the same team on a permanent basis, CFT members will generally return to their functional area or move to other teams after a project is accomplished. The final and fifth

Table 1. Core Differences Between Conventional Work Teams and Cross-functional Teams

	Conventional Work Teams*	Cross-functional Teams
1. Diversity of Expertise	Low (Exclusion)	High (Inclusion)
2. Work Processing	Sequential	Parallel
3. Focus of Team Members	Single (Internal)	Dual (Internal and External)
4. Time Duration	Permanent	Temporary
5. Organizing Principles Among Teams	Similar	Different

Note: * 1 to 3 are differences between functional work teams and cross-functional teams; 4 to 5 are differences between divisional work teams and cross-functional teams.

characteristic is diversity of organizing principles. There is generally not a common orientation among CFTs. This is in contrast with work teams within a divisional structure that share a common orientation of product, customer, or geographical region. Members comprising CFTs may not even share the same organizational affiliation. The team is created to address what top management perceives as unique problems for the organization. For example, some teams may work on new product development while other teams may work together to serve certain clientele.

Any one of these characteristics by itself is *not* sufficient to separate CFTs from conventional work teams. Nor can it be assumed that each CFT must have all five of these inherent characteristics in order to perform effectively. Our purpose in delineating these five characteristics is to explain and explore these features in order to pinpoint the type of management interventions necessary to unleash the inherent benefits of CFTs.

CONCEPTUAL MODEL AND PROPOSITIONS

Our conceptual model (see Figure 1) focuses on identifying individual-, group-, organizational-, and industry-level factors that can contribute to or block CFT effectiveness. The model derives from research on groups and teams by Hackman (1987); Mohrman, Cohen, and Mohrman (1995); Schwarz (1994); and Sundstrom, De Meuse, and Futrell (1990). However, while their models focus on the effectiveness of general work teams (Hackman, 1987), our model is directed specifically toward CFTs. General team-effectiveness models can certainly be applied to CFTs; for example, in both Hackman's (1987) and Sundstrom et al.'s (1990) models, certain effectiveness-enhancing strategies are applicable to both conventional work teams and CFTs. These include creating shared vision, defining team roles, and providing sufficient resources. However, Hackman (1987) suggests different teams face different critical task demands. Thus, although all effectiveness-enhancing strategies are salient to a certain extent, some strategies are particularly important to CFTs because of the distinctive characteristics of their structure. As a result, our conceptual model does not address the general conditions of work-team effectiveness. Instead, the purpose of our model is to concentrate on the core characteristics, as indicated previously in Table 1, that are most unique to the implementation of CFTs. These unique characteristics of CFTs interact with external conditions and determine when CFTs should be adopted and how internal interventions should be implemented.

Our multilevel model consists of several components: (1) industry and organizational context, (2) characteristics of CFTs, (3) benefit-enhancing interventions, (4) detriment-reducing interventions, and (5) outcomes of implementation. First, we propose that the decision to implement CFTs, because this is an inherently expensive intervention, should be a function of context. Industry and organizational context are external factors that affect the *value* of CFTs to organizations prior to

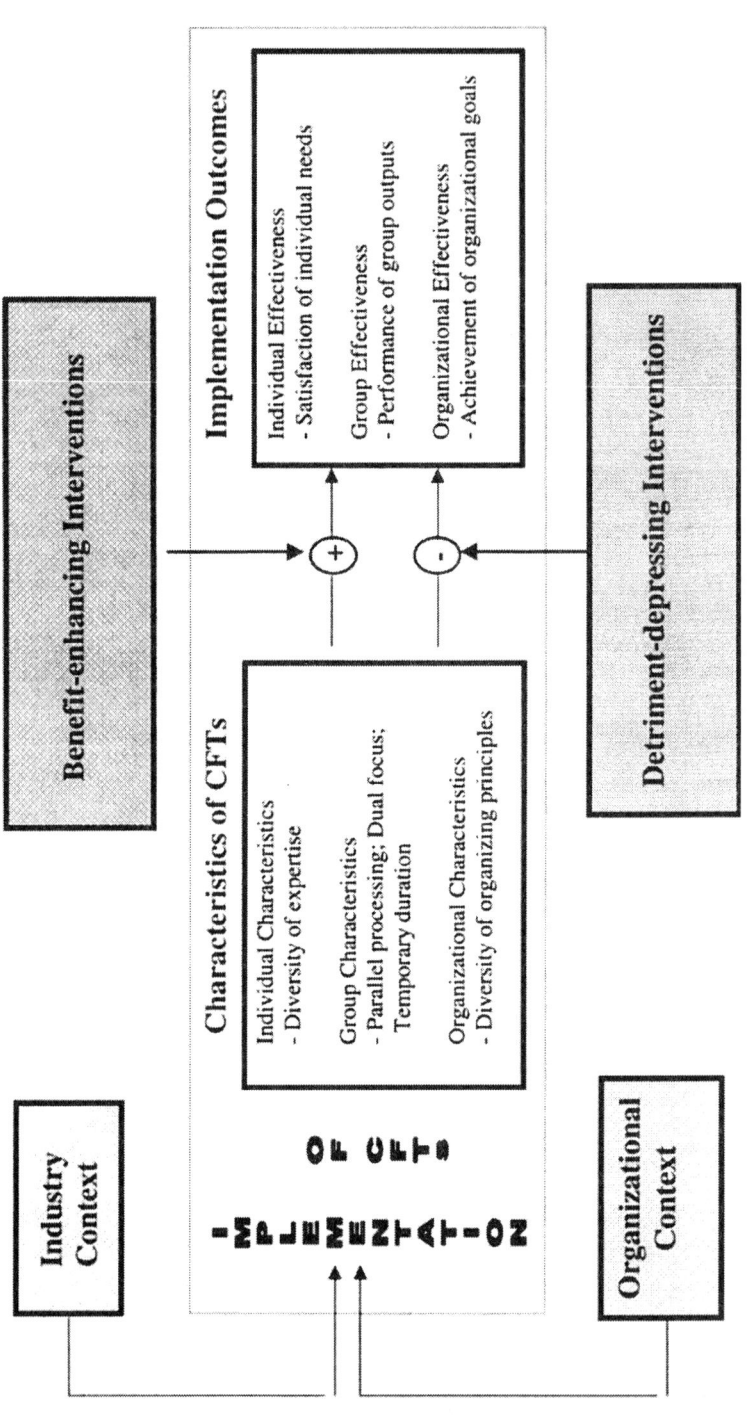

Figure 1. A Multilevel Model of Implementing Cross-functional Teams (CFTs)

their actual implementation. Second, if the decision is made to implement CFTs, organizations need to pay close attention to the core characteristics that separate CFTs from conventional work teams in functional and divisional organizations (see Table 1). An understanding of these characteristics will help in determining if the implementation will be successful. Third and fourth, organizations can adopt interventions during the CFT implementation phase in order to promote benefits or depress detriments. Interventions in this case are techniques and methods that organizations can adopt regarding the design of CFTs. We argue that each of the unique characteristics of a CFT has the potential to benefit organizations but these benefits will not be accrued automatically and must be promoted through benefit-enhancing interventions. However, some of these benefit-enhancing interventions will create unintended consequences. These negative effects must then be offset by detriment-reducing interventions in order for CFTs to be totally effective.

Our final analysis addresses the effects and outcomes on organizations after CFTs are implemented. We suggest that implementation of CFTs has the potential to benefit all three levels of effectiveness—individual, group, and organizational. However, there are situations in which improving one level of effectiveness may come at the expense of another level of effectiveness.

The Impact of Industry and Organizational Context

The implementation of CFTs is an organizational-level intervention, and specific external conditions should determine when they are adopted. Research on this topic, however, has often taken CFT implementation as a given and has rarely addressed if the organization really requires the CFT form in the first place. Therefore, the first component of the conceptual model is to analyze when and under what conditions CFTs should be implemented. Our first major proposition in this case is that:

Proposition I. The CFT design is inherently expensive; adoption should occur only given certain industry and organizational contexts.

CFTs are a more complex design for coordination than other mechanisms available to an organization. Galbraith (1994) claims that organizations may be better off using conventional mechanisms, such as telephone calls and top-management intervention, instead of CFTs. Other methods that are both less expensive and less complex include personal contacts, electronic mail, and interdepartmental meetings. When implementing CFTs, organizations expend significant resources. This includes requests that participants move outside of their traditional job responsibilities—responsibilities that still need to be performed in the organization. Thus, departments incur additional expenses in finding replacement workers and paying extra wages and benefits. There are also burdens on individual employees. Some workers in the earlier cited Honeywell case felt the system exerted too much pressure on them and they became frustrated on how to split their time among multiple projects.

> We have to make a decision on the deployment of resources. When it comes to choosing between things to do, the answer above is, "Do both"—with no added resources. (Margolis & Donnellon, 1991, p. 4)

> The (new) system is heavily loaded, especially when we are learning a new way of working. There are many things to do with little headcount and no relief with the project schedule. (Margolis & Donnellon, 1991, p. 4)

Although CFTs are more expensive and complex to implement, they should also produce more in terms of improved information-processing and decision-making capability (Galbraith, 1994; Galbraith & Nathanson, 1978; Lawrence & Lorsch, 1967). Thus, CFTs become beneficial when contextual factors suggest organizations need to equip themselves with such capabilities. In other words, the decision to implement CFTs is a function of fit or necessary alignments between contextual requirements and organizational design mechanisms (Nadler & Tushman, 1997). In this case, the choice among different design mechanisms should match the level of coordination and information-processing needs. The failure to achieve the match will lead to excessive coordination (i.e., inefficiency) or inadequate integration among functional departments (i.e., ineffectiveness; see Souder & Sherman, 1993).

Industry Context

Assessment of industry conditions is an essential first step in determination of the necessity of CFTs (Galbraith, 1994; Hitt et al., 1993). According to Daft, Bettenhausen, and Tyler (1993), organizations facing turbulent and complex environments have two problems regarding information-processing requirements. First is the amount of information that organizations need to process from the environment. Second is the ambiguity of information, the uncertainty and equivocality of the meaning of information. The "information load" of organizations is said to be high when the environmental context is characterized by voluminous and ambiguous information. This situation occurs when industries are undergoing discontinuous changes with rapid technological or environmental developments—at times from almost every sector in their industry.

Changing conditions forced the Honeywell Building Control Division to face the first and only year when their division lost money. John Bailey, general manager of the Division explained the impact of industry environment.

> In the early 1980s the move to electronics and microelectronics was accelerating, and we were having a hard time dealing with that by using engineering and manufacturing techniques that had evolved over one-hundred years and were slightly toward a really slow-moving industry and slow-moving technology.... Our competitive environment was changing, technology was chang-

ing, and our customers were demanding a different set of requirements from us. So there was no alternative but to change. (Margolis & Donnellon, 1991, p. 3)

It was under this industry and environmental context that Honeywell began experimenting with the CFT form. A product team was formed in the mid-1980s to develop a new motor (code name Mod IV) used in heating, ventilation, and air-conditioning. The team was comprised of representatives from manufacturing, marketing and sales, and engineering. Their charter was to work together "to guide the project from the conceptual stage all the way through final production" (Margolis & Donnellon, 1991, p. 3). Top management paid close attention to the Mod IV project and hoped the new product would account for 30 percent of the division's profit.

Compared with conventional work teams, CFTs have the information-processing advantage because of diversity of expertise and dual focus. Diversity of expertise enables CFTs to have more knowledgeable experts as sources of information. Having different specialists with diverse backgrounds encourages debate and clarification, which serves to resolve information ambiguity. Dual focus also allows CFTs to look for outside help from other functional departments when team members do not have the information internally. Thus, although CFTs are more costly to implement, the change effort is justified when the industry conditions demand additional information-processing capabilities.

Proposition I.1. Organizations should implement CFTs when the industry context experiences strong demands for information load. Alternatively, organizations should rely on more conventional coordination mechanisms when the industry context is characterized by low information load.

Organizational Context

In addition to better information-processing capability, research has indicated that CFTs are capable of making decisions faster than conventional teams. An understanding of competitive strategy in describing organizational context is helpful when faster decision making becomes necessary. While there are multiple ways to classify strategy types, a majority of competitive strategies is based on one of the following characteristics—efficiency, innovation, and fast response (Dess & Miller, 1993; Miles & Snow, 1978; Porter, 1985).

The emphasis behind an efficiency strategy is to achieve low costs compared to rivals in the industry. Because CFTs are most complex and expensive to implement, firms in this case may be better off relying on conventional work teams and coordination mechanisms. However, both innovation and fast-response strategies require organizations to adopt practices that reduce time between development and commercialization of new products. One Honeywell executive explained how new technology forced the company to drastically reduce the product development time in order to meet a new product life cycle that had shrunk from thirty to three years.

He further commented that "just being biggest isn't going to do it anymore, and so speed of response is one of the things we've nailed to the ground" (Margolis & Donnellon, 1991, p. 13). This is the type of organizational scenario that captures the benefits of CFT implementation (Lutz, 1996; Takeuchi & Nonaka, 1986).

CFTs are able to make decisions faster than conventional work teams because of parallel processing (Hitt et al., 1993). When new products are developed sequentially among functional teams, delays tend to occur when problems are discovered during the latter stages of the process. In the case of Honeywell, the old system of product development was described as "a game of tossing the bear over the wall" (Margolis & Donnellon, 1991, p. 2).

> When you completed your particular piece of the project, you tossed it over the wall to the next group, not caring what took place on the other side. If you had problems with work done at previous stages, you made your changes and tossed the design back to the previous group for them to adjust their work. The process was slow and costly. (Margolis & Donnellon, 1991, pp. 2–3)

The implementation of CFTs allows problems to be resolved more rapidly when a particular project is handled simultaneously by multiple functional specialists. It was the hope of Honeywell's management that through CFT implementation the Building Control Division would be able to abandon the old product development process and "transform itself into an agile organization capable of maneuvering competitors through *faster* development" (Margolis & Donnellon, 1991, p. 2, italics added). After seeing evidence of the reduced product development time (from 38 to 15 months), the general manager of the Building Control Division was so convinced of the *speed* benefit of CFTs that he believed that this would be the weapon to reclaim the company's competitive position. Honeywell's experience exemplifies the importance of organizational context.

> **Proposition I.2.** Organizations should implement CFTs when innovations or fast responses are part of an organizational strategy. Alternatively, organizations should rely on conventional work teams when efficiency is the dominant organizational strategy.

The Role of Benefit-enhancing and Detriment-reducing Interventions

The decision to implement CFTs is a function of industry and organizational contexts, but this is not the sole criterion. Even though the specific contextual environment suggests CFT implementation, there is no guarantee that the implementation will be successful. The second part of our conceptual model examines each unique characteristic of CFTs vis-à-vis conventional work teams and explores the role of benefit-enhancing and detriment-reducing interventions. Our second major proposition is stated as follows.

Proposition II. Organizations must provide interventions to make CFTs effective. Interventions that unleash the benefits of CFTs may have unintended consequences and additional interventions may be necessary to curb these negative effects.

Diversity of Expertise

As mentioned earlier, the first difference between CFTs and conventional work teams is their diversity of expertise. Compared with other types of work teams, CFTs are cognitively capable of resolving problems of greater complexity because of their diversity of skills, orientations, and functional training. For example, CFTs are believed to reduce the uncertainty of innovation because multifunctional knowledge can be applied to the problem (Hitt et al., 1993). CFTs also are viewed as more innovative because the disagreements that arise out of various functional perspectives can protect against the problems of groupthink (Eisenhardt, Kahwajy, & Bourgeois, 1997; Nonaka, 1991). Indeed, in the banking industry, top management teams that were more functionally diverse were found to be more innovative (Bantel & Jackson, 1989). When new product teams and research and development laboratories were studied, functional diversity was found to relate positively to innovation, creativity, and inventiveness (Ancona & Caldwell, 1992c; Hull & Azumi, 1989).

However organizations cannot expect CFTs to be effective by casually assembling a diverse group of specialists (Walsh, Henderson, & Deighton, 1988). Without proper interventions, the diversity of expertise may hamper instead of help CFT effectiveness. To capture the potential benefits of diversity, the first significant managerial intervention is selecting functional specialists who *need* to be represented on the team. For example, one of the authors assisted a CFT that consisted of health-care providers and state specialists who were focused on delivering seamless care to at-risk newborns. Group members expressed frustration over the lack of contribution made by obstetricians who tended to show up intermittently at meetings and seemed to have their own agenda. Once it was recognized that mothers, not newborns, were the obstetricians' patients and the appropriate physicians, pediatricians, were included on the team, member satisfaction increased and the team could focus more effectively on their objective.

A second step includes determining a set of selection criteria to evaluate whether an individual has not only the technical skills essential to the CFT's task, but also the interpersonal skills necessary for collaboration with other team members (Galbraith, 1994; Hackman, 1987). Of five functions that Walton and Hackman (1986) identified as common to groups, four focused on the interpersonal relations among group members. The failure to establish teams with the requisite interpersonal skills also has been shown to be related to intragroup conflict that can derail effectiveness *in toto* (Jehn, 1995; McGrath, 1984).

Proposition II.1a. To unleash the benefit of diversity of expertise on CFT effectiveness, organizations should consider employing a formal selection process focused on finding CFT members who possess the requisite interpersonal skills and functional expertise.

However, too much diversity may negatively affect CFT effectiveness. Because team members come from different functional backgrounds, their understanding of the CFT's goal may not be identical. The unintended consequence in this case is that increasing diversity through selection will induce goal conflicts within the team and inability to arrive at shared goals. As a result, members of 45 new product teams reported lower group performance ratings as functional diversity increased (Ancona & Caldwell, 1992b, 1992c). The earlier example of newborn care illustrates this goal conflict. Obstetricians have a different functional specialization than pediatricians, because their primary objective is maternal, not newborn, care. Because the stated objective of the team was to optimize newborn care, it is clear where conflict arose. In the Honeywell case, the objective of the marketing department was to increase revenues whereas the engineering's objective was to reduce costs. Such difference in functional orientation led to questions about why the Mod IV project team needed to be formed in the first place. A marketer complained:

> Mod IV is replacing our bread and butter for no market-driven reason. Sure it's a cost reduction and a quality improvement, but our motors already are very high quality and provide high margins, so from a marketing standpoint, it didn't have to be done. The customer-benefits derived from Mod IV, including modules, could be developed for our present motor lines. (Margolis & Donnellon, 1991, p. 8)

The resentment experienced by the Honeywell marketing group is a classic example of the lack of a shared objective or goal. Even though the original intention of Mod IV was to make the new motor more price-attractive so that more customers would buy and sales revenue would increase, offering a lower priced item did not clearly indicate "increased profits" to marketing. Mohrman and Mohrman (1997) also concluded that after studying 200 teams in 11 corporations, poorly chartered teams were a critical problem when designing team-based organizations. Teams that are said to be poorly chartered are those that have ambiguous or inappropriate responsibilities.

To resolve conflicts relating to competing objectives, it is necessary to focus team resources on clarifying the core objective of the team. This could consist of specific training for the group, or preliminary facilitation. The earlier example of at-risk newborn care is a prime example; one of the authors was brought in to facilitate due to recognized conflict and impasses, but after facilitating at several meetings the skills of the facilitator had been learned by the team and she was no longer necessary. Given these potential problems, we recommend another set of interventions be put in place in addition to the formal selection process.

Proposition II.1b. Interventions that increase diversity of expertise can also lead to goal conflicts among CFT members. To reduce such unintended consequences, individual team members should address team charter through facilitation and training in effective group processes.

Parallel Processing

CFTs engaging in parallel processing can improve group effectiveness by accelerating the decision-making process and increasing both time and cost savings. A major U.S. automobile manufacture reported almost one billion dollars of savings on new-car development after CFTs were used (Lutz, 1996). Similarly, Takeuchi and Nonaka (1986) indicated in case studies of Fuji-Xerox, Canon, and Honda that the implementation of functionally diverse teams accelerated new product development. Compared with functional teams working sequentially, CFTs can complete a task faster because they eliminate the burden of serial processing (Hitt et al., 1993; Parker, 1994). For example, in a CFT consisting of both design and manufacturing personnel, product designs can be made more compatible with specific manufacturing specifications even before products are actually manufactured (Dean & Susman, 1989).

Despite the benefits of parallel processing, there is surprisingly little CFT research on how parallel processing can be directly facilitated. There is an implicit assumption that CFTs are naturally geared toward the work mode of parallel processing. However, in the absence of managerial interventions, one possibility is that group members may simply divide the whole task by functional expertise instead of seeking creative ways to handle tasks simultaneously. In order to enhance the opportunities for parallel processing, CFTs may consider increasing physical proximity, sharing project leadership, and focusing on self-leadership.

Minimizing physical distance can also have positive impacts on internal group processes. As one project member of Fuji-Xerox explained:

> When all the team members are located in one large room, someone's information becomes yours, without even trying. You then start thinking in terms of what's best or second best for the group at large and not only about where you stand. If everyone understands the other person's position, then each of us is more willing to give in, or at least to try to talk to each other. Initiatives emerge as a result. (Takeuchi & Nonaka, 1986, p. 140)

Increased physical proximity (e.g., the same building or floor) can lead to more face-to-face communication among team members and improve the likelihood of parallel processing (Hitt et al., 1993; Sundstrom et al., 1990). Top management at Honeywell went further by creating a new building that had adequate space for people from engineering, marketing, sales, and manufacturing to work in the same location (Margolis & Donnellon, 1991). Empirical work by Pinto et al. (1993) also confirmed that physical proximity fostered greater intragroup cooperation within 62 hospital CFTs.

The literature on CFTs has addressed the importance of project leadership (Donnellon, 1993; Mohrman et al., 1995), although little has focused on how this role should be structured in order to facilitate parallel processing. We believe that in order for the task to be handled simultaneously from multiple functional perspectives, the project leadership of CFTs should be shared instead of appointed. A designated project leader has the tendency to analyze a problem from her own functional perspective, ignoring the viewpoints of other team members. At Honeywell, a marketing person was assigned as the team leader of Mod IV and was described as the "field marshal" of the project, "responsible for the product's smooth introduction to the marketplace" (Margolis & Donnellon, 1991, p. 9). However, members of a process development team are likely to express dissatisfaction when their team is assigned a technical expert as the project leader. Certain negative comments from other functional units in the Honeywell case can be attributed to that decision. Consider a comment by one engineer:

> The marketing people have changed since the project began while the engineers have been the same since the beginning. Marketing decisions changed each time the marketing people changed. We had to do two rounds of marketing research. This has had a negative psychological effect. It leaves the impression that the rationale developed in marketing is only as good as the people who developed it. (Margolis & Donnellon, 1991, p. 7)

To avoid this problem, Denison and colleagues (1996) suggest that "an effective team has to have flexible leadership—leadership and expertise need to change according to the issues involved and get passed from person to person as required" (p. 1012). This shared leadership role is consistent with both the job characteristics approach to work design (Hackman & Walton, 1986) and the sociotechnical systems approach to work design, the cornerstone of which is the self-managing team (Cummings, 1981; Manz & Sims, 1987).

In addition to shared project leadership, self-leadership is exercised when CFTs are given the authority and autonomy to decide how to execute tasks and monitor and manage their own performance (Hackman, 1987; Manz, 1990; Manz & Sims, 1987; Sundstrom et al., 1990). In terms of self-influence, self-leading teams have more control than self-managing teams. Manz (1990) argued that self-management allows team to decide how to finish projects but that top management continues to establish and control the standards of team performance. Self-leading teams, on the other hand, "possess and exercise a greater level of influence on decisions regarding what they do and why they do things as well as how they do them" (Manz, 1990, p. 286). In other words, self-leading teams are more empowered than self-managing teams because they have controls over performance standards (i.e., what they do and why they do it). Accordingly, the use of self-leadership is appropriate for CFTs because projects assigned to CFTs are normally complex in nature, ambiguous in standards, and nonroutine in processes. Self-leadership also promotes CFT members' feelings of responsibility for the group outcome, thus making them more likely to collaborate in creative ways to resolve tasks simultaneously (Campion, Medsker, & Higgs, 1993; Uhl-Bien & Graen, 1992). As Takeuchi and Nonaka (1986) explained, "the team begins to operate like a start-up company—it takes initiatives and risks, and develops an independent agenda" (p. 139). Characteristics of autonomy also have been identified in highly effective CFTs at Honda, IBM, and NEC (Takeuchi & Nonaka, 1986).

> **Proposition II.2a.** Benefits of parallel processing will be unleashed when there are greater intragroup communication, cooperation, and autonomy in the CFT. These can be achieved if organizations intervene by increasing the physical proximity among team members and using shared- and self-leadership.

However, similar to diversity of expertise, there is also an unintended consequence associated with the use of parallel processing. Unlike sequential processing, in which feedback is more readily available once an intermediate step is completed, performance-related information in cross-functional work is more difficult to obtain because multiple tasks are considered and implemented simultaneously. For example, in the Honeywell case, one team member of the Mod IV project suggested "because we do many things at the same time in crossfunctional teams, the discrete checkpoints get lost. Definite meanings don't exist" (Margolis & Donnellon, 1991, p. 11). To reduce the drawback associated with parallel processing, CFTs need to be provided with appropriate tools for both individual and team performance feedback (Sundstrom et al., 1990). Performance feedback tools allow the team to create some of its own measures of effectiveness, because there simply may not be a precedent for the work processes that they are performing. Doing so allows the team to change its course of action in order to meet organizational objectives and expectations (Schwarz, 1994). In addition, creating these tools can provide a basis for both individual and group performance evaluations (Campion et al., 1993).

> **Proposition II.2b.** Interventions that facilitate parallel processing will benefit CFT effectiveness if team members are provided with performance feedback tools such that they can assess their own project development.

Dual Focus

Groups often need to interact with other parts of the organization to request resources, information, and feedback on their performance. CFTs, by their inherent design, have multiple contacts with the rest of the organization through members' strong functional affiliations. The increased functional diversity that defines CFTs also implies more external communication with production, marketing, finance, and other functional areas (Sundstrom et al., 1990). This external access gives CFTs the characteristic of a dual focus—not only is the team project important to CFTs (internal focus), but also how contacts with outside constituents are managed is instrumental to the team's success (external focus). By soliciting inputs from the team's external constituents, the quality of outputs have a higher likelihood of being approved and accepted by external users (Walsh et al., 1988).

However, team members may be so fully occupied by the project that no one steps beyond the team boundary. Mohrman, Cohen, and Mohrman (1995) emphasized the importance of constructing linkages between the team and its external environment. Extensive channels of external communication should be provided so that CFTs are able to solicit knowledge, support, and feedback from the rest of the organization (Ancona, 1990; Ancona & Caldwell, 1992a). It has been shown that CFTs perform better when they maintain active communication with external constituents such as top management, customers, suppliers, and other functional units (Ancona & Caldwell, 1990a). Consistent with this finding, Ancona (1990) found that state education department CFTs that lacked external communication tended to have lower achievement over the longer term.

Proposition II.3a. To unleash the benefit of dual focus on CFT effectiveness, organizations should install formal channels of communication between CFTs and outside constituents.

However, opening channels of external communication also has its own unintended consequences. By overemphasizing the importance of external focus, CFTs can become susceptible to the influences of those functional interests. CFT members, whose loyalties may be divided and rooted first and foremost in their area of functional expertise, are sometimes swayed by their functional departments to maximize their coalition's interests rather than to serve the larger team and organizational goals (Ancona & Caldwell, 1992c; Hitt et al., 1993). Denison et al. (1996) presented the dilemma of a project leader whose CFT consisted of members who were specialists from other functional units. They often came to meetings to assure that their departmental interests would not be compromised, making the CFT a magnet for departmental politics. Another drawback of dual focus is related to the "two-table problem" (Gray, 1989). Even though team members may have the desire to work collaboratively within the group, they still face the challenge of convincing external constituents that functional interests have been fulfilled by the CFT project.

As described previously in the Honeywell case, team members of the Mod IV project were also aware of their departmental interests: marketing was evaluated annually based on sales revenue, engineering's evaluation was based on operative costs, manufacturing's evaluation was based on quality and defects of the product. This division of interest had partially contributed to the inability of the Mod IV team to meet its design and production schedule. As documented by the case, internal problems were still evident right before the completion deadline of Mod IV.

> The marketers were busy preparing a training video for Mod IV users. The engineers had their hands full with a noise problem (with the new motor). And the manufacturing engineers were ordering parts and arranging the assembly lines. (Margolis & Donnellon, 1991, p. 7)

When external functional interests took over, team members placed less emphasis on the internal need of collaboration. Thus team members of the Mod IV project were still concerned about their functional issues despite the problem that an important task of the Mod IV project, the design and purpose of control modules for the new motor, had yet to be resolved by the team.

The drawback of external focus suggests there is a need for a management intervention to create an alternative reward system. This system would allow team members to take advantage of their external focus but still remain committed to their project goal (Ancona & Caldwell, 1992b; Hitt et al., 1993). Hackman (1987) also argues that for team members to contribute their share of individual effort to achieve the project goal, the implementation of group-based reward systems is necessary. Group incentive programs include the use of both physical outputs and nonphysical measures, such as adherence to deadlines and magnitude of cost savings (Welbourne & Gomez-Mejia, 1991).

The problem of dual focus can be further minimized if organizations provide sufficient distance between CFTs and functional units. Earlier we mentioned the importance of minimizing the distance among CFT members; this should be coupled with maximizing the distance between team members and their previous functional areas. This buffering strategy should enable CFTs to concentrate on fulfilling the need for interfunctional collaboration rather than becoming tools for political maneuvering. The innovation literature has indicated that product development teams tend to perform better when they are isolated from the rest of the organization (Galbraith, 1982). Kazanjian and Drazin (1986) also documented the advantages of physically isolating the location of an aerospace project team, thus allowing the team to minimize interruptions and break free from the mainstream of the organization.

In addition to physical isolation, CFTs can be buffered from internal politics among departments if the team reports its results directly to top managers who have the authority and power to circumvent internal politics among functional units (Mohrman et al., 1995). Doing so can minimize the control of functional units over the operations of CFTs (Donnellon, 1993). For example, Kazanjian and Drazin (1986) explained how a vice president of an aerospace firm buffered the attempt of the data processing and the production department to manipulate a product development team. In the case of Honeywell, the Mod IV might perform better if the general manager of the Building Control Division intervened to resolve the departmental differences on project completion.

Proposition II.3b. Interventions that increase the external focus of CFTs can lead to conflicts of functional interests, a decrease of intragroup collaboration, and an inability to meet project goals. To reduce such unintended consequences, organizations should adopt group-based reward systems, isolate CFTs physically from other functional units, and arrange for CFTs to report to top management.

Temporary Duration

Most CFTs are temporary; once the project task is accomplished, the group is disbanded and team members move on to other projects. Because of the temporary status of CFTs, a sense of urgency is normally instrumental in achieving time and cost savings, one of the many benefits that CFTs are believed to deliver (Ancona & Caldwell, 1992a, 1992b). To create a greater sense of urgency for members, organizations can consider the establishment of individual and group deadlines. For example, Fuji-Xerox's top management gave its FX-35 project team two years to develop a radically different copier with one-half the production costs of older machines (Takeuchi & Nonaka, 1986). The sense of urgency, as one executive remarked, is "like putting the team members on the second floor, removing the ladder, and telling them to jump or else" (Takeuchi & Nonaka, 1986, p. 139). Mohrman et al. (1995), in addition to deadlines, have also emphasized the importance of intermediate milestones (e.g., phase completion timetables, project review points) to help members develop a sense of accomplishment and manage project deadlines (also see Eisenhardt & Tabrizi, 1995).

Proposition II.4a. To unleash the benefit of temporary duration of CFT effectiveness, organizations should establish project deadlines and intermediate milestones for CFTs in order to create a sense of urgency.

The temporary status of CFTs, however, creates its own dilemma. The pressures of project deadlines and milestones may cause team members to de-emphasize interpersonal relationships within the team. This problem caused "antagonism" and "finger-pointing" among functional groups in the Honeywell case. One marketer commented,

> From a schedule standpoint, engineering's credibility was no good. They were telling us dates that just weren't getting met. We tried to arrange shared goals and objectives but it was like pulling teeth from engineering. They said they had their own milestones. The first shared deadline they suggested weren't valid since we needed things from them well before that. (Margolis & Donnellon, 1991, p. 6)

An engineer offered his version of the story:

> (Regarding a particular design issue), we felt the changes would require more time than the schedule allowed. We went to the head of marketing for our position. We said we were making progress but did not feel we would make our introduction date and needed more time. He said we had to stick to the dates we had. It's his prerogative to demand that the target dates be met, so the target dates were not changed, even though the team knew we weren't going to make it. Insisting that a date not change, though, can lead to a project problem. I'm not sure what's accomplished by insisting on unrealistic dates. (Margolis & Donnellon, 1991, p. 6)

With the pressure to meet project deadlines, CFTs may be highly successful in terms of task accomplishment. However, individual members may be less willing to participate in future cross-functional projects, because unresolved interpersonal issues contributed to an unpleasant experience with the CFT. If this occurs, it may endanger future implementation of CFTs. To reduce this detriment to long-term effectiveness, members need to have sufficient time to address intra-team problems and develop into a functional and effective group. In other words, organizations should place equal emphasis on task completion and positive group experiences. We, therefore, suggest that sufficient time should be allowed for normal group processes so that members are able to revise performance strategies, address working relationship issues, and evaluate their team and its development (Schwarz, 1994).

Proposition II.4b. Temporary duration has drawbacks for CFT effectiveness. To alleviate negative outcomes from project deadlines and milestones, organizations should provide extra time for team members to create solutions leading to positive group processes and individual achievement.

Diversity of Organizing Principles

Inside the same organization, different CFTs may be responsible for basic research, product development, customer satisfaction, and quality improvement. This diversity of organizing principles allows organizations more flexibility in deciding how CFTs should be formed and when they should be used. One criterion for organizations to develop a team-based structure is to acquire self-design capabilities (Mohrman & Mohrman, 1997). That is, organizations must decide how to change and how to form teams on their own. Diversity of organizing principles can, therefore, be considered part of this self-design concept, particularly when there is no common basis among the forming principles of CFTs, as organizations create CFTs based on an assessment of their own immediate needs.

Despite these benefits, without a guiding set of organizing principles, CFTs may proliferate for projects that actually require little inter-functional integration. For example, if Honeywell's objective was to install a new accounting or human resource system instead of product development, the need for inter-functional integration may be significantly less necessary. An alternative approach is to rely on one functional department (accounting or human resources in this case) to assume the leadership role and use conventional coordination mechanisms (e.g., phone calls, electronic mails, personal contacts) to solicit inputs from other functional units.

To decide under what basis CFTs are formed, Mohrman et al. (1995) suggest that managers first review underlying work processes that transcend the entire organization and determine if CFT formation is appropriate and how the team should then be constituted. In essence, work processes become the foundation for forming CFTs. If work processes require consensus and complex interdependence, they can rationally be delegated to CFTs (Nadler & Tushman, 1997). For work processes with little need for consensus or for those that require sequential interdependence, they likely should be performed by functional departments through less extensive coordinating mechanisms (Galbraith, 1994; Sundstrom et al., 1990).

Proposition II.5. Organizations should undergo internal audits of work processes in order to determine the basis for forming CFTs and capture the benefits from the diversity of organizing principles.

Outcomes of CFT Implementation

Evaluation is a key component of any organizational change—and formation and implementation of CFTs is no exception. After CFTs are implemented, how are the effects on organizations measured? Is it plausible that positive results can occur simultaneously at the individual, group, and organizational level? Or is it more likely that one or more of these areas gets sacrificed for gains in another? In prior research of CFTs, the criteria of effectiveness are often a function of the research discipline. For small-group researchers, measures of CFT effectiveness that have been used include group cohesiveness and team member satisfaction (e.g., Ancona 1990; Cohen & Bailey, 1997; Ford & Randolph, 1992; Gladstein, 1984). For studies in product development, CFT effectiveness is often measured by the quality and the efficiency of innovations (e.g., Ancona & Caldwell, 1992a, 1992b, 1992c; Hull & Azumi, 1989; Katz & Tushman, 1979). For research on top management teams, organizational-level measures such as profitability or other indicators of financial performance are often used (e.g., Korn, Milliken, & Lant, 1992; Simons, 1995). Researchers have the general tendency to adopt a single level of analysis when analyzing the effectiveness of CFTs. Thus, less is known regarding the multilevel outcomes of CFTs and whether or not outcomes occurred at one level will affect outcomes at another level. For example, when implementing CFTs, do organizations improve their financial performance at the expense of team member needs? Is individual satisfaction traded off against or positively related to the quality of an innovation? To address these issues, our third major proposition in this case is related to the multilevel effects to organizations after CFTs are implemented. We propose that:

Proposition III. The implementation of CFTs has the potential to benefit individual, group, and organizational effectiveness, but these benefits may not be achieved simultaneously. There are situations in which one level of effectiveness is increased to the detriment of another level of effectiveness.

Effects On Individual Effectiveness

Following Hackman's (1987) model of work group effectiveness, we define individual effectiveness as how much the CFT satisfies rather than frustrates the personal needs of group members. Working in a more diverse group allows team members to enhance their personal knowledge (Ancona, 1990; Ford & Randolph, 1992; Pinto, Pinto, & Prescott, 1993). For example, Epson's engineers became more knowledgeable about design and marketing after functional specialists collaborated on miniprinter development (Takeuchi & Nonaka, 1986). Because problems are resolved concurrently rather than sequentially, team members are also able to determine how different functions operate, gain a more macroscopic view of how organizations are run, and understand how the same task can be addressed and achieved in different ways (Banner, Kulisch, & Perry, 1992; Hitt et al., 1993; Hitt & Tyler, 1991; Lawler, Mohrman, & Ledford, 1992). By gaining more knowledge and better understanding on how certain tasks are accomplished, individuals achieve higher personal satisfaction afterwards. In a large auto company, Denison et al. (1996) stated that CFT members reported gains in job satisfaction through personal growth and team development. Serving on CFTs also led employees to gain positive perception of their jobs and organizations (Hull & Azumi, 1989).

> **Proposition III.1a.** The implementation of CFTs can benefit individual effectiveness by adopting interventions geared toward increasing diversity of expertise and gaining a greater understanding of parallel processing.

Effects On Group Effectiveness

The second effect of CFT implementation is related to performance of group outputs. Prior research in this area has identified CFT performance in terms of task completion (Denison et al., 1996; Kazanjian & Drazin, 1986), time and costs savings (Hitt, Hoskisson, & Nixon, 1993; Katz & Tushman, 1979; Kazanjian & Drazin, 1986), creativity of the solution, and quality of the output (Ancona & Caldwell, 1992a, 1992b, 1992c). In the case of Honeywell, the use of CFTs reduced the product development time by more than 50 percent and the Mod IV project was expected to produce an innovative design that produced substantial savings for the customers (Margolis & Donnellon, 1991). These CFTs were able to realize time and costs savings on project completion because of parallel processing and temporary duration. As discussed previously, parallel processing eliminates the delays when functional units coordinate sequentially. Temporary duration further reduces time and costs by creating a sense of urgency. Creativity and quality of output is improved because team members have greater diversity of skills and expertise and the ability to solicit support and information from other functional units.

> **Proposition III.1b.** The implementation of CFTs can benefit group effectiveness by adopting interventions geared toward parallel processing, dual focus, and temporary duration.

Effects On Organizational Effectiveness

To understand how CFTs can benefit organizational effectiveness, we need to restate two previous arguments. First, organizations should evaluate their industry and organizational context before implementing CFTs. If the information-processing and decision-making capabilities of CFTs fit the contextual requirements, organizations should benefit by their implementation. In Honeywell's case, the general manager of the Building Control Division had stated repeatedly that the company needed to become quicker and more agile (Margolis & Donnellon, 1991), thus the context was ripe for CFT implementation. Second, multiple CFTs might be responsible for different objectives within the same organization. Thus, diversity of organizing principles allows organizations to become more flexible to address their immediate problems. Following these arguments, we can define organizational effectiveness as the ability of organizations to achieve their goals and objectives, especially if these goals and objectives can be achieved through speed, flexibility, and stronger information-processing capacity.

The effectiveness of CFT implementation must be evaluated based on similar criteria. For example, CFTs have been judged to be effective if companies become more flexible and adaptable after they are adopted (Ancona & Caldwell, 1990b; Ford & Randolph, 1992; Galbraith, 1994; Katz & Tushman, 1979). Another related criterion is financial performance, which measures how effectively CFTs help an organization to achieve goals and objectives (Simons, 1995; Smith et al., 1994).

Proposition III.1c. Organizational effectiveness can be enhanced by the implementation of CFTs if achieving organizational goals calls for improved information-processing and decision-making capabilities (by evaluating industry and organizational context) and more flexibility (by adopting interventions geared toward the diversity of organizing principles).

Multilevel Interaction Effects

The interaction among multiple levels of effectiveness after CFTs are implemented must also be considered. This is important to implementers of CFTs because, unless all three levels of effectiveness are considered simultaneously, one level may gain at the expense of another. We will examine three situations in which these interactions among individual, group, and organizational effectiveness exist.

The first situation is related to individual and group effectiveness. To allow team members to gain increased cross-functional knowledge of the projects, individual effectiveness is said to increase when there are interventions geared toward diversity of expertise. However, too much diversity can lead to conflicts and other group problems, resulting in poor group effectiveness. Three types of internal group processes are likely to be affected negatively by diversity of expertise. First, group conflicts tend to escalate because of the difficulty of merging divergent cognitive styles and values (Eisenhardt & Tabrizi, 1995). Although Eisenhardt, Kahwajy, and Bourgeois (1997) argue the benefits of task conflict for work group innovation, other types of conflict can be detrimental to team performance. Second, goal setting can become difficult because of differences in functional priorities and experiences. Third, internal communication may suffer when members encounter unfamiliar mental frameworks and terminology. These problems were evident in team meetings of the Mod IV project, as this comment revealed:

> We've had a lot of problems with language on this project. We've had a lot of semantic issues with the differences between what's literally required in the application versus what's required in order to position the product in the distribution channel. What's been hard about it is that *we haven't known we've disagreed.* (Margolis & Donnellon, 1991, pp. 10–11, italics added)

Thus we postulate that:

> **Proposition III.2.** While individual team members of CFTs may benefit from greater diversity of expertise, overall group effectiveness may suffer owing to internal conflicts, problems in goal setting, and poor communication.

The second situation occurs when improvements in group effectiveness come at the expense of organizational effectiveness. Within CFTs, group members retain dual focus between internal projects and external units of functional groups. Group effectiveness increases when team members actively solicit inputs from other functional departments such that the team is able to produce better outputs. The drawback of dual focus is that if too much external influences are present, CFTs may become arenas of departmental politics. For example, the project manager of a Mod IV team explained, " (the project) is somewhat a power struggle because this has been engineering's baby. Engineering saw marketing as just getting in the way" (Margolis & Donnellon, 1991, p. 10). If interdepartmental politics escalate, the ability of organizations to achieve their goals is jeopardized.

> **Proposition III.3.** While CFT group effectiveness may benefit from emphasizing dual focus, overall organizational effectiveness may suffer owing to escalation of departmental conflicts.

The third scenario occurs when organizational effectiveness becomes the priority, and the interests and needs of individual employees suffers accordingly. The formation of CFTs (an organizational-level intervention) help organizations to become flexible, but can reduce individual effectiveness when employees are overburdened by the new work requirement. James Reiner, Honeywell's CEO, placed the company's bottom line over other issues to make the firm less likely to become a takeover target.

> How will Honeywell fend off the poachers? Says Reiner: "Get the share price up. There is no other answer." To do that, he must boost performance in the core business, which may require a less-than-gentle touch. Asked if he is, as one analyst described him, more of a "butt kicker" than his predecessor, Reiner cracks a rare smile and says, "Sometimes it is necessary. It's been necessary with my kids. I have eight, so I know something about it." (Henkoff, 1989, p. 141)

Honeywell's objective was to become more competitive by implementing CFTs so that returns-on-investment would grow continuously. While CFTs promised greater speed in product development, some employees experienced difficulty handling the ambiguity and pressures when their work routines were redesigned.

Team members now had to attend team meetings whether relevant to their functional areas or not. A manufacturing engineer, for example, had to attend team meetings even if the project was only at a design stage. Since people were accustomed to simply completing a task and passing the project on, they felt team meetings stole time from doing actual work and added to total workload. As people gradually adapted to parallel development teams, they continued to struggle with their expanded roles and responsibilities. (Margolis & Donnellon, 1991, p. 4)

This leads us to propose that:

Proposition III.4. Organizations may benefit from CFTs in terms of speed, flexibility, and improved financial performance, but overall individual effectiveness may suffer because of new work requirements and expanded job responsibilities.

As mentioned earlier, interventions related to CFT implementation can often have unintended consequences. If one level of effectiveness is overly emphasized, there may be negative spillover to another level. To capture these multilevel interactions, we suggest:

Proposition III.5. Organizations need to consider how certain interventions may benefit CFT effectiveness at one level, but result in unintended consequences detrimental at another level of effectiveness. If this occurs, there is a need to implement another set of interventions to reduce these unintended detriments.

DISCUSSION AND IMPLICATIONS

We have developed a multilevel perspective on the implementation of effective CFTs. As Figure 1 illustrates, we examined and evaluated the conditions that promote the success of CFT implementation, built on existing models of work group effectiveness, and created a conceptual model that concentrates on the core characteristics of CFTs. This paper contributes to a number of areas that are related to CFT implementation and effectiveness. First, organizations need to consider industry and organizational context in determining when, and if, CFTs should be implemented. This step allows for the consideration of other possible mechanisms of coordination; further, it highlights some of the distinctions between CFTs and functionally or divisionally based work teams. CFTs are distinct from conventional work teams because of their diversity of expertise, parallel processing, dual focus, temporary duration, and diversity of organizing principles. While research has implicitly endorsed the distinct nature of CFTs, rarely have their characteristics been formally discussed.

Second, if external contextual requirements do call for CFTs, organizations should begin to activate certain internal interventions for their successful implementation. Internal interventions—classified as either benefit-enhancing or detriment-reducing—must be assessed and evaluated at the individual, group, and organizational level. Most research on CFTs tends to focus on the positives to the exclusion of the negatives. Toward that end, Figure 2 delineates the role of these

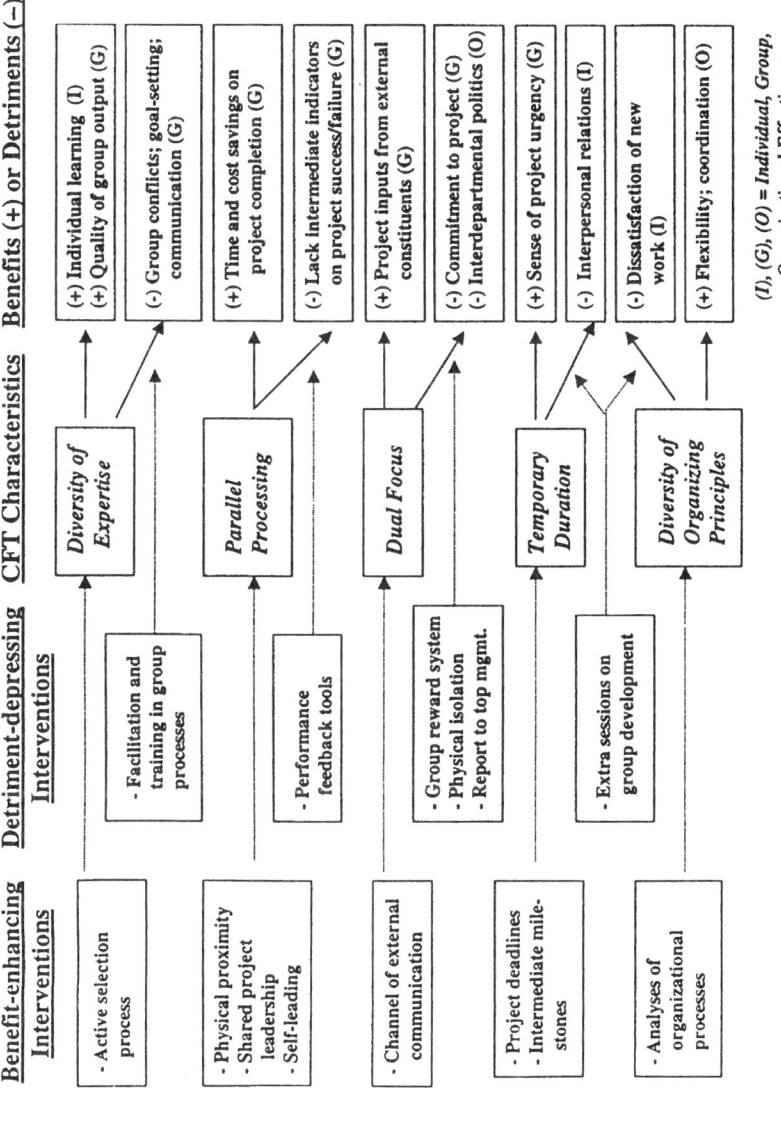

Figure 2. The Role of Management Interventions in the Implementation of Cross-functional Teams (CFTs)

internal interventions by the distinct characteristics of CFTs and illustrates what may be potential benefits, detriments, and cross-level interactions associated with each of these characteristics. As the figure indicates, CFTs consist of many benefits and detriments, with each factor either contributing or damaging individual, group, or organizational effectiveness. By understanding the role of benefit-enhancing and detriment-depressing interventions, managers will be better able to understand how to implement CFTs effectively.

Third, regarding outcomes of CFT implementation, prior research has not focused on the multilevel consequences of interventions intended to affect a single level. For example, Denison and colleagues' (1996) model does not address organizational-level outcomes, while Cohen and Bailey (1997) do not address effectiveness at the individual level. An important contribution of our approach is reflected in the identification of the multilevel consequences of single-level interventions. Figure 2 provides examples of how this multilevel view can aid in understanding how CFTs function. For example, through our multilevel analysis we can observe that channels of external communication (a group-level intervention) may help a CFT to perform better at the group level, but at the same time may inhibit organizational effectiveness by escalating departmental politics. Although not every aspect of CFT implementation involves two or more levels, such a perspective enables us to develop propositions that involve multilevel antecedents and consequences and to develop a more comprehensive view of when and how CFTs should be implemented.

Research Implications

Our theoretical framework helps to integrate the CFT literature, which is currently fragmented by research and researcher orientation. In our model, each of the characteristics of CFTs represents a partial explanation of why CFTs become effective or ineffective. Our work provides an impetus for synthesizing the divergent approaches of CFT research, as noted in our literature review. Without sound theoretical constructs, suggestions to improve CFT effectiveness may never come to fruition. This is similar to criticism leveled by Brown and Eisenhardt (1995) toward product development research—it is atheoretical because it mentions every plausible indicator of new product successes without focusing on the possibility of negative effects. While a limitation of our model may be that it does not exhaust all plausible intervening variables at different levels, we hope our work can provide some theoretical antecedents for future research on CFTs.

We have argued that the implementation of CFTs requires multilevel interventions. However, it remains unknown empirically whether or not interventions at certain levels are more important than those at other levels. For example, is an individual-level intervention (e.g., selection) more influential than an organizational-level intervention (e.g., analyses of organizational processes)? Our theoretical framework provides background for such comparative research on CFTs. By analyzing multiple factors in a single study, researchers may be able to analyze which level of analysis will exert the greatest influence on the effectiveness of CFTs. The result of such research can thereby provide directions to organizations in terms of deciding which aspects of management interventions (and research design) should receive the highest priority.

There are also questions relating to the interaction and effectiveness of these differing interventions. For example, both increasing channels of communication between CFTs and the members' functional departments *and* increasing physical separation between CFTs and functional areas have been cited by researchers as ways to increase CFT effectiveness. Testing these interactions concurrently, and assessing the impact of these interventions at each level, will provide increased knowledge about how to implement the most effective CFTs. Because almost all characteristics of CFTs share the double-edged property, our model illustrates the importance of monitoring the interactive effects of different sets of interventions during the implementation of CFTs.

The research design adopted by most CFT studies tends to be limited to multiple CFTs within one organization. With this design, researchers are able to analyze what may be the most influential group-level or individual-level indicators of effectiveness. However, what may be ignored are organizational and environmental factors that can also affect CFT effectiveness. Furthermore, by confining research to a single organization it is difficult to analyze if the adoption of CFTs actually improves organizational performance. Our multilevel analysis, therefore, suggests that future research should involve multiple organizations in different industries.

We have chosen not to address interactions between team design constructs (i.e., role allocation, size, nonfunctional heterogeneity of membership, specific functionality representation within CFTs) primarily because of the extensive prior research and knowledge in the group and team literatures on these topics. Furthermore, we believe that the level of complexity that the inclusion of these constructs would add to our propositions and framework would not substantially clarify understanding of CFTs. Future theoretical and empirical developments are needed to determine if the inclusion of these more developed constructs have unique implications for CFTs.

Managerial Implications

Our conceptual model also aids the implementation of CFTs in the corporate sector. For managers who are interested in CFTs, our propositions provide some directions for creating and evaluating effective teams in terms of their stages of implementation (Hackman, 1987; Tuckman, 1965). These stages consist of a number of steps throughout the history of a CFT. First, before CFTs are implemented, there should be conscious efforts by the host organization to consider if CFTs are necessary in the first place. It is important to understand that a CFT is not the only option for improving coordination. Less expensive options (e.g., top management interventions) or more comprehensive mechanisms (e.g., integrating departments) are also feasible (Galbraith, 1994). The basis for our proposition on industry and organizational characteristics provides the external conditions for managers to use in deciding whether or not CFTs should be adopted.

Second, during the forming stage of CFTs, we believe it is important to determine the basis that should guide group formation—especially selection and initial training. Therefore, interventions relating to the diversity of organizing principles and to the diversity of expertise are critical activities.

Third, while CFTs are performing their task, it is important that the group meets or surpasses its project goals. Interventions related to parallel processing, dual focus, and temporary duration should be encouraged, because maintaining these characteristics appropriately can be instrumental in accomplishing project goals. Our propositions related to these characteristics may improve effectiveness of these critical activities during the performing stage.

Finally, because CFTs are usually temporary, group members will eventually return to their functional areas. In this case, it is important that they have gained positive experiences from working in CFTs so that their departments may be more willing to collaborate in the future. Some of our discussions (e.g., group-based reward systems, group sessions on developing positive group processes) have provided directions for facilitating a positive CFT experience for members.

We believe that CFTs have much to offer organizations. The proper implementation of CFTs requires that managers understand the inherent advantages and probable disadvantages, as well as the multilevel effects of CFTs on individuals, groups, and organizations. We have developed a framework and a number of propositions that may shed some light on future conceptual and empirical efforts in the area of CFT implementation and effectiveness.

ACKNOWLEDGMENTS

We would like to thank Michael Beyerlein, Warren Brown, Max Elden, Alan Meyer, Nicole Steckler, and the coeditors, Bill Pasmore and Richard Woodman, for their valuable and helpful comments. A previous version of this paper was presented in the Western Academy of Management Meeting, Portland, Oregon, 1998.

REFERENCES

Ancona, D.G. (1990). Outward bound: Strategies for team survival in an organization. *Academy of Management Journal, 33*, 334–365.

Ancona, D.G., & Caldwell, D.F. (1990a). Beyond boundary spanning: Managing external dependence in product development teams. *High Technology Management Research, 2*, 119–135.

Ancona, D.G., & Caldwell, D.F. (1990b). Improving the performance of new product teams. *Research Technology Management, 33*, 25–36.

Ancona, D.G., & Caldwell, D.F. (1992a). Bridging the boundary: External activity and performance in organizational teams. *Administrative Science Quarterly, 37*, 634–665.

Ancona, D.G., & Caldwell, D.F. (1992b). Cross-functional teams: Blessing or curse for new product development? In T.A. Kochan & M. Useem (Eds.), *Transforming organizations* (pp. 154–168). New York: Oxford University Press.

Ancona, D.G., & Caldwell, D.F. (1992c). Demography and design: Predictors of new product team performance. *Organization Science, 3*, 321–341.

Ashkenas, R.N., & Jick, T.D. (1992). From dialogue to action in GE work-out: Developmental learning in a challenge process. In W.A. Pasmore & R.W. Woodman (Eds.), *Research in organizational change and development* (Vol. 6, pp. 267–287). Greenwich, CT: JAI Press.

Banner, D.K., Kulisch, W.A., & Perry, N.S. (1992). Self-managing work teams (SMWT) and the Human Resource Function. *Management Decisions, 30*, 40–45.

Bantel, K.A., & Jackson, S.E. (1989). Top management and innovations in banking: Does the composition of the top team make a difference. *Strategic Management Journal, 10*, 107–124.

Brown, S.L., & Eisenhardt, K.M. (1995). Product development: Past research, present findings, and future directions. *Academy of Management Review, 20*, 343–378.

Campion, M.A., Medsker, G.J., & Higgs, A.C. (1993). Relations between work group characteristics and effectiveness: Implications for designing effective work groups. *Personnel Psychology, 46*, 823–846.

Cohen, S.G., & Bailey, D.E. (1997). What makes teams work: Group effectiveness research from the shop floor to the executive suite. *Journal of Management, 23*, 239–290.

Cummings, T.G. (1981). Designing effective work-groups. In P.C. Nystrom & W. Starbuck (Eds.), *Handbook of organizational design* (pp. 250–271). Oxford, UK: Oxford University Press.

Daft, R.L., Bettenhausen, K.R., & Tyler, B.B. (1993). Implications of top managers' communication choices for strategic decisions. In G.P. Huber & W.H. Glick (Eds.), *Organizational change and re-design: Ideas and insights for improving performance* (pp. 112– 146). New York: Oxford University Press.

Dean, J.W., Jr., & Susman, G.I. (1989, January–February). Organizing for manufacturable design. *Harvard Business Review, 67*, 28–36.

Denison, D.R., Hart, S.L., & Kahn, J.A. (1996). From chimneys to cross-functional teams: Developing and validating a diagnostic model. *Academy of Management Journal, 39*, 1005–1023.

Dess, G.G., & Miller, A. (1993). *Strategic management*. New York: McGraw-Hill.

Donnellon, A. (1993). Crossfunctional teams in product development: Accommodating the structure to the process. *Journal of Product Innovation Management, 10*, 377–392.

Eisenhardt, K.M., Kahwajy, J.L., & Bourgeois, L.J., III. (1997). How management teams can have a good fight. *Harvard Business Review, 75*(4), 77–85.

Eisenhardt, K.M., & Tabrizi, B.N. (1995). Accelerating adaptive processes: Product innovation in the global computer industry. *Administrative Science Quarterly, 40*, 84–110.

Finkelstein, S., & Hambrick, D.C. (1990). Top management team tenure and organizational outcomes: The moderating role of managerial discretion. *Administrative Science Quarterly, 35*, 484–503.

Ford, R.C., & Randolph, W.A. (1992). Cross-functional structures: A review and integration of matrix organization and project management. *Journal of Management, 18*, 267–294.

Galbraith, J.R. (1982, Winter). Designing the innovating organization. *Organizational Dynamics, 10*, 5–25.

Galbraith, J.R. (1994). *Competing with flexible lateral organizations* (2nd ed.). New York: Addison-Wesley.

Galbraith, J.R., & Nathanson, D.A. (1978). *Strategy implementation: The role of structure and process*. St. Paul, MN: West.

Gladstein, D.L. (1984). Groups in context: A model of task group effectiveness. *Administrative Science Quarterly, 29*, 499–517.

Glick, W.H., Miller, C.C., & Huber, G.P. (1993). The impact of upper-echelon diversity on organizational performance. In G.P. Huber & W.H. Glick (Eds.), *Organizational change and re-design: Ideas and insights for improving performance* (pp. 176–214). New York: Oxford University Press.

Gray, B. (1989). *Collaborating: Finding common ground for multiparty problems.* San Francisco, CA: Jossey-Bass.

Hackman, J.R. (1987). The design of work teams. In J.W. Lorsch (Ed.), *Handbook of organizational behavior* (pp. 315–342). Englewood Cliffs, NJ: Prentice-Hall.

Hackman, J.R., & Walton, R.E. (1986). Leading groups in organizations. In P.S. Goodman & Associates (Eds.), *Designing effective work groups* (pp. 72–119). San Francisco, CA: Jossey- Bass.

Hambrick, D.C., & Mason, P.A. (1984). Upper echelons: The organization as a reflection of its top managers. *Academy of Management Review, 9,* 193–206.

Henkoff, D. (1989, May 22). Butt kicking at Honeywell. *Fortune,* p. 141.

Hitt, M.A., Hoskisson, R.E., & Nixon, R.D. (1993). A midrange theory of interfunctional integration, its antecedents and outcomes. *Journal of Engineering and Technology Management, 10,* 161–185.

Hitt, M.A., & Tyler, B.B. (1991). Strategic decision models: Integrating different perspectives. *Strategic Management Journal, 12,* 327–351.

Hull, F.M., & Azumi, K. (1989). Teamwork in Japanese and U.S. labs. *Research Technology Management, 32,* 21–26.

Jehn, K. (1995). A multimethod examination of the benefits and detriments of intragroup conflict. *Administrative Science Quarterly, 40,* 245–282.

Katz, R., & Tushman, M. (1979). Communication patterns, project performance, and task characteristics: An empirical evaluation and integration in an R & D setting. *Organizational Behavior and Human Performance, 23,* 139–162.

Kazanjian, R.K., & Drazin, R. (1986). Implementing manufacturing innovations: Critical choices of structure and staffing roles. *Human Resource Management, 25,* 385–403.

Korn, H.J., Milliken, F.J., & Lant, T.K. (1992). *Top management team change and organizational performance: The influence of succession, composition, and context.* Paper presented at the annual meeting of the Academy of Management, Las Vegas, NV.

Lawler, E.E., Mohrman, S.A., & Ledford, G.E. (1992). *Employee involvement and total quality management: Practices and results in Fortune 10 companies.* San Francisco, CA: Jossey-Bass.

Lawrence, P.R., & Lorsch, J. (1967). *Organization and environment: Managing differentiation and integration.* Boston, MA: Harvard University.

Lillrank, P., Shani, A.B., Kolodny, H., Stymne, B., Figuera, J.R., & Liu, M. (1998). Learning from the success of continuous improvement change programs: An international comparative study. In R.W. Woodman & W.A. Pasmore (Eds.), *Research in organizational change and development* (Vol. 11, pp. 47–71). Stamford, CT: JAI Press.

Lutz, R.A. (1996). Implementing technological change with cross-functional teams. In M.D. Aldridge & P.M. Swamidass (Eds.), *Cross-functional management of technology: Cases and readings* (pp. 77–82). Chicago: Irwin.

Manz, C.C. (1990). Beyond self-managing work teams: Toward self-leading teams in the workplace. In W.A. Pasmore & R.W. Woodman (Eds.), *Research in organizational change and development* (Vol. 4, pp. 273–299). Greenwich, CT: JAI Press.

Manz, C.C., & Sims, H.P., Jr. (1987). Leading workers to lead themselves: The external leadership of self-managing work teams. *Administrative Science Quarterly, 32,* 106–128.

Margolis, J.D., & Donnellon, A. (1991). *Mod IV product development team* (Case #9-491-030). Boston, MA: Harvard Business School.

McGrath, J.E. (1984). *Groups: Interaction and performance.* Englewood Cliffs, NJ: Prentice-Hall.

Miles, R.R., & Snow, C.C. (1978). *Organizational strategy, structure, and process.* New York: McGraw-Hill.

Mohrman, S.A., Cohen, S.G., & Mohrman, A.M. (1995). *Designing team-based organizations: New forms for knowledge work.* San Francisco, CA: Jossey-Bass.

Mohrman, S.A., & Mohrman, A.M. (1997). Fundamental organizational change as organizational learning: Creating team-based organizations. In W.A. Pasmore & R.W. Woodman (Eds.), *Research in organizational change and development* (Vol. 10, pp. 197–228). Greenwich, CT: JAI Press.

Nadler, D.A., & Tushman, M.L. (1997). *Competing by design: The power of organizational architecture.* New York: Oxford University Press.

Nonaka, I. (1991, November–December). The knowledge-creating company. *Harvard Business Review, 69,* 96–104.

Parker, G.M. (1994). *Cross-functional teams: Working with allies, enemies, and other strangers.* San Francisco, CA: Jossey-Bass.

Pinto, M.B., Pinto, J.K., & Prescott, J.E. (1993). Antecedents and consequences of project team cross-functional cooperation. *Management Science, 39*(10), 1281–1297.

Porter, M. (1985). *Competitive advantage.* New York: The Free Press.

Schwarz, R.M. (1994). *The skilled facilitator: Practical wisdom for developing effective groups.* San Francisco, CA: Jossey-Bass.

Simons, T. (1995). Top management team consensus, heterogeneity, and debate as contingent predictors of company performance: The complementarity of group structure and process. In D.P. Moore (Ed.), *Academy of management best papers proceedings 1995* (pp. 62–66). Vancouver, BC: Academy of Management.

Smith, K.G., Smith, K.A., Olian, J.D., Sims, H.P., Jr., O'Bannon, D.P., & Scully, J.A. (1994). Top management demography and process: The role of social integration and communication. *Administrative Science Quarterly, 39,* 412–438.

Souder, W.E., & Sherman, J.D. (1993). Organizational design and organizational development solutions to the problem of R&D-marketing integration. In R.W. Woodman & W.A. Pasmore (Eds.), *Research in organizational change and development* (Vol. 7, pp. 181–215). Greenwich, CT: JAI Press.

Sundstrom, E., De Meuse, K.P., & Futrell, D. (1990). Work teams: Applications and effectiveness. *American Psychologist, 45*(2), 120–133.

Takeuchi, H., & Nonaka, I. (1986, January–February). The new product development game. *Harvard Business Review, 64,* 137–146.

Tuckman, B.W. (1965). Developing sequences in small groups. *Psychological Bulletin, 63,* 384–399.

Uhl-Bien, M., & Graen, G.B. (1992). Self-management and team-making in cross-functional work teams: Discovering the keys to becoming an integrated team. *The Journal of High Technology Management Research, 3*(2), 225–241.

Walsh, J.P., Henderson, C.M., & Deighton, J. (1988). Negotiated belief structures and decision performance: An empirical investigation. *Organizational Behavior and Human Decision Processes, 42,* 194–216.

Walton, R.E., & Hackman, J.R. (1986). Groups under contrasting management strategies. In P.S. Goodman & Associates (Eds.), *Designing effective work groups* (pp. 168–201). San Francisco, CA: Jossey-Bass.

Welbourne, T.M., & Gomez-Mejia, L.R. (1991). Team incentives in the workplace. In M.L. Rock & L.A. Berger (Eds.), *The compensation handbook* (2nd ed., pp. 236–247). New York: McGraw-Hill.

Yang, D.J., Oneal, M.D., & Troy, S. (1990, January 9). How Boeing does it: America's export machine is ready to bet big on its embryonic 777. *Business Week,* pp. 46–50.

COLLABORATIVE ORGANIZING
AN "IDEAL TYPE" FOR A NEW PARADIGM

Peter J. Robertson

ABSTRACT

Signs of increased collaboration within and between organizations are readily apparent. An emerging "new paradigm" literature suggests that this trend is part of a fundamental transformation toward a more collaborative global culture. In this chapter, I first outline the key premises of an alternative "theory of reality" articulated by this literature, which integrates evidence and ideas from a variety of fields of human knowledge and experience. Next, I contrast the basic assumptions of the current "competition paradigm" with those of the new, emerging "collaboration paradigm." Based on the foundation provided by these new paradigm ideas, I then provide a description of "collaborative organizing." This new model of organization is described in terms of three categories of features pertaining to the purpose, design, and functioning of these organizational systems. While most of these characteristics are currently receiving considerable attention in both theory and practice of organizations, the collaborative organizing model reflects an initial attempt to integrate these features into a coherent whole that constitutes a new "ideal type" of organization. As an ideal type, I argue that this model provides a better fit with contemporary environmental conditions than the Weberian bureaucratic hierarchy. As such, it can serve as a "desired future state" to guide organizations as they undertake the transformation to a more collaborative mode of operating. The chapter concludes with additional comments regarding the possibility of a global transformation to the collaborative paradigm.

INTRODUCTION

Collaboration has become a prominent focus in the study and functioning of organizations. First, there has been increased attention to the need for more effective collaboration *within* organizations of all types. This has been demonstrated by the amount of scholarly and practitioner-oriented writing devoted recently to such topics as, for example, building effective teams (e.g., Cohen, 1993; Hackman & Walton, 1986), utilizing win–win approaches to problem resolution (e.g., Fisher & Ury, 1981; Levine, 1998), reengineering organizations to ensure greater coordination (e.g., Hammer & Champy, 1993), and developing high-involvement organizations to better capitalize on collective capacity (e.g., Lawler, 1992). Considerable attention has also been devoted to the importance of collaboration *between* organizations. Again, recent scholarly as well as practitioner-oriented literature has frequently addressed the complexities of interorganizational alliances and networks (e.g., Gray, 1985; Powell, 1990). Collaborative relationships have been recognized as valuable in both the private and the public sectors, and considerable resources in each are currently allocated to some form of cooperative venture. At their most extreme level of complexity, significant interorganizational systems exist across sectors, including the growing nonprofit or "third" sector, and/or across levels of government (e.g., Serageldin, Barrett, & Martin-Brown, 1995; Weiss & Tigue, 1997).

Collaborative systems have been created in reaction to our growing collective awareness that effective response to society's greatest challenges requires significant collaboration among a diverse array of organizations and institutions. Recognition of this fact is also driving a move toward greater international collaboration (e.g., Lipman-Blumen, 1996; Starr, 1997). The various governments of the world increasingly recognize their need to work together to develop and implement useful policies pertaining to the economic, social, military, and environmental arenas. In fact, all of the above trends can be found in nations and cultures around the world. This international move toward collaboration, at all levels of analysis, may be viewed as an indicator of significant global transformation in the institutional arrangements utilized to conduct collective action.

Many organizations are currently in the throes of this transformation process. The multifaceted shift toward collaborative arrangements noted above appears to reflect wide-scale organizational adaptation to transformative pressures existing in the global environment. Many of the changes being implemented with greatest frequency in organizations around the world can be viewed as changes needed to make organizations more successful in a collaborative context. There is reason to believe that this adaptation toward greater collaboration is simply one facet of a much more comprehensive transformation taking place on the planet. An intriguing set of ideas has emerged quite recently concerning a fundamental transformation in human civilization's dominant paradigm (cf. Woodhouse, 1996). This "new paradigm" literature is attempting to establish the foundational ideology that might serve as the basis for a new, global worldview and culture. This literature is based on research evidence and source ideas from a wide variety of fields of knowledge, and it contains powerful arguments that a paradigmatic transformation is necessary, possible, in-process, and even an inherent aspect of the natural pattern of evolution unfolding in the universe.

The dominant paradigm at the close of the millennium is that set of ideas and practices which form the basis of what can variously be referred to as the modern era, the Industrial Age, and Western civilization. It is grounded in Greek philosophy, Judeo-Christianity, the Copernican Revolution, the Newtonian-Cartesian mechanistic worldview, and the insights of the Enlightenment. It is currently manifest most directly in what has come to be known as the "Washington consensus" (Burki & Perry, 1998; Williamson, 1990), the political economic policy agenda directed by the U.S. government in collaboration with the other major industrialized nations, who exert their influence on the governments of "developing" countries to adopt this philosophy and its corresponding practices. Operationally, this is reflected by the pressure felt by countries around the world to move toward a free-market economic system supported by a democratic political system. It is widely accepted, certainly among the U.S. public, that this policy orientation is a good one, and that adoption of the Washington consensus by more and more countries around the world reflects a positive step forward for human civilization.

However, the global free-market capitalistic system, which has been generated through the implementation of this policy orientation, is certainly not without its critics. Insightful and incisive analyses of the inherent flaws and dysfunctional consequences of this system (e.g., Ayres, 1998; Greider, 1997; Henderson, 1996; Korten, 1995; Mander & Goldsmith, 1996; Terry, 1995) indicate that this system is essentially leading us toward eventual self-destruction. These observers argue that human civilization's current path of "development" is literally destroying the natural environment (our essential "life-support system"), exacerbating already significant inequalities in the distribution of resources among the world's people, and causing massive social fragmentation in societies around the world. Likewise, governmental institutions designed to address collective concerns are demonstrating their inability to respond to the multiple interdependent problems (e.g., homelessness, drug abuse, alienation, violent crime, ethnic conflict, war, pollution, etc.) generated by these various system dysfunctions.

A key premise of the new paradigm literature is that all of these problems are rooted in the underlying worldview and the fundamental beliefs on which the dominant paradigm rests. Based on an impressive array of evidence from the natural sciences, this literature explains how the dominant assumptions of the modern mechanistic worldview are no longer valid. It also identifies an alternative set of assumptions that are not only supported by data from a variety of scientific disciplines, but are also congruent with core beliefs common to most of the world's spiritual traditions. If these new paradigm assumptions were collectively recognized as viable and valuable, it would have profound implications for all facets of human civilization. The problem, of course, is that the new assumptions look neither viable nor valuable from the perspective of the old paradigm.

I would argue, however, that the new paradigm literature does in fact provide the ideological foundation for a shift to a more collaborative global culture. It provides evidence, concepts, theories, metaphors, images, frameworks, and practical applications that can be adopted by humans, individually and collectively, to facilitate a shift away from our existing competition-oriented culture toward a new collaborative paradigm. Therefore, I begin this chapter by outlining key premises of an alternative "theory of reality" articulated by the new paradigm literature, followed by a comparison of the core assumptions underlying the competition and the collaboration paradigms. While these ideas are undoubtedly interesting and provocative, I recognize that they may be too controversial and, for many, unbelievable. I present them here with the hope that the reader will not immediately discount them as unrealistic or invalid, but rather will consider the possibility that these beliefs constitute an improvement over the existing dominant paradigm.

I summarize the new paradigm ideas in some detail because this new worldview serves as the necessary foundation for the primary focus of the chapter. This focus is a description of a new model of organization I call "collaborative organizing." This model is grounded in the collaboration paradigm assumptions. To the extent that the new theory of reality and its key premises constitute an accurate and/or valuable representation of the world, collaborative organizing reflects an "ideal type" that is the logical derivative of these new starting assumptions. In other words, just as the competition paradigm quite naturally gave rise to the Weberian bureaucracy, collaborative organizing is naturally congruent with the assumptions of the new paradigm. If the new assumptions are not taken seriously, however, collaborative organizing is likely to be viewed not as "ideal" but as idealistic. While this will inevitably be the case for some, I would encourage the reader to keep an open mind and to consider the ideas presented below in their entirety. For those who find my argument interesting yet remain skeptical of the core premises, the new paradigm sources cited below provide thorough and convincing analyses of various aspects of the new theory of reality.

In describing the specific characteristics of collaborative organizing, I cite a considerable range of organizational literature that has already identified trends in this direction and/or has recommended the adoption of these features by organizations in both the private and the public sectors. The fact that organizations are already demonstrating changes toward a more collaborative mode of operation supports the contention in the new paradigm literature that this transformation is already in progress. Of course, the skeptical reader might rightfully argue that it will be exceedingly difficult for organizations to adopt a collaborative orientation while they continue to fight for survival in a competition-based political economy. Therefore, I conclude the chapter with a brief summary of the argument that a broader transformation to a more collaborative political economy and human culture is starting to become apparent.

A NEW THEORY OF REALITY

A paradigm shift in science away from the Newtonian-Cartesian mechanistic worldview began about a century ago with the initial findings that gave birth to quantum physics. Physicists' explorations of the fundamental nature of reality has led to rather remarkable findings that are inherently contradictory to mechanistic assumptions (Capra, 1991; Woodhouse, 1996; Zohar & Marshall, 1994). In simplistic terms, quantum physics has determined that everything known to exist in the universe is a manifestation of an underlying unified energy field known as "the vacuum." The vacuum is not empty, but rather constitutes the basic energic "raw material" of which everything is made and which physicists perceive as various forms of subatomic particles and waves. This energy is both particle and wave simultaneously, and while particles appear to have an autonomous existence, they are in fact inseparable from the waves which are themselves inherently interconnected. In other words, there are no separate "parts" of this unified field. The nature of this interconnectedness is such that the unified field transcends the concepts of space and time. In essence, there can be an instantaneous connection between any two points on the entire "space-time continuum," with time and space recognized as merely human "concepts" that are grounded in our *perception* of the universe. Furthermore, the unified energy field or vacuum is essentially a sea of potential. In other words, the energy itself is rather indeterminate, and becomes more "fixed" only under certain conditions. For example, physicists' intentions to measure the energy seem in turn to generate particular characteristics of the energy they are measuring.

These and other aspects of quantum reality contradict some of the mechanistic paradigm's basic assumptions regarding the nature of reality. In contrast to the reductionistic orientation of a mechanistic worldview (i.e., the belief that a system can best be understood by analyzing its parts and the forces acting on these parts), quantum physics emphasizes a holistic and ecological perspective. Capra (1996) explains how quantum findings in turn gave rise to "systems thinking" in the sciences more generally. A system is viewed as an integrated whole comprised of inherently interdependent parts, with systemic qualities that emerge from the pattern of relationships among the parts. From a systems perspective, the essential properties of an organism are properties of the whole, not of the parts. This emphasis on systems holism—on the properties of connectedness, relationships, and context—is further supported by findings from the natural sciences, especially research into the essential qualities of "life." In addition to identifying the common properties of living systems, this research has demonstrated that the living world consists of systems nested within systems, that is, the parts of a system are themselves less inclusive systems, and systems are themselves parts of more inclusive systems. In general, more inclusive systems are more complex, demonstrating properties that do not exist in lower level systems.

One key property of living systems identified through this research is the notion that they are self-organizing systems. Capra (1996, p. 85) defines self-organization as "the spontaneous emergence of new structures and new forms of behavior in open systems far from equilibrium, characterized by internal feedback loops and described mathematically by nonlinear equations." The self-organizing quality of a living system is made possible through the pattern of interactions among its various parts. This pattern of organization common to all living systems has been labeled autopoiesis (Maturana & Varela, 1980), defined as a network of production processes in which the function of each component is to participate in the production or transformation of other components in the network. Research associated with what is known as the Gaia hypothesis (Lovelock, 1988) indicates that the planet as a whole constitutes a single, living, self-organizing system.

Increased understanding of the nature of systems has in turn given rise to the study of chaos and complexity. The emergence of chaos and complexity theory has been facilitated by the development of new mathematical concepts and techniques that focus on relationships rather than objects, quality rather than quantity, and pattern rather than substance (Capra, 1996). The new mathematics has enabled scientists to discover qualitative patterns of behavior in complex systems, or a level of order underlying the apparent chaos. One such pattern has been labeled a dissipative structure (Prigogine & Stengers, 1984), which is another key property of living systems. A dissipative structure is an open system through which energy and matter flow continually, creating (somewhat paradoxically) a stable form through a self-organizing, autopoietic process. Some dissipative structures contain self-amplifying feedback loops that can push the system farther and farther away from equilibrium until it reaches a bifurcation point, a point of instability at which the system undergoes a sudden transformation. At this point, it can spontaneously self-reorganize such that a new form of order emerges, resulting in the development and evolution of the system.

These findings suggest that systems evolve, at least in part, by undergoing relatively sudden transformations to more complex forms, with these transformations taking place after the system has reached a threshold of instability that gives rise to a bifurcation point. Evidence regarding the evolution of life on the planet indicates that the evolutionary process has unfolded following just this pattern—long stable periods with very little evolutionary change punctuated by relatively short periods in which sudden and dramatic transformations occur (Capra, 1996). Combined with data supporting the Gaia hypothesis, considerable evidence is now congruent with the contention that the Earth, as a single unified system, reflects the key properties inherent in the underlying quantum vacuum and in all living systems which are manifestations of this universal energy field. Likewise, ecosystems and even human social systems can display these core systemic features (cf. Jantsch, 1980).

This perspective provides an alternative interpretation of the evolutionary process unfolding on the planet. In the dominant paradigm, evolution is viewed as resulting from a set of random mutations that are selected by the environment for their adaptive qualities. In contrast, the new paradigm perspective suggests that evolution results from an inherent characteristic of all living systems, namely, the creative tendency to constantly generate new configurations leading to ever-increasing diversity and complexity. These new configurations emerge through a process of continuous structural coupling with the environment, that is, changes in an organism in response to its environment in turn generate changes in that environment, in an ongoing process of coevolution. For example, considerable evidence suggests that symbiogenesis—the creation of new forms of life through permanent symbiotic arrangements (e.g., between bacteria and the larger cells in which they live)—is the principal avenue of evolution for all higher organisms (Margulis, 1998). More generally, recent theory and evidence regarding evolution indicate that collaboration within and among species is essential for survival and, hence, for further evolution (Augros & Stanciu, 1988). In contrast to the dominant paradigm's perspective that evolution results through competitive dynamics, this new view of evolution clarifies the important role that continual cooperation and mutual dependence have played in the unfolding of life on the planet, including the evolution of human consciousness (Elgin, 1993a). "Life is much less a competitive struggle for survival than a triumph of cooperation and creativity... evolution has proceeded through ever more intricate arrangements of cooperation and coevolution" (Capra, 1996, p. 243).

Clearly, these scientific findings are laying the foundation for fundamentally rethinking the nature of physical reality. However, further scientific evidence suggests the need for an even more revolutionary transformation in our understanding of the nature of the universe. In particular, data and insights from a number of fields of knowledge are now triangulating around the perspective that there is a fundamental connection between physical reality—grounded in the universal energy field—and consciousness. In other words, it is increasingly clear that there is an inherent linkage between thought and reality, between mind and body. The dominant paradigm, based on the Newtonian-Cartesian worldview and the mechanistic science it generated, has reflected the fundamental assumption of dualism, that is, the inherent distinction between mind as subject/observer and the physical world as object/observed. For thousands of years, philosophers have debated the validity of a dualistic worldview, relative to the merits of various alternatives. One such alternative is monism, which posits that the essential underlying reality is neither consciousness nor energy/matter, but rather is a single unified Source—the unity of energy-consciousness (Woodhouse, 1996), or the implicate order (Bohm, 1980). While modernity has adopted a dualistic stance toward the nature of reality, the transformation envisioned by the new paradigm literature includes recognition of the underlying unity of all things and the inseparability of our inner and outer worlds (Lemkow, 1990).

Scientific evidence which supports this notion comes from the natural sciences, the neurosciences, and the noetic sciences (i.e., studies of consciousness). For example, Zohar (1990; Zohar & Marshall, 1994) explains how the brain's neural activity demonstrates properties that are very similar to a particular type of quantum system identified by physicists, arguing that this similarity can help explain the holistic nature of our cognitive functioning. More broadly, some physicists and some neuroscientists have concluded that their two realms of investigation share in common the fact that they both demonstrate holographic properties (cf. Bohm, 1980; Grof, 1993). In synthesizing the findings from these two fields, Talbot (1991) develops an impressive argument concerning the essential similarity and apparent interconnection between consciousness and the universal energy field, which together constitute the monistic notion of a "holographic universe."

From a different perspective, research on living systems indicates that cognition is an inherent property of life itself, with cognition defined as the activity involved in the self-generation and self-perpetuation of autopoietic networks (Maturana & Varela, 1980). In other words, living systems at all levels of complexity display cognition, manifested in the autonomous and ongoing adaptation of the organism to its environment. Through this cognitive process of structural coupling and its coevolutionary consequences, a living system literally "brings forth" a world. Because individual systems are part of each others' worlds, their communication and coordination with each other—reflecting mutually coherent acts of cognition—generate the world they collectively inhabit. Among humans, this process of cognition is manifested in what we experience as consciousness (i.e., reflective self-awareness) which is made possible by the human capacity for abstract thinking and the use of language. Because human consciousness is integrally tied to the systems of shared beliefs, meanings, and interpretations created through language and the social context in which it is embedded, consciousness is best thought of as an inherently social phenomenon (Capra, 1996).

A third perspective on the relationship between consciousness and the external world is emerging from research on the link between mind and body. Considerable evidence from studies in many fields exploring a wide variety of issues demonstrates that there is a functional connection between cognition and physical reality. For example, research into the nature of consciousness combined with findings from medical research provides compelling support for the premise that physical health and recuperation from illness are a function, at least in part, of our own thoughts and feelings as well as the nature of the support we get from those around us (cf. Harman, 1998; Schlitz, Taylor, & Lewis, 1998). On one hand, evidence regarding the biochemical processes involved has helped to clarify the nature of the physiological mechanisms through which the mind–body link is manifest (Pert, 1997). Other explanations derive from the notion that each human being constitutes a complex energy system, and that manifestations at the physical level (e.g., of sickness) are related to energic imbalances and dysfunctions at other levels (e.g., mental or emotional) of the system (Myss, 1996; Schultz, 1998). This notion of an energy system connected to physical well-being is foundational to Chinese medicine, and the efficacy of acupuncture—which is based on the premise of an "energy grid" in the human body—provides support for its existence and relevance. Rapid growth in the use of various forms of "alternative medicine" in the United States— many of which are grounded in the premise of a mind–body link—suggest growing public recognition of the validity of this new paradigm premise.

Taken together, this research establishing the inherent connection between consciousness and the material world contradicts the basic Newtonian-Cartesian perspective of an essential distinction and separation between these two realms of reality, and it undermines the more contemporary belief that matter (i.e., the brain) gives rise to mind. The type of evidence reviewed above is generating a critical change in the dominant scientific paradigm toward a worldview in which consciousness is viewed as "causal." Harman (1998, p. 30) explains that this perspective "finds the ultimate stuff of the universe to be consciousness. Mind (or consciousness, or spirit) is primary, and matter-energy arises in some sense out of the mind. The physical world is to the greater mind as a dream image is to the individual mind. Ultimately the reality behind the phenomenal world is contacted, not through the physical senses, but through deep intuition. Consciousness is not the end-product of material evolution; rather, consciousness was here first!"

This notion of consciousness as causal reality has profound implications not just for science, but for the whole of human civilization. At a minimum, it results in a significant revision of the philosophy of science underlying almost all scientific endeavor, in both the natural and the social sciences. Harman (1998) explains how the perspective of consciousness as causal—what he calls a "transcendental monism" metaphysic (mind giving rise to matter)—is radically different from the "dualism" (matter plus mind) and "materialistic monism" (matter giving rise to mind) metaphysics which dominate contemporary science. He also clarifies how this new metaphysic yields theoretical frameworks and logical conclusions that can account for existing scientific data and human experiences which *cannot* be explained in the context of the dominant paradigm assumptions. In addition to evidence supporting the role of consciousness in health, illness, and healing, scientific findings and anecdotal information regarding various "psychic" (e.g., extrasensory perception and psychokinesis) and paranormal (e.g., out-of-body and near-death experiences) phenomena can be readily explained, given new paradigm assumptions (cf. Grof, 1993; Talbot, 1991; Woodhouse, 1996). Ultimately, this new metaphysic—or set of ontological and epistemological assumptions—replaces the objectivist, positivist, and reductionist assumptions of Newtonian-Cartesian science with a more holistic and phenomenological epistemology (cf. Harman & de Quincey, 1994, cited in Harman, 1998) coupled with what can be called a more "spiritual" ontology.

Ultimately, the most transformative feature of this new philosophy of science is the link it provides between science and spirituality (cf. Wilber, 1998)—between a relatively limited scientific worldview that discounts the validity and thus the reality of nonmaterial dimensions, and a more expansive spiritual worldview which holds that reality is composed of different levels reaching from the lowest, most dense, and least conscious to the highest, most subtle, and most conscious. This spectrum of consciousness (Wilber, 1977) is what has been referred to in spiritual writings as "the great chain of being." At one end of this continuum of being is matter, and at the other end is what has been called "spirit," the "godhead," or the "superconscious"—the all-pervading ground of the entire spectrum. This conception of the nature of reality is a key premise of the Perennial Philosophy (Bailey, 1949, 1954; Blavatsky, 1974; Hall, 1972; Huxley, 1944) which reflects the common core of the world's primary religions and spiritual traditions (Miller, 1992).

The emergence of the new paradigm is being fueled, at least in part, by growing recognition of the remarkable similarities and parallels between the description of reality generated through scientific investigation and the description of reality provided by the Perennial Philosophy (cf. Capra, 1991; Harman, 1998; Woodhouse, 1996). These include the basic notions that: there is an underlying ground of reality (i.e., the vacuum, the universal energy field, the Source, the Creator, God) from which the entire universe is manifest; this underlying ground is a unified whole; the many diverse manifestations of this underlying unity are inherently interconnected; the apparent separation among presumably distinct entities is an illusion resulting from human perception; consciousness is a property of the energy field itself; and conscious thought is the causal force determining the nature of physical reality manifested from the underlying Source.

These consistencies between contemporary scientific findings and ancient spiritual premises in turn lend credence to the validity of other ideas comprising a spiritual or transpersonal worldview. These include key ideas regarding the nature of the Source, the nature of human beings, and the essential purpose of life itself (Miller, 1992). In addition to being conscious (the quality of omniscience typically ascribed to God), the energy field is also seen as inherently creative and loving, and it has free will to do and to be whatever it desires. Because humans are simply manifestations of this energy field, these qualities are inherent in us as well. While "normal" human consciousness is clearly not omniscient, the Perennial Philosophy teaches that humans have the capacity to gain access to higher levels of consciousness, including what has traditionally been thought of as our "soul" or higher self. For example, intuition is recognized as the mechanism through which normal consciousness receives information and insights from this higher level (e.g., Naparstek, 1997; Zukav, 1989). Likewise, mystics and spiritual adepts throughout the ages have described an experience—frequently labeled "enlightenment"—of total and complete connection with Source, the ground of all being. In contemporary society, various types of spiritual or meditative practices are recognized by their practitioners as enabling spiritual growth, that is, enhancing the connection to higher levels of consciousness.

Extensive scientific research on human potential has already demonstrated that, through a wide variety of practices (e.g., hypnosis, biofeedback, mental imaging, fitness training, somatic education, martial arts, meditation, and spiritual healing) individuals demonstrate a widespread ability to transcend what are usually seen as normal human limitations (Leonard & Murphy, 1995). People frequently display dramatic alterations of mind and body, exhibiting extraordinary versions of a number of essential human attributes, including perception of external events, kinesthetic awareness, communication skills, vitality, movement abilities, bodily processes, capacities to influence the environment directly, pain and pleasure, cognition, volition, individuation, and love. All in all, these findings demonstrate that humans have extraordinary capacities for change and suggest that people have a wider range of latent metanormal capacities than is generally believed. What is more, Leonard and Murphy (1995) point out, these capacities can best be developed intentionally through transformative practices that explicitly utilize a holistic, integrated approach that focuses on the whole person—mind, body, heart, and soul.

Taken together, these ideas and findings indicate that inherent human potential is much greater than is typically assumed, and that this potential is a reflection of the true nature of our being. Specifically, the new paradigm holds that the human being is a multidimensional energic being, with each dimension reflecting one facet of the underlying energy field. The energy system comprising a human being is regulated through a number of energy vortexes (i.e., identified in yoga practice as the chakras), which function more effectively if we intentionally "take care of" them (through, for example, the kinds of practices identified above along with our daily patterns of eating, working, playing, sleeping, etc.). This energy system is part of the mechanism through which the mind–body link discussed earlier functions, and the general public in recent years has demonstrated increased interest in and use of a number of alternative medical therapies that focus directly on manipulating and improving this energy system (e.g., Gerber, 1988). Ultimately, this perspective makes it clear that our physical and our spiritual well-being are intimately connected, and that our general lifestyle and daily practices can facilitate or constrain our spiritual development (Miller, 1992).

Individuals who proactively pursue a path of spiritual growth frequently experience an increase in those qualities typically espoused as "good" in most societies—love, compassion, wisdom, peace, and so forth. Because these are recognized in the new paradigm as qualities of the Source and thus of our souls, it is only logical that these characteristics would become more evident as people enhance the quality of their connection to their higher selves, improving the linkage between normal consciousness and soul consciousness. In fact, the Perennial Philosophy teaches that this process of spiritual development is the essential purpose of human existence—the whole process of life provides us with an opportunity to pursue our reconnection with the Source, to reestablish conscious awareness of the underlying unity of all things. In other words, our own personal evolutionary journey is a process of spiritual development that concludes with our arrival at the "enlightened" state in which we experientially realize our inherent oneness with Spirit. The Perennial Philosophy teaches that this process unfolds across many human lifetimes through the process of reincarnation, and contemporary research evidence provides intriguing support for the viability of this premise (e.g., Almeder, 1992; Stevenson, 1987; Wambach, 1978).

While spiritual development is a process of individual growth along the spectrum of consciousness, another facet of the new paradigm perspective is that the true evolutionary saga being played out in the unfolding of creation is not merely the emergence of more complex physical life forms, but rather an evolution toward higher levels of consciousness (de Chardin, 1975; Wilber, 1995). In other words, the material world—as a manifestation of the underlying spiritual ground—is seen as evolving up the spectrum of consciousness toward ultimate realization of its oneness with the Source. It appears that this evolutionary path has included a number of discreet "steps" in which a qualitative transformation in the nature of consciousness occurred. In simplistic terms, for example, single-celled life forms evolved into plants with more complex "cognition," that is, in Maturana & Varela's (1980) terms; life evolved next into animals that demonstrate a more sophisticated level of consciousness; and humans then evolved with a self-reflective consciousness that differentiates them from other animals. Given the assumption of a complete spectrum of consciousness, it is only reasonable to conclude that life on this planet has the potential, if not the purpose, to evolve to the next higher level of consciousness. In fact, evidence suggests that the human nervous system is actually evolving so as to enable us to achieve higher states of consciousness (Bentov, 1988).

The argument of many new paradigm theorists is that human civilization is on the verge of a transformation reflecting the next step forward in the evolution of our collective consciousness (Devereux, 1989; Eisler, 1987; Elgin, 1993a; Harman, 1998; Hubbard, 1998; Russell, 1995; Snyder, 1995; Tarnas, 1998). Whether meant literally or metaphorically, it is frequently suggested that Gaia—Mother Earth, a planetary being—is "waking up" through the collective consciousness of the human race. Regardless of the exact nature of this awakening, by definition it implies greater awareness of the true, spiritual nature of our being and of our essential interconnectedness and interdependence with all of life. This awareness is the foundational belief, I would argue, for a cultural transformation away from the current competition paradigm to a new collaboration paradigm.

THE COLLABORATION PARADIGM

The new theory of reality articulated in the new paradigm literature lays the foundation for a different set of "starting assumptions" on which to base the re-creation of the various institutions that shape the development and evolution of human civilization. Currently, our primary institutions, grounded in the paradigmatic assumptions of the dominant paradigm, are taking global society down a path of development that leads to an unsustainable future (cf. Clark, 1995; Gladwin, Kennelly, & Krause, 1995). As we enter the new millennium, the global political economic system is clearly unstable. The intractable problems associated with population growth, unequal distributions of wealth, resource depletion, environmental destruction, and mass urbanization, combined with the threat of economic collapse, irresponsible use of weapons of mass destruction, and/or technological crises, have generated a world that seems almost ready to disintegrate into anarchy and chaos. To the extent that these problems are the natural consequences of the dynamics inherent in the existing system, potential "solutions" based on the starting assumptions of this system have no hope of resolving the problems.

It is therefore necessary, I would argue, to rethink our starting assumptions so that we can identify new, creative institutions and solutions that reverse the trend toward global destruction and instead enable human civilization to pursue a path of development and evolution that improves the quality of life for ourselves and for future generations (Anand & Sen, 1992; Olson, 1995). Rather than passively suffering the consequences of the global deterioration that continues unabated all around us, humanity must proactively choose to adopt a new cultural paradigm that yields dramatically new approaches for addressing the fundamental needs and problems of the global community. Once we recognize—collectively, publicly—that we and the rest of the planet constitute a single, interdependent system, we will realize the necessity and value of working together for our collective and individual well-being, rather than fighting each other—as individuals, organizations, interest groups, ethnic communities, and nations—to get whatever we can for ourselves. Only then do we have any hope for ensuring a sustainable future.

Successful transformation in this direction will require an alternative set of starting assumptions that provide the foundation for a collaboration paradigm. In this section, I articulate a set of these assumptions, contrasting them with the parallel assumptions of the competition paradigm. These two sets of assumptions are juxtaposed in Table 1. I have divided them into four categories, namely, the underlying motivating force, fundamental orientations, core premises, and primary objectives. Two general comments about these assumptions are worth noting here. First, each set of assumptions constitutes an interconnected, internally consistent set of beliefs. In other words, the basic tenets of either paradigm reinforce each other to support a way of thinking that is logically coherent. From "within" either worldview, therefore, the alternative set of assumptions looks entirely unrealistic and/or unreasonable. Second, the collaboration paradigm assumptions are not to be thought of as the "opposites" of the competition paradigm assumptions. More often than not they reflect the *integration* of the competition paradigm assumptions and their opposites, that is, the new transcends the old in the sense that the new also *includes* the old. This "holarchical" quality reflects the nested "systems within systems" feature of the universe and the pattern of evolution it reflects (cf. Harman,

Table 1. The Transformation in Progress

The Competition Paradigm	to →	The Collaboration Paradigm
The underlying motivating force:		
fear	→	love
Fuels two fundamental orientations:		
competition	→	collaboration
control	→	freedom
Supported by four core premises:		
individualism	→	connectivism
science	→	metaphysics
rationality	→	creativity
patriarchy	→	equalitarianism
Reflected in two primary objectives:		
economic growth	→	conscious evolution
consumerism	→	quality of life

1998; Wilber, 1996). The fact that the collaboration paradigm transcends the competition paradigm further suggests that this transformation is congruent with the evolution of consciousness unfolding on the planet.

Underlying Motivating Force

It is useful to begin with a discussion of the underlying motivating force because it gets most directly at the critical distinction between the competition and the collaboration paradigms. At the heart of it all, the competition paradigm is driven by fear, while the collaboration paradigm is driven by love. In the new paradigm, it is recognized that the true nature of human beings is to love, which we fail to do to the extent that our thinking and behavior is driven by fear. In other words, fear rather than hate is identified as the true "opposite" of love. Fear generates hate, as it does all of the dysfunctional behavior reflected in human civilization. One of the more useful analytical devices provided by the new paradigm is this recognition that every assumption, intention, action, and institution can be diagnosed in terms of the extent to which it is driven by fear or by love.

In the current worldview, fear is reflected in the basic assumptions that the most important resources are scarce, and that our efforts to acquire these resources are threatened by the existence of competitors whose pursuit of the same resources serves as an inevitable impediment to our ability to accomplish our own objectives. These fears have shaped the dominant interpretation of the nature of life and the evolutionary process, just as they have shaped the formation of our dominant political and economic systems. Rooted in fear-based assumptions, the political systems in the world's representative democracies have been designed to generate adversarial dynamics which ultimately impede effective decision making that best serves the interests of the collective. Free-market economics, likewise, is grounded in the fears of scarcity and competition, with the latter even identified as the way we *should* behave.

Fear in the dominant paradigm is also associated with our collective uncertainty regarding the nature of "life after death." Science—because of its materialist assumptions—has chosen, essentially, not to even dignify the question by undertaking an investigation as to whether or not there *is* some form of life that continues after physical death. In the absence of any scientific "proof," the world's religions each articulate their own version of the possibility of "eternal life," with adherents then competing for whose definition is best, most righteous, or most just. The primary religions in the dominant paradigm too readily reflect a tendency to instill fear in their followers through the premises that they are sinful, there is some price to be paid to or for God because of these sins, and this may affect the nature of their life after death. These basic assumptions, especially when religious beliefs are backed by the power of the state, have provided justification—historically and currently—for such activities as holy wars, inquisitions, witch burnings, the conquering of foreign lands and the enslavement of their peoples, the domination of women by men, a punitive legal system, restrictions on the expression of sexuality, and laws or social norms against behavior deemed immoral. Fear, rather than love, is easily recognized as the motivating force behind such actions.

In contrast, the collaboration paradigm rests on the primary motivating force of love. According to the Perennial Philosophy, our failure as humans to act out of love is not because we are inherently selfish or sinful, but because of our misperception of the true nature of reality. Perceiving ourselves as separate and distinct from the Source and from each other generates fear, for example, of being alone, disconnected, and/or unloved. The new paradigm, however, affirms that each of us is inherently connected to a loving Source, that we are *made* of that Source, and thus that the essential nature of our being is to love. Scientific advances, as noted above, confirm that the underlying ground of existence is entirely interconnected and interdependent. More and more people are undergoing psychological and/or spiritual growth that enables them to transcend their fears and recognize more fully the true nature of their being. Together, these scientific and spiritual advances are changing our collective consciousness, reshaping our worldview so that we can choose to begin acting more out of love and less out of fear. Love for others fuels our desire to interact collaboratively, while love for self drives us toward self-actualization (Maslow, 1954), that is, the freedom to fully express who we want to be and to create whatever we want to create. In these ways, love identifies the direction in which to change the fundamental orientations of global society.

Fundamental Orientations

In the competition paradigm, the primary orientation is, of course, competition. This reflects the basic assumption that competition is the dominant orientation in living systems, that is, that life itself is an ongoing competition for scarce resources and, hence, for survival. This competitive orientation is believed to exist within as well as between species, at any point in time as well as across the eons of evolution. Competition, in essence, is seen as inherent and inevitable. As a result, we have designed our various institutions to take this competition into account. Collectively, our political economic institutions promote competition, regulate it, address its negative consequences, and make some of it illegal. Underlying all of these approaches is the basic belief that competition is a "given" and, thus, we should take advantage of it where we can and restrict it where it is problematic.

This leads naturally to the second fundamental orientation of the competition paradigm. Because competition is often harmful, to the individuals involved as well as to the broader collective, it becomes imperative to control the forms of competition that are believed to bring undue harm. Control is then viewed more generally in the competition paradigm as the most effective mechanism through which to limit, reduce, or eliminate behaviors and activities considered to be harmful, bad, or wrong. While the decision regarding what to control is vested in the state, religious institutions have, over the years, exerted considerable influence on our collective definitions of what should be promoted and what should be prohibited. Ultimately, this control orientation permeates families, schools, organizations, and the governments of the world. The control systems utilized in all these institutions tend to be based on fear (i.e., of punishment, of failure, etc.) and, thus, tend to generate fear and competition among those who are subject to the controls. In this way, competition and control are both rooted in fear and in turn give rise to fear, in a mutually reinforcing cycle. More specifically, one can conclude that the competition- and control-based political economy of the dominant paradigm is actually *creating* the very form of selfish, competitive, dominating behavior which it then argues is inherent and inevitable.

In contrast, the new paradigm's fundamental orientation toward collaboration reflects the basic premise—supported now by the evidence and theory regarding evolution and the very nature of life mentioned earlier—that all living systems are inherently collaborative. Need theories of motivation (Alderfer, 1972; Maslow, 1954; McClelland, 1961) have long recognized that humans are inherently social and have intrinsic "relatedness" needs, and most people display a vast array of spontaneously cooperative behavior in many of their daily activities (e.g., driving, working, leisure). Based on this assumption, institutions in the collaboration paradigm should be designed, first and foremost, to promote collaborative rather than competitive behavior. Because the social sciences have clearly determined that human behavior is readily shaped by the incentives and constraints embedded in the environment, creating collaborative institutions would undoubtedly increase the occurrence of collaborative behavior and reduce the level of problematic competitive behavior generated by competition-oriented institutions.

This in turn would enable us to replace our dominant orientation toward control with a renewed emphasis on freedom. The notion of individual rights emerged with the Enlightenment to enable individuals to stake out a claim of "private" space against the intrusive control of the state and religious institutions. These rights were meant to expand the amount of personal freedom individuals could enjoy. Yet control-oriented governments in place around the world today are more naturally inclined toward restriction than expansion of freedom. Government controls, as indicated above, are necessary because of the competitive orientation of individuals and society, an orientation which in turn is actually promoted by the government. By promoting collaboration and reducing competition, the institutions and organizations in a collaborative society could be designed to promote greater freedom in terms of potential and acceptable human behavior, rather than focusing substantially on the need to exert control, as they do currently. Expansion of freedom is worrisome to those in charge of a control system, especially when behavior tends to be competitive. But as human activity becomes more collaborative, there will be less need to restrict others' behavior and instead people can be encouraged to seek their own happiness however they see fit. Free will is recognized in the new paradigm as a fundamental property of all life, of the energy field itself, and in the context of collaborative assumptions, people will be inclined to pursue self-interest in such a way as to also improve collective well-being. This kind of "win–win" orientation in our interactions with others is the cornerstone of a collaborative society.

Core Premises

Four core premises generate and reinforce the dominant paradigm's emphasis on competition and control. The first of these is individualism, which reflects a number of more specific beliefs about the nature of reality. First, the individual human being is presumed to be the highest life form in the universe (with the possible exception of some kind of Divine Being, whose existence remains disputed in the competition paradigm). Second, humans are viewed as willful and intentional beings with the freedom to decide our own course of action, with the further assumption that humans use this free will, first and foremost, to pursue our own self-interest. Third, individuals (and all other sentient beings) are assumed to be separate and distinct and thus independent of each other, unless or until they choose otherwise.

These beliefs have a number of relevant consequences. The inherent connections and interdependencies among individuals or between humans and other entities are frequently ignored, and human collectives and multi-entity ecosystems are not viewed as living systems. Furthermore, self-interest is viewed as conceptually distinct from, if not opposed to, collective interest, with greater importance attached to the former than to the latter. Likewise, human interests carry greater weight than those of all other life forms, such that everything on the planet is viewed as available to be used for human purposes. Ultimately, pursuit of human self-interest is given the highest priority in the competition paradigm, and the world's institutional context is designed to promote and reward this behavior. This results, too readily, in a disregard for the collective interest, or at least the lack of any felt responsibility on the part of the individual to help take care of or even improve collective well-being. In fact, the dominant paradigm argues that we will maximize our collective well-being economically if everyone pursues their own personal self-interest. Once again, then, it is reasonable to conclude that the selfish behavior characteristic of the dominant paradigm is better viewed as a natural consequence of its starting assumptions, that is, a self-fulfilling prophecy (cf. Wolfe, 1989), rather than as evidence that these premises are valid.

The collaboration paradigm rests instead on the assumption of connectivism, which holds that the entire universe is inherently interconnected and interdependent. As indicated above, the validity of this assumption is supported by evidence from a wide range of natural sciences, which demonstrate it to be true of the underlying universal energy field as well as of living systems at all levels of analysis from molecular to planetary. The connectivism assumption is further reflected in the belief that the connections among individuals in human systems (e.g., groups, organizations, communities) are real and important. A growing literature pointing to the importance of trust (e.g., Carnevale, 1995; Fukuyama, 1995; Robinson, 1996; Shaw, 1997) and social capital (e.g., Coleman, 1988; Nahapiet & Ghoshal, 1998; Putnam, 1995) in the functioning of human collectives reflects recent acknowledgment of the significance of these important but intangible connections.

Furthermore, the collaboration paradigm does *not* assume that all individuals are primarily self-interested (in the narrow sense articulated in the dominant paradigm). Existing research already indicates that individuals, groups, and cultures vary in terms of their level of individualism (e.g., Chatman & Barsade, 1995; Earley, 1993; Hofstede, 1980), with some demonstrating a high degree of collectivism. The "individualism versus collectivism" dimension used in this research, however, seems to imply that a person or culture is oriented either toward self *or* toward the collective—that is, those that are more self-oriented are, by definition, less collective-oriented. In contrast, a logical conclusion in the new paradigm is that we can and should be both self- *and* collective-oriented. In fact, the connectivist assumption means that self- interest cannot be viewed accurately as independent from collective interest; the well-being of the part is intimately connected to the well-being of the whole. In the new paradigm, then, tending to collective well-being is a necessary prerequisite, if not a sufficient condition, for providing a context in which all individuals can effectively pursue their own self-interest. This perspective enables individuals to adopt the win–win orientation necessary for a collaborative society, replacing the win–lose orientation generated by the competition paradigm's emphasis on self-interest at the other's or the collective's expense.

The second core premise of the competition paradigm is science, or more specifically, the basic beliefs defining the nature and focus of legitimate scientific endeavor. In the dominant paradigm, knowledge is, for the most part, only accepted as true or real if it has been gained through the scientific process. But this process is based on some starting assumptions that create inherent biases which limit its scope and thus circumscribe our collective definition of reality.

First, the general purpose of science is to understand the universe, which is accomplished by identifying patterns in the activity of the subject of study (Stewart, 1995), making the subject more predictable and thus, ultimately, more controllable. A key element of this process is the identification of cause–effect relationships through an iterative process of generating hypotheses and testing them with data. Second, the only legitimate data are those that can be measured, which basically means that only evidence obtained through the five senses (including technological "extensions") is accepted by science. In essence, this limits the focus of legitimate scientific study to the material world, which means in turn that only the material world is defined as real in the competition paradigm. Third, the scientific process is grounded in a mechanistic worldview, in which scientific efforts to explain the nature of reality reflect reductionist thinking (i.e., the notion that "wholes" are best understood only in terms of their "parts") which in turn leads to fragmentation (i.e., the whole is divided into smaller and smaller parts for more and more specialized study). Fourth, science is inherently skeptical and thus conservative, in that new ideas, theories, and explanations are not accepted as valid until they have been adequately confirmed through testing and/or the accumulation of data.

In the collaboration paradigm, the biases inherent in the assumptions of science are recognized, allowing our collective definition of reality to be expanded to a metaphysical understanding of the universe. First, the key notion of cause-and-effect is brought into question. Quantum physics has determined that subatomic reality demonstrates the principle of nonlocality (Capra, 1991). This means that the behavior of any part of the universal energy field is determined by its instantaneous, nonlocal connections to the whole field. Because these connections are immediate and thus transcend the presumed constraints of time and space, they defy the notion of cause-and-effect that undergirds scientific explanations of the nature of reality. Second, nonmaterial realms of existence are acknowledged as real in the collaboration paradigm. Scientific evidence mentioned above points to the essential unity of material reality and consciousness. Coupled with the premise of a spectrum of consciousness and a great chain of being, the new paradigm accepts as real a higher, spiritual realm that is inherently connected to the material world. With this as a starting assumption, contemporary experiences of and interactions with a nonmaterial or spiritual realm (e.g., Carey, 1991; Castaneda, 1968; Dennis, 1997; Monroe, 1971; Moody, 1975, 1993; Walsch, 1995) can be taken seriously and learned from rather than discounted or even ridiculed.

As a result of this expanded definition of reality, the mechanistic worldview of the competition paradigm is recognized as an inadequate model for understanding the universe. The new paradigm science is slowly leading the scientific community to the realization that holistic analyses of integrated systems are necessary to understand them effectively. This shift in thinking is best reflected in the areas of systems theory, complexity theory, and chaos theory (e.g., Prigogine & Stengers, 1984; Von Bertalanffy, 1968; Waldrop, 1992). Ultimately, the scientific and mathematical breakthroughs derived from these and other fields of inquiry are providing the foundational knowledge the human race needs to transition successfully to the collaboration paradigm. When the emerging theory and accumulating evidence become sufficiently compelling, the resistance caused by science's inherent conservatism can be overcome and the new theory of reality can be accepted as legitimate.

The third underlying assumption of the competition paradigm is rationality, which can be broadly defined as the reliance on human reasoning as the appropriate basis for analysis and assessment. This assumption holds that a rational approach is the normative model for reaching judgment or making decisions. In practice, the concept of rationality has taken on three interrelated connotations that further shape its meaning and thus constrain its use. First, rationality reflects an emphasis on data-based analysis, with data defined as measurable, quantifiable information. Taken to the extreme, this leads to the notion of a rational decision maker choosing a course of action to maximize self-interest based on a cost-benefit analysis of data presented in terms of a common metric, which is typically the monetary value of each potential alternative. Second, rationality refers to an objective or dispassionate analysis, which further implies that emotions and other subjective (and thus, presumably, *non*rational or even *ir*rational) factors should be omitted from the analysis. This ignores the fact that the goals driving a rational decision-making process, derived from self-interest, reflect the decision maker's "wants," which are inherently subjective and frequently emotional. Third, because analytical reasoning is essentially a "left-brain" intellectual process, "right-brain" activities such as the use of intuition are discounted and thus not viewed as making a legitimate contribution to the decision-making process. For example, our formal education system places much greater emphasis on the development of analytic skills than it does on the development of intuition and nonanalytic modes of thinking.

A broad effect of this imbalance is that rational decision-making processes underestimate the value of creativity and inhibit its occurrence. Given the focus on monetary cost-benefit analysis, another effect is that rationality overemphasizes the efficiency criterion, that is, maximizing the benefit-to-cost ratio, relative to other important criteria. Finally, the competition paradigm's assumptions regarding the importance of rationality and efficiency tend to dictate the design and management of organizations striving to be "rational actors" (e.g., Taylor, 1911; Thompson, 1967; Weber, 1947). The result is the bureaucratic hierarchy, which has long been criticized (e.g., Argyris, 1957; McGregor, 1960) for the demotivating and dehumanizing effects of a top-down control system designed to promote organizational rationality and efficiency (i.e., the interests of those individuals controlling the system) at the expense of individual well-being.

Creativity replaces rationality as the normative orientation in the collaboration paradigm. Because considerable literature makes it clear that decision making by individuals, groups, and organizations is typically *not* rational (e.g., Cohen, March, & Olsen, 1972; Janis, 1971; Lindblom, 1959; Pfeffer, 1977a; Staw, 1980), the normative status of the rational model in practice is already questionable. The rational model, in other words, is more of an "espoused theory" than a "theory-in-use" (Argyris, 1977). To enhance creativity, however, the various limitations of the rationality assumption must be overcome.

First, a full range of costs and benefits, especially those that cannot be translated into monetary values, should be incorporated into decision-making processes. While efficiency need not become irrelevant as a criterion, other subjective values such as happiness, beauty, and fairness should be considered as well. Second, creativity will be enhanced by greater development of and reliance on intuition (Adams, 1979), which again is simply the mechanism through which our normal consciousness receives valuable information from higher levels of consciousness (Naparstek, 1997). Creative activity frequently results from an intuitive idea that makes subsequent rational analysis feasible and meaningful (Agor, 1984). Because intuitive capacity can be developed (Ray & Myers, 1989), educational activities should be designed to promote such development. Third, the creativity assumption in the collaboration paradigm requires that the values and interests of *everyone* affected by a given decision must be taken into account in an effort to reach a consensus decision that is satisfactory to all concerned. The process of striving for consensus often requires, and frequently generates, creative solutions to the problem at hand (cf. Shuster, 1990). Thus, consensus decision making should replace top-down decision-making processes that are inherently biased toward the interests of those with the most power. Ultimately, a key advantage to organizations of emphasizing creativity rather than rationality is that it enables them to become more innovative, which is replacing efficiency as the dominant criterion for organizational success and survival (Alter & Hage, 1993).

The fourth assumption of the competition paradigm is patriarchy. A patriarchy is a cultural system in which men have more power and control than women. This is reflected in the fact that men hold most of the positions of power in the world, and thus have control over most of the world's material resources. More deeply, however, patriarchy reflects the fact that the dominant paradigm and its key orientations and assumptions are based on a male-oriented or masculine worldview. This is because men have been the source of almost all of the ideas on which this paradigm and its dominant institutions have been based. Patriarchy is reinforced by the additional assumption that the inequality between men and women inherent in a patriarchy is natural, inevitable, and/or desirable. This notion is based on the premise that human civilization has always reflected this imbalance, and thus generates the further belief that it always will. The most common assumption as to why men have more power and control than women is because men are physically stronger, which has enabled them to exert and maintain their power and control at will. Pervasive subordination and degradation of women is a natural consequence of the patriarchal assumption.

Whereas patriarchy reflects the deeper assumption that one group of humans has the right to exert control over another group of humans, the collaboration paradigm rests instead on the assumption of equalitarianism. Reflecting the premise that all *humans* are created equal, equalitarianism recognizes that men and women are equal partners in the development and maintenance of an effective human civilization. Furthermore, it assumes that equality between men and women, and likewise between all the peoples of the world, is a viable option for humanity. This assumption is based on considerable archaeological evidence which indicates that early human societies (i.e., up until about 5,000 years ago) operated according to the principles of a "partnership" cultural model rather than the "dominator" model inherent in a patriarchal system (Eisler, 1987). These equalitarian, partnership societies were ultimately overrun and destroyed by invading patriarchal, dominator societies. However, given that humans have already demonstrated their capacity to build and thrive in an equalitarian culture, the competition paradigm's assumption that patriarchy is inevitable can be rejected as invalid. Furthermore, because human civilization has now moved into the Information Age, where knowledge and intelligence rather than physical strength become the primary basis for power and authority, the justification for male domination over women disappears. The continued role of the feminist movement for bringing about greater equality between the sexes reflects slow but steady progress back to a partnership model of society rooted in the equalitarian assumption.

As indicated earlier, the four premises of the collaboration paradigm do not constitute the opposites of the dominant premises, but rather transcend them. A typical feature of analysis in the dominant paradigm is to utilize "either/or" thinking, that is, to identify apparent contrasts and then assume that only one or the other can be "true." It was very difficult for early quantum physicists, for example, to accept the fact that light was neither a particle nor a wave but was in fact both. "Both/and" thinking is the hallmark of the new paradigm. If we build a society that values both individualism and collectivism, we have connectivism. If we build a society that accepts both scientific and spiritual knowledge as valid and legitimate, we have metaphysics. If we build a society that utilizes both left-brain rationality and right-brain intuition, we have creativity. Also, if we build a society that honors and empowers both men and women, we have equalitarianism. To adopt the collaboration paradigm, therefore, it is not necessary to deny the value and validity of individual self-interest, scientific progress, rational analysis, or masculine energy. We simply have to recognize the equal value and validity of community well-being, spiritual growth, intuitive insights, and feminine energy, and thus transcend the limitations that naturally result from giving primacy to only one-half of the whole.

It should be clear that, for each paradigm, the four core premises are mutually reinforcing of each other and of the fundamental orientations. A focus on self-interested rational actors in pursuit of finite material resources generates competition, while patriarchy's natural inclination toward domination leads to power-seeking behavior that is intended to facilitate one's pursuit of self-interest through the control of available resources (including other people). Fear of others, fear of failure, fear of losing control, and many other fears are pervasive in this system. In contrast, the new paradigm holds forth a vision of creative men and women working collaboratively to improve the collective well-being so as to provide a context in which each individual has more freedom to pursue their spiritual development and/or whatever else makes them happy. Love of self, love of others, and love of community would be pervasive in this system. Very simply, the primary objectives in the collaboration paradigm would be quite different from those in the competition paradigm.

Primary Objectives

In the dominant paradigm, "progress" for human civilization is seen as being achieved through the process of development, which is defined almost exclusively as economic growth. This is based on the assumption that improvement in the economic or material standard of living of individuals and communities is the most important or direct means of enhancing their quality of life. With economic growth thus widely viewed as the key indicator, if not definition, of our collective well-being, the world's political economy is designed to promote such growth. This growth mentality in turn drives the actions of nations, communities, organizations, and individuals. With the fall of communism, the dominant paradigm has declared that free-market capitalism is clearly the best system through which to achieve economic growth and personal advancement. Thus, the policy agenda of the developed world is to open up the markets of developing countries in order to expand operations into these countries and thus spur their domestic economic development. This unquenchable thirst for economic growth continues to be the primary objective in the dominant paradigm even in the face of mounting evidence that it is causing irreparable damage to the natural environment and threatening global sustainability.

The second primary objective in the competition paradigm is consumerism, which can be thought of as the aggregate set of ideas and activities associated with the development, production, marketing, purchase, and use of consumer goods and services. Consumerism is the primary modality through which economic growth can occur. Thus, the most important roles individuals play in the dominant paradigm are as producers and as consumers. As consumers, we are encouraged by the corporate media-delivered mass-marketing system of the free-market economy to maximize our purchases and expenditures and to increase our debt, if necessary, to do so. As producers, on the other hand, the incentives of capitalism are designed to keep wages as low as possible so as to improve bottom-line indicators of profit and growth. This profit is then distributed to the suppliers of capital, that is, the same institutions (and the individuals behind them) that provide the consumer credit which supports the "demand" side of the equation needed to spur economic growth—all in all, a rather vicious cycle. Essentially, people become simply the means through which to pursue the growth objective and thus ensure the continued wealth accumulation and power maintenance of the richest people on the planet. What is apparently overlooked in the dominant paradigm, however, is that any competitive system has losers as well as winners, costs as well as benefits—by definition, not everyone can win, and not all "growth" improves societal well-being. Unfortunately, under current economic policies and accounting methods, resources allocated to addressing the personal and social costs of competition are simply tallied as additional economic growth. As Hawken (1997, p. 48) points out, "where economic growth is concerned, the government uses a calculator with no minus sign."

The first primary objective in the collaboration paradigm is conscious evolution (McWaters, 1982), which can be broadly defined as an intentional process of growth and development by individuals, communities, and societies. Hubbard (1998) points out that we must, from here on, choose a path of conscious evolution in which we explicitly and collectively decide the nature of our development and the direction of our evolution. We cannot afford to simply let development happen *to* us, with whatever consequences emerge from the dynamics of the free market as shaped by state interventions that attempt to guide it in one direction or another. As indicated earlier, a key premise in the new paradigm is that the true evolutionary saga being played out in the unfolding of creation is the emergence of higher and higher levels of *consciousness*, rather than simply more and more complex physical life forms (Wilber, 1995). This perspective suggests that human civilization should place a higher priority than it currently does on facilitating the development of consciousness, individually and collectively. To support this objective, the dominant institutions of society should be redesigned to encourage and enable this process. At a minimum, this will require a shift in attention away from economic growth toward activities that contribute more directly to human growth and development defined much more holistically. A further logical conclusion is that education, broadly defined as those processes which contribute to one's pursuit of self-actualization, becomes the most critical activity in the collaboration paradigm.

The second primary objective in the new paradigm is quality of life (cf. Olson, 1995), which must be assessed in terms of the full range of criteria that are important to people as individuals and as collectives. For example, such noneconomic "goods" as peace, justice, beauty, health, and happiness are viewed as critical to our individual and collective quality of life. In other words, our needs and interests are not simply material in nature, but reflect those associated with our whole person— body, mind, heart, and soul. Thus, the competition paradigm's myopic overemphasis on consumer goods and services as the primary determinant of quality of life is replaced by a collective focus on ensuring that everyone can maintain a minimum standard of "health" in each of these areas and has a meaningful opportunity to improve themselves however they so choose. Many individuals already have consciously adopted a lifestyle of "voluntary simplicity" (Elgin, 1993b) based on their belief and experience that quality of life is actually diminished by the work-and-spend pattern inherent in the consumerism mentality. Interestingly, the key dimensions and determinants of quality of life in the new paradigm are not "scarce resources" and are likely to be achieved only through collaborative efforts. Thus, a competition-based system for pursuing improved quality of life under these circumstances is both unnecessary and counterproductive.

Conclusion

The collaboration paradigm assumptions outlined above are justifiable given the new theory of reality and the empirical evidence which supports it. They would also seem to be preferable to the old assumptions in that they point the way to a better future civilization than the one generated by the competition paradigm. Undoubtedly, these assumptions will readily be perceived as unrealistic, which in fact they are in the context of existing competition-based assumptions. This is why a planetary transformation to the collaboration paradigm is as difficult, but equally as easy, as a change in our consciousness, that is, through the adoption of the new paradigm assumptions. If the new theory of reality's ontological premise is correct—that consciousness precedes reality (Havel, 1990)—society will turn toward collaboration if and when enough people adopt a new perspective and thus change their beliefs, attitudes, and actions. Evidence suggests that millions of people (Ray, 1996) are already adopting a new orientation or lifestyle that moves away from the competition paradigm. Likewise, myriad social, economic, and political innovations congruent with the themes of the new paradigm are emerging around the world. Such transformations are certainly visible in the world of organizations. Although the broader political economy appears to reflect a state of "hypercompetition" (D'Aveni, 1994)—an issue I will return to in the conclusion of the chapter—the organizational world is restructuring itself to facilitate internal as well as external collaboration. In the next section, a number of these changes are synthesized into a new model of organization that is grounded in and supports the collaboration paradigm assumptions.

COLLABORATIVE ORGANIZING

The new paradigmatic assumptions outlined above provide the foundation for a fundamentally different organizational form than the bureaucratic hierarchy. The bureaucratic model is based on and reflects key assumptions of the competition paradigm. It was articulated as an ideal type of rational organization (Weber, 1947), and it has incorporated key principles from "scientific management" (Taylor, 1911). Managers' emphasis on the implementation of appropriate control systems has reflected their "Theory X" (McGregor, 1960) fears that employees dislike work, have little ambition, wish to avoid responsibility, prefer to be directed, and must be coerced to put forth adequate effort toward the achievement of organizational goals. Weber viewed the bureaucracy and its rational-legal authority structure as a valuable improvement over more traditional patrimonial systems. He associated the bureaucratic form with the growth of rationality in Western civilization, and viewed its characteristics as congruent with the requirements of a society that was becoming increasingly modernized and industrialized. The fit between the bureaucratic model and modern society was a good one, such that it has become institutionalized throughout this century as the dominant organizational form (Zucker, 1983).

However, the bureaucratic hierarchy is *not* congruent with the requirements of a postmodern society that has entered the Information Age. A transformation to the collaboration paradigm is being made possible by growing awareness that intellectual capital is now the critical resource in the global economy (cf. de Geus, 1997). Economic success is increasingly recognized as being a function of the extent to which organizations effectively acquire, create, and/or utilize new knowledge (e.g., Shukla, 1997; Wick & León, 1993). The fact that knowledge is inherently a positive-sum or infinite resource (Halal, 1998) undermines the basis for and value of competing for it. Instead, significant benefits to society could accrue if knowledge were readily disseminated and used for collaborative win–win purposes, rather than fought over in a futile attempt to control it for private gain. The emergence of the internet as a highly decentralized mechanism for the distribution and exchange of information is already facilitating the development of new kinds of organizations, business strategies, and economic systems that are contributing to a trend toward collaboration.

Just as the bureaucratic model served as a valuable organizational form for a society transitioning into the modern era, the time has come to specify a new "ideal type" organization that is congruent with the societal changes taking place as part of the present paradigmatic transformation. The new theory of reality outlined above gives rise to a new set of collaborative assumptions which, when taken as "givens," provides a basis for rethinking the essential characteristics of an ideal organizational form. In this section, I outline what might be thought of as a "blueprint" for such an ideal type—a set of guidelines regarding the purpose, design, and functioning of "collaborative organizing." In an effort to identify a broad range of characteristics of this organizational form, the discussion provides only a skeletal framework of the model and avoids extensive elaboration of relevant details pertaining to many of the issues raised. Because there is clear indication that organizations are already exhibiting evidence of most of the espoused features of this model, I cite relevant literature throughout the discussion below to support the premise that this new organizational form is in fact already emerging on a broad scale.[1]

Purpose

In the collaboration paradigm, every organization should have as its purpose the improvement of society, which includes contributing to both individual and collective well-being (cf. Harrison, 1984; Maynard & Mehrtens, 1993; Weisbord, 1987). Rather than viewing people as existing to serve the interests of organizations, organizations should exist to serve the interests of people. This requires attending to the overall well-being of the planetary ecosystem, and also to future generations' need for an equal or better quality of life than is available to the present generation. These are, of course, two fundamental facets of a sustainable development strategy (Purser, Park, & Montuori, 1995; Starik & Rands, 1995), and in the collaboration paradigm, it is imperative that all organizations enhance rather than detract from our collective ability to pursue growth and development in a sustainable manner.

Organizational effectiveness is defined, then, in terms of the extent to which an organization improves society. Societal improvement will be maximized only to the extent that organizations adopt a collaborative, win–win approach in their various interactions, and abide by "norms of reciprocity" so as to maintain equitable exchange relationships over time. In contrast, the use of win–lose strategies in the dominant paradigm generates various "costs of competition" that inherently detract from our ability to maximize societal well-being. Because it is often easier for diverse actors to agree on which outcomes are least desirable (i.e., impose the greatest costs) than it is for them to agree on which outcomes are most desirable, organizational effectiveness can be assessed operationally in terms of the problems it causes for society (cf. Cameron, 1984). From this perspective, "harm minimization" can replace "utility maximization" as the primary criterion by which organizational effectiveness, and social justice more generally, is evaluated (Keeley, 1984).

Organizational survival should in turn be a function of its effectiveness. Organizations that contribute to societal well-being while minimizing the negative effects of their activities should have more opportunity to continue to operate, while those generating undue harm and/or not adding value to society should not be allowed to continue. This link between effectiveness and survival would be mediated by the notion of "surplus." In systems terms, organizations acquire inputs from the environment, transform inputs into outputs, and dispose of outputs into the environment in exchange for additional inputs. The more an organization's activities generate a surplus of resources (human, material, financial, etc.), the greater its capacity to continue adding value to the world. By successfully engaging in win–win exchanges that improve outcomes for all the parties involved in the exchange, more effective organizations will, by definition, generate greater surpluses. Society as a whole would benefit both from the consequences of these win–win interactions as well as from the reduction in negative consequences resulting from a careful assessment of the costs generated by the organization's activities.

In the competition paradigm, this premise of greater returns accruing to organizations operating more effectively is found in the concept of "profit" (at least in the case of for-profit organizations). Profit is presumably a measure by which society, through the market, determines the overall value of the organization's activity. (This is more true in theory than in practice, however, because accounting practices can help disguise how much profit was made and because other indicators of an organization's value, such as stock price or price/earnings ratio, are viewed by the financial community as a more direct measure of an organization's worth.) There are three key differences between the concept of profit and the alternative concept of surplus. First, surplus is explicitly defined more broadly than simply the monetary value of "revenues minus costs." Surplus is an indicator of *all* the resources accruing to an organization as a result of its contributions to society, including those which cannot be readily valued in monetary terms. Second, whereas profit maximization is the primary *purpose* in the competition paradigm, surplus generation in the collaboration paradigm is a *consequence* of the effective pursuit of an organization's primary purpose to improve society (cf. Collins & Porras, 1994). Profit maximization strategies lead, quite naturally, to a competitive, win–lose exchange orientation. Ultimately, the focus on win–win, reciprocity-based interactions in the collaboration paradigm even reduces the importance of making a profit or surplus through the process of exchange (as is currently the case in nonprofit organizations).

A third key difference is the decision rule defining how surplus, rather than profit, should be "distributed" among organizational stakeholders. According to the rules of capitalism in the competition paradigm, profits are to be distributed primarily, if not exclusively, to the organization's "owners," defined as those who provide its financial resources. In the collaboration paradigm, surplus is distributed more equitably to a broader range of key contributors to the organization's capacity. Furthermore, because the organization's primary purpose is to promote individual and collective well-being, the surpluses generated by the organization's activity should be distributed in such a way as to further these two objectives. For example, surplus could be distributed to all individuals who contributed a resource to conduct the organization's activity, it could be utilized to develop the community in which the organization operates, it could be allocated to provide valuable benefits for society-at-large, and so forth. Guided by the principle of improving societal well-being, it is unlikely that surplus would be distributed only to those stakeholders who already have access to or control over an inordinate share of the organization's financial resources, which is the default outcome under the rules of capitalism.

This discussion of surplus raises the issue of how the level of surplus, and thus an organization's survivability, will be ascertained. In the competition paradigm, two primary mechanisms are utilized to determine the level of resources an organization can accrue to carry out its activities. These are the market and the state, or government. Generally speaking, the market is viewed as the preferred mechanism for allocating resources among organizations. Reliance on the market is based on the assumption that organizations providing socially desirable goods and/or services will be able to exchange these outputs for the inputs needed to continue their activities. Those that provide their goods and/or services at an efficient price will be able to make a profit, which in turn will attract additional resources needed to survive and possibly grow. The state, on the other hand, is primarily used to counteract various difficulties associated with the market mechanism. Public organizations exist, for example, to provide goods and/or services for which there are "market failures," to restrict various undesirable features of market activity, and to respond to various problems caused by market activity. Allocation of resources among the various public organizations is determined through the political processes of government.

The market and the state, then, are viewed as two distinct mechanisms used to make society's collective decisions regarding resource allocations among organizations. Because these two mechanisms operate according to different decision rules, the private and public sectors are often characterized as being distinct and quite different from each other. In reality, of course, the two sectors are not independent, and the boundaries between them are becoming increasingly blurred (Bozeman, 1987; Perry & Rainey, 1988). For most organizations, the resources they accrue, and thus their level of profit or surplus, are a function of the dynamics or "decisions" of both the market and the state. For example, a private organization's revenues are determined largely through market dynamics of supply and demand, while the state imposes some costs on the organization through various types of regulatory activity. Similarly, a public organization's operating budget is usually determined through legislative action, but the costs associated with acquisition of needed inputs are primarily dictated by market forces. Thus, at the organizational level of analysis, the amount of resources available to any given organization is likely to be jointly determined by the market and the state.

At the societal level of analysis, resource allocation decisions are also jointly determined by the market and the state. But the conceptual distinction between the private and the public sectors, including assumptions about the different functions or purposes of these two spheres of activity, creates a significant complication for our collective pursuit of societal well-being. In particular, the dominant paradigm's political economic system operates such that profits are largely privatized while many of the costs generated by the negative consequences of profit-seeking activity are instead socialized (Hansen, 1995; Vansina & Taillieu, 1996). In other words, private organizations have considerable freedom to engage in a broad range of profit-seeking activities, and any surplus resources an organization accrues become private property. While the state regulates these activities to limit some of their potential negative externalities, other externalities—both direct and indirect—are not regulated effectively or at all. As a result, public money is used to fund the organizational activities required to address the various problems caused by these unmitigated negative externalities. In this way, many private organizations do not pay the full costs of their activities (cf. Cairncross, 1991; Hawken, 1997), and thus they are able to accumulate profits that do not accurately reflect their overall contribution (or lack thereof) to individual and collective well-being.

To put all this another way, by separating the market from the state and the functions of the private sector from those of the public sector, the competition paradigm's political economy creates a dysfunctional system in which the benefits and the costs of an organization's activities are frequently not considered *simultaneously* in our collective decisions as to which organizations are best contributing to societal well-being. Benefits are presumed to be adequately assessed through market dynamics. In the marketplace, individual participants have direct input into the process by virtue of their freedom to "vote with their feet," with the result that resources flow to those organizations best responding to the preferences of these participants. As a collective decision-making process, the market is an "aggregative" mechanism in that the collective outcome is simply a sum of the decisions and actions of the individual participants. In contrast, the state attempts to provide a more "integrative" collective decision-making mechanism, through which a reasonable synthesis of the diverse preferences of many individual participants can be identified and implemented. Individual participants do not have direct input into the state's decisions regarding how to minimize and address the various costs of organizational profit-seeking behavior, as they influence this process only indirectly through their elected representatives. The worst consequence of this separation is that organizations can continue to survive and prosper because of market-generated profits, while the state is unable or unwilling to require these organizations to be accountable for the full range of costs associated with their activity. Ultimately, decisions regarding organizational survival are thus fragmented, such that people do not have the opportunity to decide explicitly whether the costs associated with an organization's activities outweigh the benefits to collective well-being provided by that organization.

In the collaboration paradigm, the mechanism for assessing the benefits and costs of an organization's activities should not be bifurcated in this manner. In other words, if surplus is to be used as an indicator of the effectiveness of an organization's activities, the mechanism through which surplus is determined should simultaneously factor in both the societal benefits and costs of these activities. To the extent possible, participants in the system should have direct rather than indirect input into these decisions, such that they collectively decide which organizational activities to support and which to discontinue. Likewise, a mechanism that encourages integrative decisions would be preferable to an aggregative system that does not effectively take into account the specific interests of the collective (as opposed to the sum of the interests of the individuals in the collective). What is needed, then, is an organizational governance mechanism that builds on the strengths of both the market and the state, without suffering from the problems inherent in each of these as well as in their separate use to make collective decisions. A design for collaborative organizing that reflects such a governance mechanism is described next.

Design

The label collaborative *organizing* is used to reflect the fact that these organizational systems are best thought of in dynamic rather than static terms (cf. Weick, 1969). (For convenience, however, these systems are also sometimes referred to below as organizations.) These systems are defined as ongoing sets of activities (cf. Pfeffer & Salancik, 1978), characterized in terms of patterns of interactions rather than in terms of time and/or space constraints. A critical feature of these systems is that they are self-organizing. The notion of self-organizing systems has been gaining attention in the organizational literature recently (e.g., Leifer, 1989; Stacey, 1996; Stamps, 1997; Wheatley, 1992). Grounded in research on chaos theory and complexity theory (e.g., Prigogine & Stengers, 1984; Waldrop, 1992), the key element of a self-organizing system is its ability to reconfigure itself to perform more effectively at a higher level of complexity. When environmental turbulence becomes too great for an organization to manage adequately, it can undergo a relatively spontaneous and sudden transformation to a new way of operating. This pattern of transformation is reflected in a punctuated equilibrium model of organizational change (e.g., Tushman & Romanelli, 1985). After a self-organizing transformation occurs, an organization is able to generate a higher level of internal activity due to its improved ability to cope with environmental complexity. In managing the environment more effectively, the organization is also better able to seek out and attract enough resources and skills to offset the potential disintegration of the system.

The "attractor" which draws together the resources—that is, the energy (cf. Ackerman, 1984)—needed to conduct the activities of a particular collaborative organizing system is its mission. Mission refers to the organization's specific intentions as to how it will contribute to societal well-being. The various energy sources needed to pursue this mission self-organize, creating the necessary activities, and then they maintain an ongoing process of self-organizing as long as interest remains in pursuing this mission. The requisite energy sources (e.g., material, financial, human, intellectual, and social capital) come together in a collaborative process of interaction and exchange reflecting their shared interest in the mission, with individuals, groups, and organizations providing and receiving the goods and services needed to pursue this mission. Thus, the mission and the visions and values cultivated to support it serve as the guiding framework for what the organization will do and how it will conduct its activities. Because participation in the system is voluntary, acceptance of this normative framework can serve as an adequate "control" function in the organization (cf. Walton, 1985), as suggested by the literature on organizational culture (e.g., Trice & Beyer, 1993). When a culture is mutually agreed upon, its normative control function is quite useful, especially when the culture itself reinforces collaborative interactions.

A collaborative organizing system thus consists of a self-organizing network of individuals, groups, and organizations that contribute to the overarching purpose and mission of the organization. The notion of a network form of organization has been identified in the literature as a new mechanism for collective activity that is distinct from both hierarchy and market (Powell, 1990). In essence, a network model changes the key axis of organizational analysis from the vertical to the horizontal. This shift parallels widespread calls for reducing or eliminating hierarchy (e.g., Drucker, 1988; Handy, 1990; Kanter, 1983; Osborne & Gaebler, 1992; Peters, 1988) as well as various organizational redesign practices intended to enhance "horizontal" capacity (Galbraith, 1994; Hammer & Champy, 1993; Linden, 1994). In the absence of a vertical authority structure through which key decisions get made, network organizations require horizontal interactions among partners with varying levels of formal and informal power. These interactions inherently require more collaboration than competition, as competitive behavior impedes the development of trust and social capital required for effective interaction among "peers." Effective interaction among individuals is critical to the development of successful teams and collaborative alliances, which are increasingly recognized as useful tactics for improving innovative and adaptive capacity (Mohrman, Cohen, & Mohrman, 1995; Powell, Koput, & Smith-Doerr, 1996). More generally, innovativeness and adaptability are essential qualities of collaborative organizing, as the improvement of society requires innovative (i.e., creative) responses to the evolving preferences of individuals and collectives.

In essence, then, collaborative organizing constitutes a network of voluntary participants engaging in processes of collaborative exchange oriented toward pursuing the organization's mission while also fulfilling participants' preferences. All participants are members of the system, and the system is member-owned. The notion and practice of employee ownership has existed for a number of years (cf. Rosen, Klein, & Young, 1986), and Greider (1997) suggests that a broad-scale movement toward greater labor ownership is a necessary step toward overcoming the dysfunctions inherent in the existing global capitalist economy. In collaborative organizing, all member-owners would determine how any surplus created by the organization would be distributed. As part of this process, the criticality of all types of capital, not just financial capital, should be recognized and rewarded. Likewise, the interests of the broader collective within which the organization functions should also be taken into account, not just in terms of how surplus is distributed, but how the organization conducts its activities more generally.

These various properties of collaborative organizing are all consistent with the "cellular" form of organization recently described by Miles, Snow, Mathews, Miles, and Coleman (1997). The notion of a cellular system reinforces the basic premise that organizations, as human collectives, are themselves living, adaptive systems (cf. de Geus, 1997). Cellular organizations are comprised of individual cells, which can be conceptualized as individuals, teams, profit centers, business units, or entire organizations (or, for that matter, as organizational alliances, communities, or even nation-states). Cellular organizations form when independent, autonomous cells self-organize into more complex, interdependent, higher-level systems that can engage in more sophisticated activities and accomplish more difficult objectives. At the individual level, for example, this self-organizing quality is reflected in such practices as individuals choosing which projects to pursue, which teams to work on, and which tasks they can perform such that their capabilities are used most effectively. The individual cells are held together through various types of connecting mechanisms such as shared interests, information technology, opportunities for external professional interactions, and so forth. These connecting devices are also self-organizing, reflecting needs and opportunities recognized by the members of the system. Furthermore, protocols jointly defined by system members replace bureaucracy as a primary mechanism for coordination and control.

In addition to its self-organizing property, collaborative organizing is also self-managing (cf. Zeleny, 1990). Each cell is self-managing in that "management" functions (e.g., planning operations, ensuring output quality, interfacing with other cells, and responding to external demands) are the joint responsibility of all the cell's members. More generally, each cell in a cellular organization is responsible for its own performance, and all the cells are collectively responsible for the long-term success of the organization as a whole. In this way, cellular systems as a whole are also self-managing, with "clusters of self-organizing components collaboratively investing the enterprise's know-how in product and service innovations for markets that they have helped create and develop" (Miles et al., 1997, p. 12). At the interorganizational or macro-system level, the notion of boundaries between separate and distinct organizations starts to lose its meaning (reinforcing the blurring of boundaries between organizational sectors previously noted), with most collective activity being accomplished through a collaborative effort among networked cells at various levels of analysis. This collaborative activity adds value throughout the system—within and between cells, within and between organizations—improving "individual" and "collective" well-being simultaneously. As a result, collaborative organizing through a cellular model serves as a useful design through which organizations can effectively achieve their purpose of improving societal well-being.

While the concept of self-management is an essential characteristic of collaborative organizing, additional description of the governance mechanism to be used in these systems is warranted. Basically, governance refers to the "formal structure" for organizational decision making, which in the case of a dynamic system like collaborative organizing might be better thought of as design criteria. The criteria for a collaborative governance system are reflected in what I call an "inside-out" model of governance, which replaces the top-down governance structure used in the bureaucratic hierarchy. An inside-out model is designed to explicitly identify the preferences and interests of the members of a collective, with these then serving as the fundamental basis for collective decisions. The term "inside-out" refers to the notion that the decision-making process begins in small cellular systems and works its way through ever-larger cellular systems until it reaches the most inclusive level needed to successfully implement the decision. For example, individuals meet in relatively small cells, such as teams or small business units, to determine their collective interests and preferences regarding the decision at hand. These cells (or, probably, their representatives) then meet with other similar cells to determine the collective stance of this larger cellular organization. This process expands until all cells who are affected by a given decision have had their interests and preferences taken into account in the decision.

Through a consensus-based decision process, the collective decision at any given level reflects a synthesis of the ideas and preferences of the individual cells in that system. While consensus is typically viewed as an impossible ideal in the competition paradigm, it becomes much more feasible and necessary in the collaboration paradigm. Most collective decisions in the competition paradigm are either made autocratically (with or without "consultation" from others) or determined through a voting process. While an autocratic decision maker may or may not take others' interests into account, this approach to decision making (which follows logically from an overemphasis on rationality) is undoubtedly biased in favor of the decision maker's interests. A voting process, while designed to allow various interests to be addressed in the decision-making process, ultimately forces decisions to be made in either–or, win–lose terms. In both cases, then, the result is that one alternative is usually selected over other alternatives, with no effort made to creatively synthesize alternatives into a solution that responds to a broader range of interests. Consensus decision making, in contrast, is oriented toward integrating ideas, preferences, options, and so on, enabling decisions to be made which address the broadest range of individual concerns and thus more successfully reflect the collective interest.

Approaches such as the Delphi and the nominal group techniques have long been available to help small groups make more consensus-oriented decisions. More recently, information technology-based decision-making tools have become available which facilitate integrative decision making in "virtual" groups, that is, networks of people—or, a cellular system—who interact with each other and make decisions, across time and space. The newest technologies available for encouraging consensus decision making include approaches explicitly oriented toward identifying positive steps to take to create a desired future. These include appreciative or collaborative inquiry (Cooperrider & Srivastva, 1987; Torbert, 1983), future search conferences (Emery & Purser, 1996; Lippitt, 1998; Weisbord, 1992), open space technology (Owen, 1997), and other large group interventions (Bunker & Alban, 1992). These approaches can be used effectively with relatively large numbers of individuals (i.e., up to 1,000), and thus they serve as a powerful decision-making methodology for collaborative organizing.

These approaches have their roots in the change technologies developed in the field of organization development (e.g., T-groups, or laboratory training groups; intergroup conflict resolution techniques). As a result, they stress the importance of, and in turn capitalize on, participants' willingness to express their thoughts and feelings openly and honestly. This practice of bringing people together to make consensus decisions through a process in which everyone gets to "speak from the heart" is certainly not a new one. This is how collective decisions were made, historically, in many indigenous cultures around the world (cf. Newhouse & Chapman, 1996). Current examples of this practice of "calling the circle" (Baldwin, 1994) for a communal decision-making task, sometimes under conditions of considerable conflict, indicate that it can often result in creative, integrative, even surprising solutions that have widespread collective support. The political nature of competition-based decision making usually precludes the openness and honesty necessary to achieve these valuable outcomes (cf. Wisely & Lynn, 1994). According to Argyris (1985), the truthfulness of most interactions among people is limited by the nearly universal adoption of what he calls Model I assumptions, which can be thought of as natural by-products of a competition-based worldview. Argyris has long argued for the value of Model II assumptions which, if adhered to, would more readily generate the truthful interactions needed for consensus decision making to be successful.

Governance in collaborative organizing, then, follows a consensus-driven, inside-out model of decision making. Participation in any given decision process should be voluntary, in that any cell can choose not to participate; however, nonparticipation must then be tantamount to a willingness to accept any decision made. The participants should also include any cell which believes it has a stake in the decision; in other words, the process should err on the side of being inclusionary. Inclusion of all stakeholders in a decision helps to ensure that the full range of benefits and costs of a potential action is considered, so that the principle of "harm minimization" (Keeley, 1984) can be applied. Furthermore, full participation is necessary for the decision-making process to be as democratic as possible, reducing the incongruence between the principles of democracy and the functioning of a hierarchical system of governance (Bachrach & Botwinick, 1992; Elden, 1986). Collaborative governance thus provides a means through which to further realize these democratic principles in an important sector of society's collective action arena.

The trend in this direction is already clear, as considerable organizational literature and practice of late have clarified the importance of employee involvement, participation, and empowerment (Cotton, 1993; Lawler, 1992; McLagan & Nel, 1997; Quinn & Spreitzer, 1997), inclusion of customers and clients in organizational decisions (Milakovich, 1995; Osborne & Gaebler, 1992), and other strategies that expand participation and democracy in organizational decisions (Ackoff, 1994). Governmental devolution, greater involvement of citizens in municipal decision making (Berry, Portney, & Thomson, 1993; Thomas, 1995), the growth of the nonprofit or third sector (Hodgkinson & Weitzman, 1996), and the increased focus on participative community and international development (Chrislip & Larson, 1994; Henton, Melville, & Walesh, 1997; Jackson & Kassam, 1998; Organization of American States, 1997; Robertson & Speier, 1998; World Bank, 1996) reflect a similar trend taking place in society-at-large. All of these changes, in organizational as well as in governmental systems, give people more direct input into a broader range of decisions that determine the quality of their lives. As a result, collaborative governance enhances the power of people and their communities to prevent an organization from doing "harm" to which they are opposed. Given the impact of organizations on our individual and collective quality of life, this kind of empowerment is a necessary prerequisite to the "full flowering of the idea of democracy" ("Full Democracy," 1996).

Functioning

The specific operational activities engaged in by the participants of a collaborative system would be guided by the organization's purpose, with participants collectively determining (through inside-out governance) the nature of those activities and/or delegating to specific cells the authority to make these decisions. In addition to their particular operational (i.e., input, transformation, and output) activities, participants would collectively attend to a variety of "system maintenance" activities that are critical to the effectiveness of collaborative organizing. In this section, I will discuss three broad categories of activities, outlining collaboration-based guidelines in the areas of leadership and management, rewards and evaluation, and development and learning. Of course, these three categories are not clearly differentiated from each other, as the discussion below demonstrates.

A first key point about leadership is that it can be viewed as an organizational function, analogous to the classical administrative theory notion of the functions of management. In contrast to the classical perspective that the management function should be centralized, however, collaborative organizing reflects the premise that leadership is best thought of as an organizational variable (Ogawa & Bossert, 1995). In other words, rather than relying on the designation of one or a few individuals as the formal leaders of an organization, leadership is recognized as more effective when it is widely distributed among members throughout the system (cf. Barry, 1991; Senge, 1997). Ideally, all participants will exert leadership under some circumstances or conditions. More realistically, some participants will demonstrate greater aptitude or propensity for leadership; by virtue of their ability to generate voluntary followership (Kelley, 1991), they will emerge as informal leaders. While these members may be delegated responsibility and authority for particular activities, this authority would not translate into the institutionalized differences in authority and hierarchical distinctions on which the concept of bureaucracy is based. In other words, formal authority would be situation-specific, with different individuals holding greater authority over different decisions. While the feasibility of the absence of a formal hierarchy is rejected by those who believe that organizations must be "controlled," others have questioned whether formal leaders actually make any difference in terms of organizational performance (Pfeffer, 1977b). In collaborative organizing, distributed leadership—not a formal leader— is seen as the key to organizational effectiveness.

The leadership orientation most appropriate for collaborative organizing is the notion of stewardship, or servant leadership (Block, 1993; Greenleaf, 1977). This approach is based on the idea that a leader exists to serve, rather than control, his or her followers, whose willingness to follow is a response to the leader's servant nature. In collaborative organizing, this also means that leaders act in service to the organization's purpose and are willing to be held accountable for the well-being of the system as a whole. In other words, they take action after identifying the best way to proceed to serve the highest purpose of all involved (cf. Ackerman, 1984). Servant leaders bring high levels of compassion to the organization, and their values, competence, and judgment inspire trust and reinforce their credibility (cf. Kouzes & Posner, 1995). This leadership model is compatible with the emergence of a stronger "service" orientation in a wide array of organizations (Halal, 1996). It also supports broader trends toward participation and empowerment, and thus serves to promote and facilitate democracy in organizations. Servant leadership is also consistent with other recent leadership literature that recommends incorporating spirituality into the practice of leadership (e.g., Bolman & Deal, 1995; Conger & Associates, 1994; Harrison, 1984). The recent popularity of these spiritual approaches to leadership is congruent with, and may reflect (Lee & Zemke, 1993), broader social trends that have been recognized as signs of an impending spiritual renaissance in the United States (Spayde, 1998).

Given the self-managing quality of collaborative organizing, the concept of "facilitation" replaces the obsolete notion of management. This is consistent with management literature that increasingly characterizes the requirements of effective managers in such terms as facilitator, coach, mentor, and developer (e.g., Bradford & Cohen, 1984; Orth, Wilkinson, & Benfari, 1987; Waldroop & Butler, 1996). Similar to servant leadership, facilitation is primarily an enabling function, oriented toward helping participants carry out their activities and accomplish their objectives more effectively. This could include, for example, serving as a "broker" (cf. Lawless & Moore, 1989; Mandell, 1984; Selsky, 1991) by generating agreement and gaining commitment within and between cells. It could also include identifying and eliminating obstacles in the system so as to increase participants' opportunity to make a significant contribution. It could entail tracking whether the various energy sources in the system (goals, events, people, information, symbols, etc.) are helping or hindering the organization's effectiveness (Ackerman, 1984). While this is by no means a comprehensive list, the point is that facilitation is intended to enhance participant freedom and creativity, eliminating the "control" orientation prevalent in traditional approaches to management. As with leadership, therefore, there are no formal managers with "permanent" hierarchical authority. Instead, performance of the various facilitation functions becomes the collective responsibility of all participants, in the context of a "shared responsibility team" (Bradford & Cohen, 1984).

The use of power and influence in collaborative organizing is modified to reflect the fact that all participants are inherently equal. First, power is conceptualized as a positive-sum resource used to enhance productivity, rather than as a fixed-sum resource used to exert control over people (Kanter, 1979; McClelland, 1975). Power is most effective when shared, increasing the overall amount of productive power in the organization; it becomes dysfunctional when hoarded, usurped, fought over, and/or used coercively. Second, influence results from effective reliance on the principle of reciprocity, the nearly universal belief that any act of goodwill should be reciprocated in the future (Cohen & Bradford, 1989). By being influenceable, participants also become influential, leading to the development of reciprocal relationships characterized by a long-term equality between partners that constitutes an effective win–win, mutually beneficial outcome (cf. Axelrod, 1984). Third, expert power (French & Raven, 1959) is the dominant source of power in collaborative organizing. In those situations where it is useful or necessary for one or more participants to have formal authority over particular decisions or activities, power should be delegated by the system as a whole to those participants with the most expertise pertinent to the situation. This approach is similar to what Mohr (1994) calls a "voluntarist model of organizational democracy."

"Intrinsic" motivation plays an important role in collaborative organizing. With many cells participating in an organization because of a shared interest in its purpose, alignment between individual and collective interests is enhanced (Culbert & McDonough, 1985; Harrison, 1984), with participation then likely to be more meaningful and effective. This kind of organizational identification, or internalization of values and goals, is frequently a primary basis of commitment in both private and public organizations (e.g., Balfour & Wechsler, 1994; O'Reilly & Chatman, 1986). Alignment also serves as the foundation for the development of "covenantal" relationships (DePree, 1990; Graham, 1991) between a collaborative system and its constituent cells. A covenantal relationship is characterized by mutual trust, shared values, and open-ended commitment, and is based on ties that bind individuals to their communities and communities to their members (Kanter, 1968). This kind of equitable relationship between the individual and the organization may increase the frequency of the kinds of "extra-role" organizational citizenship behaviors that can enhance the organization's overall functioning (e.g., Moorman, Blakely, & Niehoff, 1998; Tsui, Pearce, Porter, & Tripoli, 1997).

In return for a cell's agreement to buy into the organization's core beliefs and to abide by any rules it has established, collaborative organizations are committed to promoting the health and development of the individual cells. The growing literature proposing the integration of spirituality into organizations and work (e.g., Briskin, 1996; McKnight, 1984; Owen, 1987) attests to the increasing popularity of the notion (cf. Galen & West, 1995) that organizations should focus on the well-being of the "whole person," reflecting norms of mutual caring and nurturing (Egri & Frost, 1991). "Work" under these conditions becomes a vocation, with individuals more readily operating in a "flow" state (Ackerman, 1984; Csikszentmihalyi, 1990) or having a peak experience (Adams, 1984a). The result is a win–win situation in which both individuals and organizations benefit, in contrast to the frequent occurrence in bureaucratic organizations of high levels of employee dissatisfaction and/or alienation.

Any given cell may also have its own distinct purposes for wanting to participate in a system, which could be independent of the organization's purpose. In particular, cells will usually expect some extrinsic rewards or inducements in exchange for their contributions to the organization (cf. Barnard, 1938; Simon, 1957). The nature of these rewards is clarified in an explicit agreement or "contract" between the cell and the system (i.e., the community of cells) when they mutually agree on the terms of the cell's participation. This could entail simply being paid in exchange for doing some type of work, but the potential range of exchange possibilities is much greater when the guiding principle is a focus on achieving win–win outcomes in the context of addressing the full range of needs associated with the development of the whole person or cell. From the system's perspective, the cell's expected contributions can be defined in terms of activities, where inducements are offered in exchange for the cell's performance of particular processes or behaviors, and/or in terms of outcomes, where the organization specifies a particular result or consequence it desires independent of the activities needed to achieve this outcome. A global trend toward increased use of outsourcing, contract employment, and temporary workers, which has been viewed as a fundamental change in the nature of the employment relationship (Capelli, Bassi, Katz, Knoke, Osterman, & Useem, 1997), reflects significant movement in this direction.

Just as the initial relationship between an organization and any given cell requires agreement by both sides on the conditions of exchange, a continued relationship is based on mutual satisfaction that the terms of the contract have been or are being fulfilled. Either the cell or the system can initiate a change in the terms of the relationship. If such a change is pursued by the cell, the community of cells would collectively decide whether they agree to the new terms. This would follow a collective evaluation of the cell's performance and contribution to the system. Alternatively, the results of such an evaluation, whether positive or negative, may provide the rationale for the organization to initiate a change in its agreement with a cell. Feedback based on a collective assessment by those participants who are interdependent with the cell can also serve as an important mechanism through which to bring about improvement in cell performance evaluated as unsatisfactory. For example, implicit or explicit pressure from a group can be a powerful influence on the behavior of group members (Keisler & Keisler, 1969). Collective evaluation makes it clear to the cell that being broadly perceived as a valuable contributor and an attractive partner is a key asset. The value of this approach is reflected in growing use by organizations of "360-degree assessments" (Tornow & London, 1998), in which an individual is collectively evaluated by managers, peers, employees, and even sometimes customers or users.

Feedback generated by collective evaluations also provides impetus for and guidance regarding subsequent efforts to improve cell competence and capacity (cf. Antonioni, 1996). Because one of the basic purposes of collaborative organizing is to enhance individual well-being, organizations should adopt a developmental approach (rather than a punitive stance) toward any cells that do not perform adequately. For example, efforts should be made to improve cell performance prior to any decision to discontinue the cell's relationship with the organization. Organizational resources should be allocated for educational and training activities that serve as an investment in human resources intended to build individual and organizational capacities for the future. These development efforts should focus in part on improving participants' skills. In collaborative organizing, three broad sets of skills are particularly valuable: (1) technical skills that enable a cell to perform its primary function(s) effectively; (2) interpersonal skills that enable a cell to manage its relationships in a collaborative manner; and (3) governance skills that enable a cell to contribute to the leadership and management of the system(s) in which it participates. Of course, it is important that development in collaborative organizing not be limited to a focus on skills but instead is conceived more broadly as a multifaceted process addressing the various aspects of the whole person. In the broadest sense, development should be oriented toward enabling individuals to meet their "higher order" growth or self-actualization needs (Alderfer, 1972; Maslow, 1954).

In addition to the development of organizational participants, collaborative organizing requires effective learning mechanisms at the system level as well. The literature on organizational learning has grown rapidly in recent years (e.g., Argyris & Schön, 1996; Huber, 1991; Senge, 1990), and a variety of practices (e.g., quality circles, total quality management, parallel learning structures) have been introduced into organizations as a way to enhance their ability to learn and undergo continuous improvement. Improved organizational learning provides a number of key benefits that are valuable for collaborative organizing. First, it enhances a system's capacity for self-management by improving its ability to take self-correcting actions that help it more effectively accomplish established goals, to solve ill-defined problems (Bushe & Shani, 1990), and to innovate more readily (Nonaka & Takeuchi, 1995). Organizational involvement in networks has been recognized as an important means through which to import, develop, and utilize new knowledge and capacity (Inkpen, 1996). Second, learning enables organizations to identify whether their goals, and even their basic purposes, are still appropriate or worthwhile, which helps them respond to changing environmental conditions (Argyris, 1977). Because most organizational environments can be portrayed as "permanent white water" (Vaill, 1989), this adaptability is a key value-adding activity in an organization, replacing efficiency as the primary prerequisite to organizational survival (Alter & Hage, 1993). Third, organizational learning can contribute to more fundamental organizational transformation (Bushe & Shani, 1990), for example, a permanent change in the consciousness of the system (Fletcher, 1990) or a fundamental shift in the "shared reality" of the organization's participants. Such a shift will be necessary, of course, for organizations to incorporate more fully and more systemically the principles and practices of collaborative organizing outlined here.

Conclusion

The description of collaborative organizing presented above provides a radically different ideal type of organization than the bureaucratic organization which has served as the normative blueprint for "modern" organizations under competition paradigm assumptions. In this sense, the model builds on previous efforts by organizational theorists to specify new or alternative models and images of organizations (e.g., Bergquist, 1993; Drucker, 1988; Handy, 1990; Lawler, 1992; Miles et al., 1997; Peters, 1988; Shrivastava, 1995). Many of the specific ideas incorporated into this template have been prescribed by organization development scholars and practitioners since the emergence of the field in the 1960s, and it is obvious now that most of these recommendations have considerable currency in both theory and practice of organizations. Thus, I would argue that collaborative organizing is simply the "logical extreme" of a variety of trends that can already be seen taking place in organizations around the world (cf. Hargrove, 1998; Maynard & Mehrtens, 1993) as we evolve through a "postmodern" transformation to a new paradigm (cf. Banner, 1987). What I hope to have accomplished here is to tie these characteristics together into a coherent whole, extending and building on them so as to ground the whole in the foundational assumptions of the collaboration paradigm. It is this extrapolation which makes collaborative organizing a significant departure from the bureaucratic model.

Because collaborative organizing is grounded in a different set of assumptions than those on which bureaucracy is based, the culture of these organizations will be quite different than the typical bureaucratic culture. The emergence of organizational culture in the early 1980s as a useful concept for explaining organizational functioning was due in large part to evidence that high-performing organizations reflected beliefs and practices that deviated from those in traditional hierarchical systems (e.g., Deal & Kennedy, 1982; Ouchi, 1981; Peters & Waterman, 1982). Recent empirical research has further demonstrated that some types of organizational cultures appear to have significant performance benefits. For example, Denison and Mishra (1995) found that a variety of objective and subjective measures of firm performance were related to one or more of four cultural traits: (1) involvement and participation that creates a sense of ownership and responsibility; (2) consistency, or the degree of normative integration that provides an implicit control system based on internalized values; (3) adaptability, or the capacity for internal change in response to external conditions; and (4) a sense of mission or long-term vision that provides purpose and meaning to the work being performed. Sheridan (1992) found that organizations with cultures based on the values of collaboration, teamwork, fairness, and tolerance were able to retain employees longer than those organizations with cultures that emphasized precision and accuracy in the work along with norms of predictability and rule orientation. The differences in retention rates were estimated as generating "opportunity losses" of approximately $6 to $9 million for those firms that emphasized work task values rather than the quality of interpersonal relationships.

A logical explanation for the effectiveness of these cultural characteristics is that they are more attractive to employees, who thus become more committed to the organization and more motivated to contribute to its success. This notion is supported by findings regarding organizational members' perceptions of an "ideal" culture which would promote both organizational effectiveness and personal satisfaction (Cooke & Rousseau, 1988). Organizations with ideal cultures were viewed as those that: (1) are managed in a participative and person-centered way; (2) place a high priority on constructive interpersonal relationships; (3) do things well and value members who set and accomplish their own goals; and (4) value creativity, quality over quantity, and both task accomplishment and individual growth. In short, ideal cultures are those that promote humanistic, affiliative, achievement-oriented, and self-actualizing thinking and behavioral styles. Clearly, the cultural orientations and values identified in all three of these studies are inherent to the collaborative organizing model. This research thus supports the premise that organizations and their members could benefit from efforts to implement the changes required to move toward the collaborative organizing model.

While many organizations are in fact implementing a variety of these changes, they are typically intended to modify rather than replace the bureaucratic hierarchical arrangements so readily viewed as essential to organizational success. To fully adopt the collaborative organizing model, organizations will have to undergo a more fundamental transformation that entails a systemic and holistic redesign based on a new set of assumptions regarding critical organizing criteria. Coincident with the emergence of the culture school in the organizational literature, the field of organizational change began to focus on the process of organizational transformation (e.g., Adams, 1984b; Levy & Merry, 1986), drawing explicit attention to the importance of changing the basic assumptions underlying an organization's design and functioning. Organizational transformation is typically viewed as requiring a paradigm shift in an organization's basic vision, management philosophy, or view of reality (Egri & Frost, 1991; Fletcher, 1990; Porras & Silvers, 1992). Not surprisingly, descriptions of the process and outcomes of organizational transformation identified in this literature are quite congruent with the principles of collaborative organizing outlined here.

An important facet of the transformation process is that organizational members should have a clear image of a "desired future state" describing the kind of organization they intend to become. The value of an image or vision of a desired future has been recognized as playing an important role in the creation of effective organizations (Collins & Porras, 1994) and in the process of organizational change in general (e.g., Beckhard & Harris, 1977; Jayaram, 1976). To facilitate the kind of paradigmatic change involved in organizational transformation, it is useful for members to envision a future state that transcends the limits of what seems realistic or feasible given existing beliefs and practices. This kind of "generative" theory is a potent force for affecting transformation in social systems (Cooperrider & Srivastva, 1987). The notion of expanding the sense of what is possible, of thinking beyond what seem to be reasonable limits, is a key component in the development of an appreciative learning culture that facilitates organizational innovation (Barrett, 1995). Thus, while the collaborative organizing model might easily be viewed as too idealistic to be of any practical value, I would argue that its efficacy derives from the fact that it constitutes an attractive vision of a desired future state that can serve as a beacon to guide the current and future efforts of all those who desire to transform their organizations into collaborative institutions that contribute to the well-being of individuals, communities, and the planet as a whole.

Olson (1995) points out that virtually all the social sciences—history, psychology, social psychology, humanistic psychology, sociology, anthropology, and economics, as well as the interdisciplinary fields of cybernetics, complexity theory, management science, and futures research—have recognized that images of the future can exert a powerful influence on the behavior of individuals and collectives. Drawing on Polak (1973), he argues that a positive image of the future is largely missing in contemporary industrial societies, which instead consist of "moment-ridden modern cultures desperately repressing fears of what tomorrow might bring, their imaginative capacities crippled by disuse and pervasive cynicism, lacking any compelling vision of human possibilities beyond riches and technological power devoted to gaining more riches and technological power" (1995, p. 17). Historical and anthropological analyses suggest that many cultures have experienced a crisis of direction when their images of the future lagged behind their social, economic, or technological development (Markley & Harman, 1982). Furthermore, when a society holds on to prior images of the future past the point of their effectiveness, these images cause more problems than they solve. The global deterioration resulting from the practices and institutions grounded in the competition paradigm bears witness to the fact that its assumptions and values have become obsolete and dysfunctional (cf. Gladwin, Newburry, & Reiskin, 1996).

The collaboration paradigm and the new theory of reality on which it is based provide a vision of the future that can guide the more extensive social transformation required to avoid the catastrophic consequences which are inevitable if such a transformation does not occur (cf. Elgin, 1993a; Henderson, 1996). While Olson (1995) proposes that "sustainability" should be the central focus of a new image of the future, I would argue that adoption of the collaboration paradigm assumptions is a necessary, if not sufficient, condition for a global transformation to a sustainable human civilization. In keeping with the key premise of the new theory of reality that "consciousness precedes being" (Havel, 1990), I hold that the collaboration paradigm constitutes the "thought" that could lead to a new "outer reality" (Harrison, 1984). Rather than being merely an unattainable ideal, then, a collaborative society can be recognized as a laudable goal toward which human civilization can choose to strive. Ultimately, by pursuing a path toward this desirable future, we will exhibit much more collective rationality than we do by simply letting the future unfold as it may through the workings of the "invisible hand of the market."

A VISION OF GLOBAL TRANSFORMATION

The collaboration paradigm assumptions are undoubtedly controversial, and many readers will be skeptical of their validity or viability. However, arguments against these premises invariably boil down to a simple reassertion of the basic assumptions of the competition paradigm, rendering these arguments ultimately tautological. In particular, skepticism regarding the possibility of a transformation to the collaboration paradigm typically reduces to two primary claims. The first is that the world presents to us a competitive environment, that is, that competition is inherent in the natural order of things. Based on this assumption, the socially constructed institutional environment that human beings have created over time is also designed to promote competitive behavior. The second claim is that human beings are also inherently competitive and self-interested, and thus they cannot or will not act collaboratively enough to allow the collaboration paradigm to work. Of course, this ignores the incredible array of collaborative behavior that nearly all human beings demonstrate on a daily basis. It also discounts the role of the institutional environment as a *determinant* of competitive, self-interested behavior. This claim thus appears to be a paradigm-level example of the "fundamental attribution error" (Ross, 1977), in which a dispositional attribution regarding behavior is made even though the behavior can reasonably be construed as having been caused by situational forces.

This paradigmatic attributional error is due, largely, to the shallow and myopic view of the human being adopted by the field of economics, which is the dominant ideology used to guide the public policies around the world which determine the very conditions of our day-to-day lives. Competition-oriented economics, for example, is based on the simplistic premises that people are inherently and inevitably self-interested, that resources are inherently limited or scarce, and thus that self-interested parties who want some portion of these resources have no option but to compete for them. This way of thinking is exacerbated by the economic system's further assumption that the primary determinants of quality of life are material goods (including money as the primary mechanism through which these goods can be acquired), such that acquisition of greater amounts of these resources, through a competitive process, becomes the primary goal of individuals, communities, and nations.

In this context, collaborative behavior is actually discouraged. Because people are assumed to be exclusively self-interested, we are led to believe that others will take advantage of us if given the chance; we also recognize that, by acting collaboratively, we increase others' opportunities to do just that. More generally, by inducing competitive self-interested behavior, the competition paradigm instills fear of others and undermines the building of trust that is a critical prerequisite for successful collaboration. Because people are rewarded for competing and discouraged from collaborating, human civilization has in fact created a very competitive society. The argument that this competitive environment is inevitable, however, is true only if other starting assumptions are taken as givens. One might argue that competition paradigm assumptions are "right," or at least more valid than the collaboration paradigm assumptions, but merely claiming this does not make it so. In other words, such claims do not constitute a meaningful argument against the viability of the collaboration paradigm.

The point here is that, from "within" the assumptions of the competition paradigm, the feasibility of the collaboration paradigm, and thus of collaborative organizing, appears to be limited. However, any set of assumptions is just that—a *taken-for granted* belief system that serves as the foundation for consequent values and practices (Schein, 1985). Ideally, of course, assumptions are supported by available evidence, and proponents of the competitive worldview can undoubtedly point to evidence which supports their assumptions. However, as suggested above, their interpretation of this evidence (e.g., regarding the cause of self-interested behavior) may be biased by their starting assumptions, leaving open to question the validity of these assumptions. Furthermore, their skepticism about the validity of the collaboration paradigm at least implicitly discounts a significant body of empirical evidence (Woodhouse, 1996) that supports key premises of the new theory of reality and its basic assumptions. Again, explicit arguments against this evidence usually require a restatement of the starting assumptions of the competition paradigm. However, because much of the new paradigm evidence also indicates that the competition assumptions are *not* valid, it is hardly appropriate to rely on these assumptions as the basis for an argument against this evidence. Ultimately, if one evaluates the new theory of reality and its supporting evidence from outside the blinders created by the assumptions of competition, the collaboration paradigm assumptions seem quite reasonable.

As suggested earlier, probably the most important of these assumptions—a premise supported by extensive scientific evidence (Capra, 1996) as well as by scientific and spiritual exploration of the nature of consciousness (Harman, 1998)—is that everything is inherently interconnected. Recognition of the validity of this single premise provides the necessary foundation for a transformation to a collaborative society. Once we realize that individual and collective interests are intimately tied together—that actions which bring harm or do damage to some parts of the system actually decrease the quality of life for the system as a whole—then the inherent dysfunctions of a competitive orientation and the natural advantages of collaboration will be readily apparent. Fortunately, global civilization's recognition of our interdependence and interconnectedness is growing almost on a daily basis. There are two distinct facets of this process, which are working together to increase our collective awareness of the necessity of a global transformation.

First, it is increasingly evident that the significant and presumably intractable problems of the world—overpopulation and resource depletion, pollution and civil unrest, widespread poverty and famine, war and crime, destabilized global climate and ecological devastation—are all intertwined and closing in on us rapidly. It is hard not to be aware of the fact that the trend line in these areas does not look good, and that when one adds all the consequences of these problems together, the future in fact looks rather bleak. In a 1992 statement called the World Scientists' Warning to Humanity issued by the Union of Concerned Scientists, 1,700 signatories (including over 100 Nobel laureates) offered the following perspective:

> Human beings and the natural world are on a collision course. Human activities inflict harsh and often irreversible damage on the environment and on critical resources. If not checked, many of our current practices put at serious risk the future that we wish for human society and the plant and animal kingdoms, and may so alter the living world that it will be unable to sustain life in the manner that we know. Fundamental changes are urgent if we are to avoid the collision our present course will bring about...A great change in our stewardship of the earth and the life on it, is required, if vast human misery is to be avoided and our global home on this planet is not to be irretrievably mutilated...A new ethic is required—a new attitude towards discharging our responsibility for caring for ourselves and for the earth...This ethic must motivate a great movement, convince reluctant leaders and reluctant governments and reluctant peoples themselves to effect the needed changes.

Informed estimates by futurists (e.g., Elgin, 1993a; Henderson, 1996; Hubbard, 1998; Snyder, 1995) suggest that this planetary crisis will reach the breaking point within one or two generations, with predictions that systemic breakdown will occur in the first half of the next century if the fundamental assumptions and practices of the modern, Western, industrialization paradigm are not revised.

The second facet of our growing awareness of interconnectedness is the emergence of mass communication technologies that enable humanity as a whole to better understand the nature and consequences of the problems we face and to recognize that, ultimately, we are all "in this together." As computers merge with television, telephones, and satellites into a single, integrated global multimedia system, a "central nervous system" for the planet is being created that is transforming virtually every aspect of life and awakening a new collective consciousness among the people of the Earth (Elgin, 1993a). It is clear that a transformation in consciousness is necessary if the human family is to learn how to live together cooperatively based on a genuine capacity for mutual understanding and appreciation. "Fortunately, this transformation in consciousness is already in progress and appears to be gaining momentum roughly in proportion to the pace with which the world moves into the full blossoming of the communications era" (Elgin, 1993a, p. 118). In other words, our technological capacity to disseminate information globally and to communicate with others anywhere in the world is literally creating the interconnectedness that in turn provides the foundation for the collaboration paradigm.

Increased global awareness of the need for a fundamental transformation in the nature of human civilization provides the basic motivation for engaging in efforts to develop and implement the requisite changes in our institutions and practices. While skeptics might agree that the need for transformation is obvious, they are typically less sanguine about the possibility of actually generating the changes required to bring forth this transformation. However, a wide variety of theoretical frameworks and practical strategies for moving toward a more collaborative, sustainable political economic system and a life-enhancing social structure are being proposed in the new paradigm literature. Likewise, this literature is documenting the myriad transformational changes already being implemented through the actions and activities of individuals, groups, organizations, and communities around the world.

In the area of business and economics, ecological economists are establishing a theory base (e.g., Krishnan, Harris, & Goodwin, 1995) and proposing numerous solutions (e.g., Callenbach, Capra, Goldman, Lutz, & Marburg, 1993; Capra & Pauli, 1995; Daly & Cobb, 1994; Hawken, 1993) to move the economy in this direction. For example, a "natural capitalism" approach (Hawken, 1997) suggests that we stop treating waste and environmental destruction as externalities—costs to be borne by society-at-large—and instead "internalize" a more accurate assessment of the costs of the natural resources used and destroyed by corporate activities. This perspective also promotes the development of new technologies that will help shift patterns of production and consumption to more closely resemble the characteristics of living systems, in which the output or "waste" of one part of a system serves as input to another part of the system in an ongoing, mutually dependent cycle. More broadly, businesses are under pressure to adopt more environmentally sustainable practices, as a result of increased consumer demand for "green" products and services and willingness to boycott organizations that use destructive or unethical practices. Likewise, considerable growth recently in the total assets invested in various socially responsible stock funds demonstrates greater willingness on the part of individuals to exert their voice as investors to broaden the range of interests taken into account by corporate leaders.

Ultimately, a fundamental transformation in the dominant economic paradigm will require a wide variety of changes in basic economic policies, accounting rules, tax laws, and other incentives and constraints used by government leaders to steer market dynamics in a preferred direction. Henderson (1996) provides a thorough overview of the wide range of policy devices currently available or being developed at the national and international levels that will help build a more collaborative, win–win economy. She also identifies many examples of economic innovations at the grassroots level that reflect individuals' dissatisfaction with dysfunctional aspects of the global economy (cf. Meeker-Lowry, 1988). For example, a number of communities throughout the United States and around the world have implemented a community-based trading network that uses a local currency to facilitate the exchange of goods and services between users. These systems help reduce the flow of money out of these communities and into the coffers of large corporations and financial institutions. Likewise, by decoupling some of their local economy from the influence of broader economic forces, these communities are retaking some control over the quality of their personal and communal well-being. Another set of mechanisms being used successfully to stimulate economic development in low-income communities are the various micro-finance systems in place around the world (Otero & Rhyne, 1994). Micro-finance, conducted through such institutions as the Grameen Bank in Bangladesh and ACCION International in Latin America, has become a worldwide movement and serves as a central tenet of socially oriented and sustainable development (Versluysen, 1999). Through innovative practices such as these, a growing number of communities and cities around the world are demonstrating an impressive capacity to use their own initiative, resources, and creativity to effectively address shared concerns and improve their collective well-being (Lean, 1995; Pye-Smith & Feyerabend, 1994). All in all, this literature makes a powerful argument that significant economic change is already in progress, with a multitude of practical applications available to replace the destructive elements of the existing economy with those more appropriate for a collaborative society.

Relative to the emphasis on economic transformation, less attention has been given in the new paradigm literature to the types of political changes needed to facilitate a global transformation. This may be due in part to the fact that there is a noticeable worldwide trend toward the adoption of democratic forms of government, and democracies are inherently more compatible with the principles of collaboration than the autocratic governments they typically replace. Even so, however, critiques of existing political arrangements have pointed out the dysfunctional consequences of democratic systems that are grounded in adversarial dynamics and which focus primarily on our material well-being at the expense of other issues and values that are critical to our overall quality of life (cf. Lerner, 1996). Williamson (1997), for example, brings a spiritual perspective to her discussion of the changes needed in American politics. Likewise, McLaughlin and Davidson (1994) analyze historical and contemporary political themes from the perspective of the Perennial Philosophy, outlining useful proposals for changing the political process to be more congruent with the values and ideals of the new paradigm.

While there is not yet much explicit evidence of any significant changes actually taking place in the U.S. political system, the emergence of the communitarian movement (Bellah, Madsen, Sullivan, Swidler, & Tipton, 1991; Etzioni, 1996) might be viewed as a social trend congruent with a political shift toward a more collaborative society. Similarly, increased citizen participation in community-level government (Berry, Portney, & Thomson, 1993; Thomas, 1995)—made easier by internet access (Ishida, 1998; Weare, Musso, & Hale, 1999)—can be viewed as spurring a transformation toward more citizen-based, collaborative political institutions (Box, 1998; Grossman, 1995; Robertson & Paredes, 1999). Finally, "green politics" (Capra & Spretnak, 1984) provides an ideological foundation that unites a number of citizens' movements (e.g., the peace, ecology, and feminist movements), and while the Green Party has not yet had a significant impact on politics in the United States, its recent rise in popularity in Europe seems to reflect widespread recognition there of the desirability of a political orientation supporting more environmentally sustainable policies.

All in all, the myriad political economic changes and trends that have already been proposed and/or implemented constitute considerable evidence that global civilization is currently showing signs of transformation in a direction consistent with the principles of collaboration. Unfortunately, the many valuable operational ideas identified in the new paradigm literature do not easily make their way into discussions among key decision makers in the existing public policy and organizational arenas. However, while the dominant institutions on the planet conspire to maintain and reinforce the status quo, people throughout the world continue to generate collaboration-oriented "social innovations" that they successfully put into practice in their homes, neighborhoods, and communities. As the number of these creative ideas and practices increases, and as the people engaged in these activities develop networks through which to share information and facilitate their diffusion, the behavioral foundation for a collaborative society is slowly but surely being put into place (Hubbard, 1998). With growing recognition of the legitimacy and value of these efforts, global society may be approaching the time when the viability of a transformation to a collaboration paradigm becomes readily apparent.

While a transformation in the global political economy and movement toward collaborative organizing are best conceptualized as different facets of the same underlying shift in collective consciousness, it is possible to analyze the link between these two trends in terms of a "natural selection" process. The world's political economy and the social and environmental conditions it generates constitute the environment of the many organizations populating the planet. The dominant perspective in the field of organization theory—from early analyses of the determinants of organizational structure (e.g., Burns & Stalker, 1961; Lawrence & Lorsch, 1967; Thompson, 1967) to the more recent resource dependence (Pfeffer & Salancik, 1978), institutional (Meyer & Rowan, 1977), and population ecology (Hannan & Freeman, 1977) frameworks—is that organizations must adapt to environmental conditions if they desire to succeed or even survive. From a natural selection perspective, there are clearly pressures in the environment that are leading organizations to strive for more effective collaboration. My argument here is that these environmental forces include: (1) growing interdependence, real and perceived; (2) more difficult problems or challenges that cannot be effectively addressed by individuals or organizations acting alone; (3) heightened competitive pressures that require effective resource-sharing among organizations in order for each to ensure its own survival; (4) a level of environmental turbulence and uncertainty that demands organizations to be innovative and adaptive; (5) an "institutional environment" that is exerting pressure on organizations to at least attempt to collaborate (e.g., research and project funding from the government or foundations now frequently requires some type of interdisciplinary or interorganizational collaboration); and (6) a global culture that is already demonstrating, especially at the grassroots level, signs of the emergence of a transformation toward collaboration.

In short, the global environment of the world's population of organizations has reached a point in which the ability to collaborate effectively has become a necessary skill for organizational survival. The pressure is clearly on organizations to become more collaborative, that is, to adopt the kinds of properties outlined above that constitute the qualities of collaborative organizing, and this pressure exists in all three sectors. Private sector organizations that are unable to adapt are likely to perform poorly and ultimately go out of business. In the public sector, government organizations do not necessarily stop functioning altogether, but they do undergo a variety of reorganizations that can be viewed as analogous to the "death" of private organizations (cf. Peters & Hogwood, 1991). Furthermore, the privatization movement reflects an effort to submit more of these organizations—often those that are perceived to be functioning poorly—to the pressures of the market, which would force them to become more adaptable, innovative, and responsive. Nonprofit organizations are typically, almost by definition, more readily aligned with many of the principles of collaboration. By their very nature, they usually have to collaborate effectively with sources of funding and with the members of the communities they are intending to serve. Failure to do this effectively would invariably reduce a nonprofit organization's long-term viability.

My basic premise here, then, is that collaborative organizing is a model of organization that better "fits" existing environmental conditions. The bureaucratic hierarchy constituted a successful fit with the modern, industrial environment emerging at the turn of the last century, and its proliferation and success are a testament to the congruence between the assumptions and values of the competition paradigm and the essential features of the bureaucracy. However, the features of collaborative organizing are more congruent with the demands and opportunities of the postmodern, information-based, globalized political economy of the coming century. Thus, a natural selection process is underway that is producing visible signs of the emergence of this new form, both through strategic adaptations by existing organizations and through the "birth" of new organizations that reflect greater resemblance to the collaborative organizing template.

This selection process, therefore, is taking place at the global ecological level-of-analysis. The planet, as a completely interdependent and interconnected system, constitutes a single environment populated by many different kinds of organizations (along with thousands of other life forms). From "inside" this system, it seems quite apparent that the overall level of competition is increasing dramatically, reaching the point of hypercompetition (D'Aveni, 1994). I would argue that the heightened competitive pressures felt in the global political economy are due, in very broad terms, to the overall deterioration of the system resulting from the dysfunctional practices generated by the dominant paradigm. More specifically, the rapid population growth of the twentieth century has been coupled with a serious depletion of key resources on the planet, brought on at least in part by the extremely wasteful nature of our collective patterns of production and consumption (Hawken, 1997). Basically, the planetary environment has nearly reached its carrying capacity, exacerbating the competition for resources required for survival. We have reached the limits of our "economic growth," especially as defined in terms of the competition paradigm's priorities.

Heightened competition also seems obvious to most people because, after all, the competition paradigm has taught us to expect, and thus to perceive, competitive behavior. But using a perceptual lens that filters our observations of the world through expectations grounded in the new theory of reality, we can also readily see that these competitive pressures are being joined by the forces eliciting collaboration. In particular, the new paradigm findings regarding evolutionary dynamics summarized earlier support an alternative perspective on the nature of the natural selection process now taking place. In addition to recognizing the legitimate and functional role that collaboration plays in the maintenance and development of all living systems, these findings indicate that a punctuated equilibrium model of evolutionary change is a normal pattern for systems at all levels of analysis. I would argue that the global political economy, and human civilization as a whole, is in the midst of a fundamental reorientation that has spawned considerable "variation"—innovation and novelty—in our organizations and other political economic institutions. While these new forms and practices are implicitly competing for survival, those which survive will increasingly reflect qualities oriented toward collaboration. The value of organizational collaboration for providing a competitive advantage has already been acknowledged (Inkpen, 1996; Powell et al., 1996). As this transformation proceeds, the viability and value of the collaborative organizing model should become increasingly clear.

Human civilization, from this perspective, is confronted with something of a paradox. On one hand, the level of competition is obviously increasing, and the social and environmental deterioration this is causing has brought us literally to the brink of global disaster. The competition paradigm and its natural by-products—war, economic injustice, environmental destruction, overpopulation, and so on—have created significant and imminent threats to the continued existence of Earth as we know it. Other telling signs of planetary disintegration are not difficult to identify—proliferation of ethnic rivalries, uncontrolled diffusion of weapons of mass destruction, widespread poverty and political alienation, the anger among our youth that leads them to shoot and kill their classmates, financial instability in a very interdependent economic system, and the impending "Y2K" computer problem, to name a few of the most obvious. It does not take much thought to conclude that our current path toward a competition-driven destiny culminates in a very undesirable future.

While disintegration is readily apparent, the fundamental premise of this chapter is that the foundation for a self-organizing global transformation to a higher level of complexity is also now in place. The new theory of reality provides an ontological and epistemological basis for a transformation to a higher level of collective consciousness; the collaboration paradigm establishes an ideological foundation for a transformation of the values, practices, and institutions which underlie and create human civilization; collaborative organizing constitutes an ideal type template to guide the transformation of arguably the dominant institution of society, the formal organization; and signs of a transformation in actual practice are visible in the host of social, economic, and political changes springing forth among individuals and communities around the world. The paradox, then, is that the levels of competition and collaboration are both increasing simultaneously (cf. Snyder, 1995).

In light of this analysis, the planetary system appears to be reaching a bifurcation point (cf. Eisler, 1987), the threshold point of instability at which a system either disintegrates or spontaneously reorganizes to a higher level of complexity and functioning. In dissipative structures, an increase in environmental turbulence and fluctuation generates the potential for a sudden transformation through a systemic self-reorganization. In planetary terms, the turbulence generated by the competition paradigm is naturally increasing the potential for a global transformation to the collaboration paradigm. I would argue that collaborative endeavors are increasing in frequency precisely *because* people are recognizing the need to confront, ameliorate, and overcome the dysfunctions of competition. There is no guarantee, however, that this transformation will come to complete fruition. The dilemma is that humanity is literally in a race with itself—and the question is whether this transformation will proceed quickly enough to stave off the inevitable disaster that will result if we do not fundamentally reorient the political economy toward collaborative, sustainable development.

In this sense, human civilization has reached a fundamental choice point regarding our collective future. For the first time ever in our relatively brief history, humanity has the opportunity—and the responsibility—to explicitly, reflectively, and collectively *choose* the direction of our future development rather than just allowing it to unfold uncontrolled and unwanted. This is the essential notion of conscious evolution (Hubbard, 1998). Ultimately, the planet's destiny is a function of what humans individually and collectively choose to think and in turn how we choose to act. Any individual who chooses to think and act collaboratively thus contributes to the transformation process and helps to ensure global sustainability. If enough individuals attune themselves to a collaborative worldview, a critical mass could be achieved that would greatly quicken the pace of transformation. Given recent findings which suggest that there are as many as 44 million people in the United States who hold "transformational values" (Ray, 1996), just such a critical mass may in fact be emerging (Elgin & LeDrew, 1997).

As we arrive at the threshold of a new millennium, then, human civilization confronts an auspicious moment in its history. The human race must choose whether to hold on to the competition paradigm and deal with the growing "costs of competition" that threaten global destruction, or to begin thinking and acting in ways more fully aligned with the collaboration paradigm on the premise that doing so is a necessary requirement for creating a more desirable future. More simply, it is a choice as to whether we want to continue holding a paradigmatic worldview that is driven by fear—of scarcity, of each other, of death—or to adopt a new paradigm based on love—for self, for community, for life. The most fascinating and optimistic premise of the new theory of reality is that we are literally undergoing a shift in collective consciousness, becoming what Hubbard (1998) calls "universal humans." As more and more individuals adopt a global mindset—thinking of themselves as "citizens of the world," for example—we increase our individual and collective capacity to make decisions that are intended to benefit the global community rather than simply localized self-interests. As more and more individuals achieve higher, more expansive levels of conscious awareness—through meditation and other forms of personal and spiritual growth—we increase our individual and collective capacity to make decisions based on love rather than fear.

How the future will unfold is really quite unpredictable, but one thing is certain—interesting times are ahead. As the transformational tendencies strive to overcome the inevitable resistance from the status quo, nothing less than the future of global civilization is at stake. Ultimately, the choice of which path to follow in pursuit of further progress comes down to a choice between two alternative belief systems. At a minimum, I hope in this chapter to have clarified the fact that such a choice even exists. I also hope to have offered a provocative argument that the resulting transformation would not only be possible but is actually in progress. Finally, I hope to have provided a thoughtful description of the kinds of collaborative organizations we could strive for if we want to facilitate this transformation. Only time will tell whether the exciting possibilities laying before us become fully realized in our conscious evolution to the collaboration paradigm.

NOTE

1. The material in this section was informed by a series of discussions with a group of doctoral students in the Public Administration program at the University of Southern California School of Policy, Planning, and Development, and was originally presented at a conference in Shanghai, China, in July 1998 (Robertson & Associates, 1998). This collaborative group of students included: Muhittin Acar, Nobuyuki Ainoya, Kwi-Hee Bae, Young-Soo Choi, Eloise Dellagnelo, Armando Garcia, Otto Paredes, Vandana Prakash, Jonathan Speier, and Joanna Yu.

REFERENCES

Ackerman, L.S. (1984). The flow state: A new view of organizations and managing. In J.D. Adams (Ed.), *Transforming work* (pp. 114–137). Alexandria, VA: Miles River Press.

Ackoff, R.L. (1994). *The democratic corporation: A radical prescription for recreating corporate America and rediscovering success.* New York: Oxford University Press.

Adams, J.D. (Ed.). (1984a). Achieving and maintaining personal peak performance. In *Transforming work* (pp. 194–207). Alexandria, VA: Miles River Press.

Adams, J.D. (Ed.). (1984b). *Transforming work*. Alexandria, VA: Miles River Press.

Adams, J.L. (1979). *Conceptual blockbusting: A guide to better ideas* (2nd ed.). New York: W.W. Norton.

Agor, W.H. (1984). *Intuitive management*. Englewood Cliffs, NJ: Prentice-Hall.

Alderfer, C.P. (1972). *Existence, relatedness, and growth: Human needs in organizational settings*. New York: Free Press.

Almeder, R. (1992). *Death and personal survival: The evidence for life after death*. New York: Rowman and Littlefield.

Alter, C., & Hage, J. (1993). *Organizations working together*. Newbury Park, CA: Sage.

Anand, S., & Sen, A.K. (1992). *Sustainable human development: Concepts and priorities*. New York: United Nations Development Programme, Office of Development Studies.

Antonioni, D. (1996, Autumn). Designing an effective 360-degree appraisal feedback process. *Organizational Dynamics, 25*, 24–38.

Argyris, C. (1957). The individual and the organization. *Administrative Science Quarterly, 2*, 1–24.

Argyris, C. (1977). Double loop learning in organizations. *Harvard Business Review, 55*, 115–125.

Argyris, C. (1985). *Strategy, change and defensive routines*. Boston: Pitman.

Argyris, C., & Schön, D.A. (1996). *Organizational learning II: Theory, method, and practice*. Reading, MA: Addison-Wesley.

Augros, R., & Stanciu, G. (1988). *The new biology: Discovering the wisdom in nature*. Boston: Shambhala.

Axelrod, R. (1984). *The evolution of cooperation*. New York: Basic Books.

Ayres, R.U. (1998). *Turning point: The end of the growth paradigm*. New York: St. Martin's Press.

Bachrach, P., & Botwinick, A. (1992). *Power and empowerment: A radical theory of participatory democracy*. Philadelphia: Temple University Press.

Bailey, A. (1949). *The destiny of nations*. New York: Lucis Publishing.

Bailey, A. (1954). *Education in the new age*. New York: Lucis Publishing.

Baldwin, C. (1994). *Calling the circle: The first and future culture*. Newberg, OR: Swan Raven and Co.

Balfour, D.L., & Wechsler, B. (1994). A theory of public sector commitment: Towards a reciprocal model of person and organization. In J. Perry (Ed.), *Research in public administration* (Vol. 3, pp. 281–314). Greenwich, CT: JAI Press.

Banner, D.K. (1987). Of paradigm, transformation and organizational effectiveness. *Leadership and Organizational Development Journal, 8*, 17–28.

Barnard, C.I. (1938). *The functions of the executive*. Cambridge, MA: Harvard University Press.

Barrett, F.J. (1995, Autumn). Creating appreciative learning cultures. *Organizational Dynamics, 24*, 36–49.

Barry, D. (1991, Summer). Managing the bossless team: Lessons in distributed leadership. *Organizational Dynamics, 20*, 31–47.

Beckhard, R., & Harris, R.T. (1977). *Organizational transitions: Managing complex change*. Reading, MA: Addison-Wesley.

Bellah, R.N., Madsen, R., Sullivan, W.M., Swidler, A., & Tipton, S.M. (1991). *The good society*. New York: Alfred A. Knopf.

Bentov, I. (1988). *Stalking the wild pendulum: On the mechanics of consciousness*. Rochester, VT: Destiny Books.

Bergquist, W. (1993). *The postmodern organization: Mastering the art of irreversible change*. San Francisco, CA: Jossey-Bass.

Berry, J.M., Portney, K.E., & Thomson, K. (1993). *The rebirth of urban democracy*. Washington, DC: The Brookings Institution.

Blavatsky, H. (1974). *The secret doctrine*. Pasadena, CA: Theosophical University Press. (Originally published 1888)

Block, P. (1993). *Stewardship: Choosing service over self-interest*. San Francisco, CA: Berrett-Koehler.

Bohm, D. (1980). *Wholeness and the implicate order*. London: Routledge & Kegan Paul.
Bolman, L.G., & Deal, T.E. (1995). *Leading with soul: An uncommon journey of the spirit*. San Francisco, CA: Jossey-Bass.
Box, R.C. (1998). *Citizen governance: Leading American communities into the 21st century*. Thousand Oaks, CA: Sage.
Bozeman, B. (1987). *All organizations are public: Bridging public and private organizational theories*. San Francisco, CA: Jossey-Bass.
Bradford, D.L., & Cohen, A.R. (1984). *Managing for excellence: The guide to developing high performance in contemporary organizations*. New York: Wiley & Sons.
Briskin, A. (1996). *The stirring of soul in the workplace*. San Francisco, CA: Jossey-Bass.
Bunker, B.B., & Alban, B.T. (1992). Editors' introduction: The large group intervention—A new social innovation? *Journal of Applied Behavioral Science, 28*, 473–479.
Burki, S.J., & Perry, G.E. (1998). *Beyond the Washington consensus: Institutions matter*. Washington, DC: World Bank.
Burns, T., & Stalker, G.M. (1961). *The management of innovation*. London: Tavistock.
Bushe, G.R., & Shani, A.B. (1990). Parallel learning structure interventions in bureaucratic organizations. In W.A. Pasmore & R.W. Woodman (Eds.), *Research in organizational change and development* (Vol. 4, pp. 167–194). Greenwich, CT: JAI Press.
Cairncross, F. (1991). *Costing the earth*. London: Business Books.
Callenbach, E., Capra, F., Goldman, L., Lutz, R., & Marburg, S. (1993). *EcoManagement: The Elmwood guide to ecological auditing and sustainable business*. San Francisco, CA: Berrett-Koehler.
Cameron, K.S. (1984). The effectiveness of ineffectiveness. In B.M. Staw & L.L. Cummings (Eds.), *Research in organizational behavior* (Vol. 6, pp. 235–285). Greenwich, CT: JAI Press.
Capelli, P., Bassi, L., Katz, H., Knoke, D., Osterman, P., & Useem, M. (1997). *Change at work*. New York: Oxford University Press.
Capra, F. (1991). *The tao of physics* (3rd ed.). Boston: Shambhala.
Capra, F. (1996). *The web of life: A new scientific understanding of living systems*. New York: Anchor Books.
Capra, F., & Pauli, G. (Eds.). (1995). *Steering business toward sustainability*. Tokyo: United Nations University Press.
Capra, F., & Spretnak, C. (1984). *Green politics*. New York: Dutton.
Carey, K. (1991). *The third millennium: Living in a posthistoric world*. San Francisco, CA: HarperSanFrancisco.
Carnevale, D.G. (1995). *Trustworthy government: Leadership and management strategies for building trust and high performance*. San Francisco, CA: Jossey-Bass.
Castaneda, C. (1968). *The teachings of Don Juan: A Yaqui way of knowledge*. Berkeley: University of California Press.
Chatman, J.A., & Barsade, S.G. (1995). Personality, organizational culture, and cooperation: Evidence from a business simulation. *Administrative Science Quarterly, 40*, 423–443.
Chrislip, D.D., & Larson, C.E. (1994). *Collaborative leadership: How citizens and civic leaders can make a difference*. San Francisco, CA: Jossey-Bass.
Clark, M.E. (1995). Changes in Euro-American values needed for sustainability. *Journal of Social Issues, 51*, 63–82.
Cohen, A.R., & Bradford, D.L. (1989). Influence without authority: The use of alliances, reciprocity, and exchange to accomplish work. *Organizational Dynamics, 17*, 4–17.
Cohen, M.D., March, J.G., & Olsen, J.P. (1972). A garbage can model of organizational choice. *Administrative Science Quarterly, 17*, 1–25.
Cohen, S.G. (1993). New approaches to teams and teamwork. In J.R. Galbraith, E.E. Lawler, III, & Associates (Eds.), *Organizing for the future: The new logic for managing complex organizations* (pp. 194–226). San Francisco, CA: Jossey-Bass.

Coleman, J.S. (1988). Social capital in the creation of human capital. *American Journal of Sociology, 94*, S95–S120.
Collins, J.C., & Porras, J.I. (1994). *Built to last: Successful habits of visionary companies.* New York: HarperCollins.
Conger, J.A., & Associates. (Eds.). (1994). *Spirit at work: Discovering the spirituality in leadership.* San Francisco, CA: Jossey-Bass.
Cooke, R.A., & Rousseau, D.M. (1988). Behavioral norms and expectations: A quantitative approach to the assessment of organizational culture. *Group & Organization Studies, 13*, 245–273.
Cooperrider, D.L., & Srivastva, S. (1987). Appreciative inquiry in organizational life. In R.W. Woodman & W.A. Pasmore (Eds.), *Research in organizational change and development* (Vol. 1, pp. 129–169). Greenwich, CT: JAI Press.
Cotton, J.L. (1993). *Employee involvement: Methods for improving performance and work attitudes.* Newbury Park, CA: Sage.
Csikszentmihalyi, M. (1990). *Flow: The psychology of optimal experience.* New York: Harper & Row.
Culbert, S.A., & McDonough, J.J. (1985). How reality gets constructed in an organization. In R. Tannenbaum, N. Margulies, F. Massarik, & Associates (Eds.), *Human systems development* (pp. 122–142). San Francisco, CA: Jossey-Bass.
Daly, H.E., & Cobb, J.B., Jr. (1994). *For the common good: Redirecting the economy toward community, the environment, and a sustainable future* (2nd ed.). Boston: Beacon Press.
D'Aveni, R.A. (1994). *Hypercompetition: Managing the dynamics of strategic maneuvering.* New York: The Free Press.
Deal, T., & Kennedy, A. (1982). *Corporate cultures: The rites and rituals of corporate life.* Reading, MA: Addison-Wesley.
de Chardin, T. (1975). *The phenomenon of man.* New York: HarperCollins.
de Geus, A. (1997). *The living company.* Boston: Harvard Business School Press.
Denison, D.R., & Mishra, A.K. (1995). Toward a theory of organizational culture and effectiveness. *Organization Science, 6*, 204–223.
Dennis, L. (1997). *The pattern.* Lower Lake, CA: Integral Publishing.
DePree, M. (1990). *Leadership is an art.* New York: Dell.
Devereux, P. (1989). *Earthmind: A modern adventure in ancient wisdom.* New York: Harper & Row.
Drucker, P.F. (1988). The coming of the new organization. *Harvard Business Review, 66*, 45–53.
Earley, P.C. (1993). East meets west meets mideast: Further explorations of collectivistic and individualistic work groups. *Academy of Management Journal, 36*, 319–348.
Egri, C.P., & Frost, P.J. (1991). Shamanism and change: Bringing back the magic in organizational transformation. In R.W. Woodman & W.A. Pasmore (Eds.), *Research in organizational change and development* (Vol. 5, pp. 175–221). Greenwich, CT: JAI Press.
Eisler, R. (1987). *The chalice & the blade: Our history, our future.* San Francisco, CA: Harper & Row.
Elden, M. (1986). Sociotechnical systems ideas as public policy in Norway: Empowering participation through worker-managed change. *Journal of Applied Behavioral Science, 22*, 239–255.
Elgin, D. (1993a). *Awakening earth: Exploring the evolution of human culture and consciousness.* New York: Morrow.
Elgin, D. (1993b). *Voluntary simplicity: Toward a way of life that is outwardly simple and inwardly rich* (rev. ed). New York: Morrow.
Elgin, D., & LeDrew, C. (1997). *Global consciousness change: Indicators of an emerging paradigm.* Sausalito, CA: Institute of Noetic Sciences.
Emery, M., & Purser, R.E. (1996). *The search conference: A powerful method for planning organizational change and community action.* San Francisco, CA: Jossey-Bass.
Etzioni, A. (1996). *The new golden rule: Community and morality in a democratic society.* New York: Basic Books.
Fisher, R., & Ury, W. (1981). *Getting to yes: Negotiating agreement without giving in.* Boston: Houghton Mifflin.

Fletcher, B.R. (1990). *Organization transformation theorists and practitioners: Profiles and themes.* New York: Praeger.
French, J.R.P., Jr., & Raven, B. (1959). The bases of social power. In D. Cartwright (Ed.), *Studies in social power* (pp. 150–165). Ann Arbor: University of Michigan, Institute for Social Research.
Fukuyama, F. (1995). *Trust: The social virtues and the creation of prosperity.* New York: The Free Press.
Full democracy. (1996, December 21). *The Economist,* pp. 3–14.
Galbraith, J.R. (1994). *Competing with flexible lateral organizations.* Reading, MA: Addison-Wesley.
Galen, M., & West, K. (1995, June 5). Companies hit the road less traveled. *Business Week,* pp. 82–85.
Gerber, R. (1988). *Vibrational medicine: New choices for healing ourselves.* Santa Fe, NM: Bear & Company.
Gladwin, T.N., Kennelly, J.J., & Krause, T.-S. (1995). Shifting paradigms for sustainable development: Implications for management theory and research. *Academy of Management Review, 20,* 874–907.
Gladwin, T.N., Newburry, W.E., & Reiskin, E.D. (1996). The usual mind as environmentally unsustainable: The unusual mind to sustain our common future. In J.B. Keys & L.N. Dosier (Eds.), *Academy of Management Proceedings,* pp. 429–433.
Graham, J.W. (1991). An essay on organizational citizenship behavior. *Employee Responsibilities and Rights Journal, 4,* 249–270.
Gray, B. (1985). Conditions facilitating interorganizational collaboration. *Human Relations, 38,* 911–936.
Greenleaf, R.K. (1977). *Servant leadership: A journey into the nature of legitimate power and greatness.* New York: Paulist Press.
Greider, W. (1997). *One world, ready or not: The manic logic of global capitalism.* New York: Simon & Schuster.
Grof, S. (1993). *The holotropic mind: The three levels of human consciousness and how they shape our lives.* New York: HarperCollins.
Grossman, L.K. (1995). *The electronic republic: Reshaping democracy in the information age.* New York: Viking.
Hackman, J.R., & Walton, R.E. (1986). Leading groups in organizations. In P.S. Goodman (Ed.), *Designing effective work groups* (pp. 72–119). San Francisco, CA: Jossey-Bass.
Halal, W.E. (1996). *The new management: Democracy and enterprise are transforming organizations.* San Francisco, CA: Berrett-Koehler.
Halal, W.E. (1998). *The infinite resource: Creating and leading the knowledge enterprise.* San Francisco, CA: Jossey-Bass.
Hall, M. (1972). *The secret teachings of all ages.* Los Angeles: The Philosophical Research Society.
Hammer, M., & Champy, J. (1993). *Reengineering the corporation: A manifesto for business revolution.* New York: Harper Business.
Handy, C. (1990). *The age of unreason.* Boston, MA: Harvard Business School Press.
Hannan, M.T., & Freeman, J. (1977). The population ecology of organizations. *American Journal of Sociology, 82,* 929–964.
Hansen, J.L. (1995). *Invisible patterns: Ecology and wisdom in business and profit.* Westport, CT: Quorum Books.
Hargrove, R. (1998). *Mastering the art of creative collaboration.* New York: McGraw-Hill.
Harman, W.W. (1998). *Global mind change: The promise of the 21st century* (2nd ed.). San Francisco, CA: Berrett-Koehler.
Harman, W.W., & de Quincey, C. (1994). *The scientific exploration of consciousness: Toward an adequate epistemology* (Report No. CP-6). Sausalito, CA: Institute of Noetic Sciences.
Harrison, R. (1984). Leadership and strategy for a new age. In J.D. Adams (Ed.), *Transforming work* (pp. 97–112). Alexandria, VA: Miles River Press.
Havel, V. (1990, March 5). The revolution has just begun. *Time,* pp. 14–15.
Hawken, P. (1993). *The ecology of commerce: A declaration of sustainability.* New York: HarperBusiness.
Hawken, P. (1997, March/April). Natural capitalism. *Mother Jones,* pp., 40–53, 59–62.

Henderson, H. (1996). *Building a win-win world: Life beyond global economic warfare*. San Francisco, CA: Berrett-Koehler.
Henton, D., Melville, J., & Walesh, K. (1997). *Grassroots leaders for a new economy: How civic entrepreneurs are building prosperous communities*. San Francisco, CA: Jossey-Bass.
Hodgkinson, V.A., & Weitzman, M.S. (1996). *Nonprofit almanac (1996-1997): Dimensions of the independent sector*. San Francisco, CA: Jossey-Bass.
Hofstede, G. (1980). *Culture's consequences: International differences in work-related values*. Beverly Hills, CA: Sage.
Hubbard, B.M. (1998). *Conscious evolution: Awakening the power of our social potential*. Novato, CA: New World Library.
Huber, G.P. (1991). Organizational learning: The contributing processes and literatures. *Organization Science, 2*, 88–115.
Huxley, A. (1944). *The perennial philosophy*. New York: Harper & Row.
Inkpen, A. C. (1996). Creating knowledge through collaboration. *California Management Review, 39*, 123–140.
Ishida, T. (Ed.). (1998). *Community computing: Collaboration over global information networks*. New York: Wiley.
Jackson, E.T., & Kassam, Y. (Eds.). (1998). *Knowledge shared: Participatory evaluation in development cooperation*. West Hartford, CT: Kumarian Press.
Janis, I.L. (1971). Groupthink. *Psychology Today, 5*, 43–46, 74–76.
Jantsch, E. (1980). *The self-organizing universe*. New York: George Braziller Publishers.
Jayaram, G.K. (1976). Open systems planning. In W.G. Bennis, K.D. Benne, R. Chin, & K. Corey (Eds.), *The planning of change* (3rd ed.) (pp. 275–283). New York: Holt, Rinehart, and Winston.
Kanter, R.M. (1968). Commitment and social organization: A study of commitment mechanisms in utopian societies. *American Sociological Review, 33*, 499–517.
Kanter, R.M. (1979). Power failure in management circuits. *Harvard Business Review, 57*, 65–75.
Kanter, R.M. (1983). *The change masters: Innovations for productivity in the American corporation*. New York: Simon and Schuster.
Keeley, M. (1984). Impartiality and participant-interest theories of organizational effectiveness. *Administrative Science Quarterly, 29*, 1–25.
Kelley, R.E. (1991). *The power of followership: How to create leaders people want to follow, and followers who lead themselves*. New York: Doubleday/Currency.
Keisler, C.A., & Keisler, S.B. (1969). *Conformity*. Reading, MA: Addison-Wesley.
Korten, D.C. (1995). *When corporations rule the world*. San Francisco, CA: Berrett-Koehler.
Kouzes, J.M., & Posner, B.Z. (1995). *Credibility: How leaders gain and lose it, why people demand it*. San Francisco, CA: Jossey-Bass.
Krishnan, R., Harris, J.M., & Goodwin, N.R. (Eds.). (1995). *A survey of ecological economics*. Washington, DC: Island Press.
Lawler, E.E., III. (1992). *The ultimate advantage: Creating the high-involvement organization*. San Francisco, CA: Jossey-Bass.
Lawless, M.W., & Moore, R.A. (1989). Interorganizational systems in public service delivery: A new application of the dynamic network framework. *Human Relations, 42*, 1167–1184.
Lawrence, P.R., & Lorsch, J.W. (1967). *Organization and environment: Managing differentiation and integration*. Boston: Graduate School of Business Administration, Harvard University.
Lean, M. (1995). *Bread, bricks, and belief: Communities in charge of their future*. West Hartford, CT: Kumarian Press.
Lee, C., & Zemke, R. (1993, June). The search for spirit in the workplace. *Training, 30*, 21–28.
Leifer, R. (1989). Understanding organizational transformation using a dissipative structure model. *Human Relations, 42*, 899–916.
Lemkow, A.F. (1990). *The wholeness principle: Dynamics of unity within science, religion, & society*. Wheaton, IL: The Theosophical Publishing House.

Leonard, G., & Murphy, M. (1995). *The life we are given: A long-term program for realizing the potential of body, mind, heart, and soul*. New York: G.P. Putnam's Sons.
Lerner, M. (1996). *The politics of meaning: Restoring hope and possibility in an age of cynicism*. New York: Addison-Wesley.
Levine, S. (1998). *Getting to resolution: Turning conflict into collaboration*. San Francisco, CA: Berrett-Koehler.
Levy, A., & Merry, U. (1986). *Organizational transformation: Approaches, strategies, theories*. New York: Praeger.
Lindblom, C.E. (1959). The science of "muddling through." *Public Administration Review, 19*, 79–88.
Linden, R. (1994). *Seamless government: A practical guide to re-engineering in the public sector*. San Francisco, CA: Jossey-Bass.
Lipman-Blumen, J. (1996). *The connective edge: Leading in an Interdependent world*. San Francisco, CA: Jossey-Bass.
Lippitt, L.L. (1998). *Preferred futuring: Envision the future you want and unleash the energy to get there*. San Francisco, CA: Berrett-Koehler.
Lovelock, J. (1988). *The ages of Gaia: A biography of our living earth*. New York: Norton.
Mandell, M.P. (1984). Application of network analysis to the implementation of a complex project. *Human Relations, 37*, 659–679.
Mander, J., & Goldsmith, E. (Eds.). (1996). *The case against the global economy: And for a turn toward the local*. San Francisco, CA: Sierra Club Books.
Markley, O.W., & Harman, W.W. (1982). *Changing images of man*. New York: Pergamon Press.
Margulis, L. (1998). *Symbiotic planet: A new look at evolution*. New York: Basic Books.
Maslow, A. (1954). *Motivation and personality*. New York: Harper & Row.
Maturana, H.R., & Varela, F.G. (1980). *Autopoiesis and cognition: The realization of the living*. Boston: D. Reidel Publishing.
Maynard, H.B., & Mehrtens, S.E. (1993). *The fourth wave: Business in the 21st century*. San Francisco, CA: Berrett-Koehler.
McClelland, D.C. (1961). *The achieving society*. New York: Van Nostrand Reinhold.
McClelland, D.C. (1975). *Power: The inner experience*. New York: Random House.
McGregor, D. (1960). *The human side of enterprise*. New York: McGraw-Hill.
McKnight, R. (1984). Spirituality in the workplace. In J.D. Adams (Ed.), *Transforming work* (pp. 139–153). Alexandria, VA: Miles River Press.
McLagan, P., & Nel, C. (1997). *The age of participation: New governance for the workplace and the world*. San Francisco, CA: Berrett-Koehler.
McLaughlin, C., & Davidson, G. (1994). *Spiritual politics: Changing the world from the inside out*. New York: Ballantine Books.
McWaters, B. (1982). *Conscious evolution: Personal and planetary transformation*. San Francisco, CA: Evolutionary Press.
Meeker-Lowry, S. (1988). *Economics as if the earth really mattered*. Philadelphia, PA: New Society Publishers.
Meyer, J.W., & Rowan, B. (1977). Institutionalized organizations: Formal structure as myth and ceremony. *American Journal of Sociology, 83*, 340–363.
Milakovich, M.E. (1995). *Improving service quality: Achieving high performance in the public and private sectors*. Delray Beach, FL: St. Lucie Press.
Miles, R.E., Snow, C.C., Mathews, J.A., Miles, G., & Coleman, H.J., Jr. (1997). Organizing in the knowledge age: Anticipating the cellular form. *Academy of Management Executive, 11*, 7–24.
Miller, R.S. (1992). *As above, so below: Paths to spiritual renewal in daily life*. New York: G.P. Putnam's Sons.
Mohr, L.B. (1994). Authority in organizations: On the reconciliation of democracy and expertise. *Journal of Public Administration Research and Theory, 4*, 49–65.

Mohrman, S.A., Cohen, S.G., & Mohrman, A.M., Jr. (1995). *Designing team-based organizations: New forms of knowledge work*. San Francisco, CA: Jossey-Bass.

Monroe, R. (1971). *Journeys out of the body*. New York: Doubleday.

Moody, R. (1975). *Life after life: The investigation of a phenomenon—Survival of bodily death*. Atlanta, GA: Mockingbird Books.

Moody, R. (1993). *Reunions: Visionary encounters with departed loved ones*. New York: Villard Books.

Moorman, R.H., Blakely, G.L., & Niehoff, P.B. (1998). Does perceived organizational support mediate the relationship between procedural justice and organizational citizenship behavior? *Academy of Management Journal, 41*, 351–357.

Myss, C. (1996). *Anatomy of the spirit: The seven stages of power and healing*. New York: Crown Publishers.

Nahapiet, J., & Ghoshal, S. (1998). Social capital, intellectual capital, and the organizational advantage. *Academy of Management Review, 23*, 242–266.

Naparstek, B. (1997). *Your sixth sense: Activating your psychic potential*. New York: HarperCollins.

Newhouse, D.R., & Chapman, I.D. (1996). Organizational transformation: A case study of two aboriginal organizations. *Human Relations, 49*, 995–1011.

Nonaka, I., & Takeuchi, H. (1995). *The knowledge-creating company: How Japanese companies create the dynamics of innovation*. New York: Oxford University Press.

Ogawa, R.T., & Bossert, S.T. (1995). Leadership as an organizational quality. *Educational Administration Quarterly, 31*, 224–244.

Olson, R.L. (1995). Sustainability as a social vision. *Journal of Social Issues, 51*, 15–35.

O'Reilly, C.A., III, & Chatman, J.A. (1986). Organizational commitment and psychological attachment: The effects of compliance, identification, and internalization on prosocial behavior. *Journal of Applied Psychology, 71*, 492–499.

Organization of American States. (1997). *Inter-American strategy for participation: Strengthening public participation in environment and sustainable development*. Washington, DC: Author.

Osborne, D., & Gaebler, T. (1992). *Reinventing government: How the entrepreneurial spirit is transforming the public sector*. Reading, MA: Addison-Wesley.

Orth, C.D., Wilkinson, H.E., & Benfari, R.C. (1987). The manager's role as coach and mentor. *Organizational Dynamics, 15*, 66–74.

Otero, M., & Rhyne, E. (Eds.). (1994). *The new world of microenterprise finance: Building healthy financial institutions for the poor*. West Hartford, CT: Kumarian Press.

Ouchi, W.G. (1981). *Theory Z: How American business can meet the Japanese challenge*. Reading, MA: Addison-Wesley.

Owen, H. (1987). *Spirit: Transformation and development in organizations*. Potomac, MD: Abbott.

Owen, H. (1997). *Expanding our now: The story of open space technology*. San Francisco, CA: Berrett-Koehler.

Perry, J.L., & Rainey, H.G. (1988). The public-private distinction in organization theory: A critique and research strategy. *Academy of Management Review, 13*, 182–201.

Pert, C.B. (1997). *The molecules of emotion: Why you feel the way you feel*. New York: Scribner.

Peters, B.G., & Hogwood, B.W. (1991). Applying population ecology models to public organizations. In J. Perry (Ed.), *Research in public administration* (Vol. 1, pp. 79–108). Greenwich, CT: JAI Press.

Peters, T. (1988). Restoring American competitiveness: Looking for new models of organizations. *Academy of Management Executive, 2*, 103–109.

Peters, T.J., & Waterman, R.H., Jr. (1982). *In search of excellence: Lessons from America's best run companies*. New York: Harper and Row.

Pfeffer, J. (1977a). Power and resource allocation in organizations. In B.M. Staw & G.R. Salancik (Eds.), *New directions in organizational behavior* (pp. 235–265). Chicago, IL: St. Clair Press.

Pfeffer, J. (1977b). The ambiguity of leadership. *Academy of Management Review, 2*, 104–112.

Pfeffer, J., & Salancik, G.R. (1978). *The external control of organizations: A resource dependence perspective*. New York: Harper & Row.

Polak, F.L. (1973). *The image of the future.* San Francisco, CA: Jossey-Bass.
Porras, J.I., & Silvers, R.C. (1992). Organization development and transformation. In M.R. Rosenzweig & L.W. Porter (Eds.), *Annual review of psychology* (Vol. 42, pp. 51–78). Stanford, CA: Annual Reviews.
Powell, W.W. (1990). Neither market nor hierarchy: Network forms of organization. In B.M. Staw & L.L. Cummings (Eds.), *Research in organizational behavior* (Vol. 12, pp. 295–336). Greenwich, CT: JAI Press.
Powell, W.W., Koput, K.W., & Smith-Doerr, L. (1996). Interorganizational collaboration and the locus of innovation: Networks of learning in biotechnology. *Administrative Science Quarterly, 41,* 116–137.
Prigogine, I., & Stengers, I. (1984). *Order out of chaos: Man's new dialogue with nature.* New York: Bantam Books.
Purser, R.E., Park, C., & Montuori, A. (1995). Limits to anthropocentrism: Toward an ecocentric organization paradigm? *Academy of Management Review, 20,* 1053–1089.
Putnam, R.D. (1995). Bowling alone: America's declining social capital. *Journal of Democracy, 6,* 65–78.
Pye-Smith, C., & Feyerabend, G.B. (1994). *The wealth of communities: Stories of success in local environmental management.* West Hartford, CT: Kumarian Press.
Quinn, R.E., & Spreitzer, G.M. (1997). The road to empowerment: Seven questions every leader should consider. *Organizational Dynamics, 26,* 37–49.
Ray, M., & Myers, R. (1989). Practical intuition. In W.H. Agor (Ed.), *Intuition in organizations: Leading and managing productively.* Newbury Park, CA: Sage.
Ray, P.H. (1996). *The integral culture survey: A study of the emergence of transformational values in America.* Sausalito, CA: Institute of Noetic Sciences.
Robertson, P.J., & Associates. (1998). *Collaborative organizing: Foundations for an "ideal type."* Paper presented at the Australian and Pacific Researchers in Organisation Studies International Colloquium, Shanghai, China.
Robertson, P.J., & Paredes, O. (1999). *Citizen participation and community development: A "collaborative organizing" approach to local governance.* Paper presented at the American Society for Public Administration National Training Conference, Orlando, Florida.
Robertson, P.J., & Speier, J.V. (1998). Organising for international development: A collaborative network-based model. *The International Journal of Technical Cooperation, 4,* 166–187.
Robinson, S.L. (1996). Trust and breach of the psychological contract. *Administrative Science Quarterly, 41,* 574–599.
Rosen, C.M., Klein, K.J., & Young, K.M. (1986). *Employee ownership in America: The equity solution.* Lexington, MA: Lexington Books.
Ross, L. (1977). The intuitive psychologist and his shortcomings: Distortions in the attribution process. In L. Berkowitz (Ed.), *Advances in experimental social psychology* (Vol. 10, pp. 173–220). New York: Academic Press.
Russell, P. (1995). *The global brain awakens: Our next evolutionary leap.* Palo Alto, CA: Global Brain.
Schein, E.H. (1985). *Organizational culture and leadership.* San Francisco, CA: Jossey-Bass.
Schlitz, M., Taylor, E., & Lewis, N. (1998). Toward a noetic model of medicine. *Noetic Sciences Review, 47,* 45–52.
Schultz, M.L. (1998). *Awakening intuition: Using your mind-body network for insight and healing.* New York: Harmony Books.
Selsky, J.W. (1991). Lessons in community development: An activist approach to stimulating interorganizational collaboration. *Journal of Applied Behavioral Science, 27,* 91–115.
Senge, P.M. (1990). *The fifth discipline: The art & practice of the learning organization.* New York: Doubleday.
Senge, P.M. (1997). Communities of leaders and learners. *Harvard Business Review, 75,* 30–32.

Serageldin, I., Barrett, R., & Martin-Brown, J. (Eds.). (1995). *The business of sustainable cities: Public-private partnerships for creative technical and institutional solutions.* Washington, DC: World Bank.

Shaw, R.B. (1997). *Trust in the balance: Building successful organizations on results, integrity, and concern.* San Francisco, CA: Jossey-Bass.

Sheridan, J.E. (1992). Organizational culture and employee retention. *Academy of Management Journal, 35,* 1036–1056.

Shrivastava, P. (1995). The role of corporations in achieving ecological sustainability. *Academy of Management Review, 20,* 936–960.

Shukla, M. (1997). *Competing through knowledge: Building a learning organization.* Hershey, PA: Idea Group.

Shuster, H.D. (1990). *Teaming for quality improvement: A process for innovation and consensus.* Englewood Cliffs, NJ: Prentice-Hall.

Simon, H.A. (1957). *Administrative behavior* (2nd ed.). New York: Macmillan.

Snyder, H.A. (1995). *EarthCurrents: The struggle for the world's soul.* Nashville, TN: Abingdon Press.

Spayde, J. (1998, January/February). The new renaissance. *Utne Reader,* pp. 42–47.

Stacey, R.D. (1996). *Complexity and creativity in organizations.* San Francisco, CA: Berrett-Koehler.

Stamps, D. (1997, April). The self-organizing system. *Training, 34,* 30–36.

Starik, M., & Rands, G.P. (1995). Weaving an integrated web: Multilevel and multisystem perspectives of ecologically sustainable organizations. *Academy of Management Review, 20,* 908–935.

Starr, H. (1997). *Anarchy, order, and integration: How to manage interdependence.* Ann Arbor: The University of Michigan Press.

Staw, B.M. (1980). Rationality and justification in organizational life. In B.M. Staw & L.L. Cummings (Eds.), *Research in organizational behavior* (Vol. 2, pp. 45–80). Greenwich, CT: JAI Press.

Stevenson, I. (1987). *Children who remember past lives.* Charlottesville: University Press of Virginia.

Stewart, I. (1995). *Nature's numbers: The unreal reality of mathematics.* New York: Basic Books.

Talbot, M. (1991). *The holographic universe.* New York: HarperCollins.

Tarnas, R. (1998). The great initiation. *Noetic Sciences Review, 47,* 24–31, 57–59.

Taylor, F.W. (1911). *The principles of scientific management.* New York: Harper.

Terry, R. (1995). *Economic insanity.* San Francisco, CA: Berrett-Koehler.

Thomas, J.C. (1995). *Public participation in public decisions: New skills and strategies for public managers.* San Francisco, CA: Jossey-Bass.

Thompson, J.D. (1967). *Organizations in action.* New York: McGraw-Hill.

Torbert, W. (1983). Initiating collaborative inquiry. In G. Morgan (Ed.), *Beyond method* (pp. 272–291). Beverly Hills, CA: Sage.

Tornow, W., & London, M. (1998). *Maximizing the value of 360-degree feedback: A process for successful individual and organizational development.* San Francisco, CA: Jossey-Bass.

Trice, H.M., & Beyer, J.M. (1993). *The cultures of work organizations.* Englewood Cliffs, NJ: Prentice-Hall.

Tsui, A.S., Pearce, J.L., Porter, L.W., & Tripoli, A.M. (1997). Alternative approaches to the employee-organization relationship: Does investment in employees pay off? *Academy of Management Journal, 40,* 1089–1121.

Tushman, M.L., & Romanelli, E. (1985). Organizational evolution: A metamorphosis model of convergence and reorientation. In L.L. Cummings & B.M. Staw (Eds.), *Research in organizational behavior* (Vol. 7, pp. 171–222). Greenwich, CT: JAI Press.

Vaill, P.V. (1989). *Managing as a performing art.* San Francisco, CA: Jossey-Bass.

Vansina, L.S., & Taillieu, T. (1996). Business process reengineering or socio-technical system redesign in new clothes? In R.W. Woodman & W.A. Pasmore (Eds.), *Research in organizational change and development* (Vol. 9, pp. 81–100). Greenwich, CT: JAI Press.

Versluysen, E. (1999). *Defying the odds: Banking for the poor.* West Hartford, CT: Kumarian Press.

Von Bertalanffy, L. (1968). *General system theory: Foundations, development, applications.* New York: George Braziller Publishers.
Waldroop, J., & Butler, T. (1996). The executive as coach. *Harvard Business Review, 74,* 111–117.
Waldrop, M.M. (1992). *Complexity: The emerging science at the edge of order and chaos.* New York: Simon and Schuster.
Walsch, N.D. (1995). *Conversations with God: An uncommon dialogue* (Book 1). New York: G.P. Putnam's Sons.
Walton, R.E. (1985). From control to commitment in the workplace. *Harvard Business Review, 63,* 67–74.
Wambach, H. (1978). *Reliving past lives: The evidence under hypnosis.* New York: Harper & Row.
Weare, C., Musso, J.A., & Hale, M.L. (1999). Electronic democracy and the diffusion of municipal web pages in California. *Administration & Society, 31,* 3–27.
Weber, M. (1947). *The theory of social and economic organization* (A.H. Henderson & T. Parsons, eds., trans.). Glencoe, IL: Free Press.
Weick, K.E. (1969). *The social psychology of organizing.* Reading, MA: Addison-Wesley.
Weisbord, M.R. (1987). *Productive workplaces: Organizing and managing for dignity, meaning, and community.* San Francisco, CA: Jossey-Bass.
Weisbord, M.R. (Ed.). (1992). *Discovering common ground.* San Francisco, CA: Berrett-Koehler.
Weiss, B., & Tigue, P. (1997). *Public-private partnerships 1997: Issues and resources for state and local governments.* Chicago, IL: Government Finance Officers Association.
Wheatley, M.J. (1992). *Leadership and the new science: Learning about organization from an orderly universe.* San Francisco, CA: Berrett-Koehler.
Wick, C.W., & León, L.S. (1993). *The learning edge: How smart managers and smart companies stay ahead.* New York: McGraw-Hill.
Wilber, K. (1977). *The spectrum of consciousness.* Wheaton, IL: Theosophical Publishing House.
Wilber, K. (1995). *Sex, ecology, spirituality: The spirit of evolution.* Boston, MA: Shambhala.
Wilber, K. (1996). *A brief history of everything.* Boston, MA: Shambhala.
Wilber, K. (1998). *The marriage of sense and soul.* New York: Random House.
Williamson, J. (Ed.). (1990). *Latin American adjustment: How much has happened?* Washington, DC: Institute for International Economics.
Williamson, M. (1997). *The healing of America.* New York: Simon & Schuster.
Wisely, D.S., & Lynn, E.M. (1994). Spirited connections: Learning to tap the spiritual resources in our lives and work. In J.A. Conger & Associates (Eds.), *Spirit at work: Discovering the spirituality in leadership* (pp. 100–131). San Francisco, CA: Jossey-Bass.
Wolfe, A. (1989). *Whose keeper: Social science and moral obligation.* Berkeley: University of California Press.
Woodhouse, M.B. (1996). *Paradigm wars: Worldviews for a new age.* Berkeley, CA: Frog.
World Bank. (1996). *The World Bank participation sourcebook.* Washington, DC: World Bank.
Zeleny, M. (1990). Amoebae: The new generation of self-managing human systems. *Human Systems Management, 9,* 57–59.
Zohar, D. (1990). *The quantum self: Human nature and consciousness defined by the new physics.* New York: Morrow.
Zohar, D., & Marshall, I. (1994). *The quantum society.* New York: William Morrow.
Zukav, G. (1989). *The seat of the soul.* New York: Simon & Schuster.
Zucker, L.G. (1983). Organizations as institutions. In S.B. Bacharach (Ed.), *Research in the sociology of organizations* (Vol. 2, pp. 1–47). Greenwich, CT: JAI Press.

THE RELATIONAL HEALING DIMENSION OF ORGANIZATIONAL DEVELOPMENT
TRANSFORMATIVE STORIES AND DIALOGUE IN LIFE-CYCLE TRANSITIONS

Gurudev S. Khalsa and David S. Steingard

ABSTRACT

This chapter offers a relationally based organizational development (OD) model for understanding the crisis period that characterizes an organization in transition between life-cycle stages. In this model, organizations are viewed as holographically comprising relationships at multiple levels—among people, groups, functions, and other organizations in the environment. During transition crises, relations at all of these levels tend to become polarized, threatening the organization, its people, and the mission it serves. By embracing these powerful "creative tensions" through a process we call "relational healing," stakeholders come to see their organization more holistically as a set of interwoven relationships evolving toward a new life stage of their choosing. Drawing upon OD approaches such as appreciative inquiry and dialogue, relational healing guides the organization to greater integrity via a five-stage "wholing" model: splitting, engagement, appreciation, release, and reintegration. The model is grounded in our research and consulting work with *JAZZ*, a not-for-profit arts organization that worked through a life-transition crisis over a two-year period. In-depth case stories from this work illustrate the fragmentation and subsequent healing of relationships at multiple levels, leading to a radically transformed and reenergized organization.

INTRODUCTION

JAZZ is a not-for-profit organization dedicated to promoting appreciation of jazz music in its region. It does so primarily through programming and educational initiatives that provide a wide variety of opportunities to hear and learn about jazz. Acting as both an arts promoter and educator, *JAZZ* also serves a socially integrating function by offering programs that attract a diverse constituency: from children making their first acquaintance with jazz to aficionados keen on hearing the best and learning more about their favorite styles; from inner-city senior citizens to downtown workers and suburban residents; and from fans of traditional jazz to connoisseurs of the most avant-garde forms.

Founded in 1978, *JAZZ* has grown from a club started by a few jazz enthusiasts to an organization serving over 900 members and the surrounding community. After many years of activity carried out by an all-volunteer board of directors working entirely out of their homes and businesses, the organization established an office and hired its first executive director in 1989. In the last few years, *JAZZ* has grown substantially in membership, funding, and scope of operations. But this apparent success, combined with unresolved issues from the 1989 change in operational structure, also led to a growing sense of organizational crisis.

When we were invited to conduct a field study of *JAZZ* in January 1992, a host of tensions were in evidence within the Board of Trustees and between the Executive Director and certain board members. Although manifest in strained relationships and a sometimes acrimonious working environment, these human tensions were mirrored by organizational tensions that pointed to life-transition issues the organization as a whole needed to confront. Especially in a volunteer-supported social organization, where passion for the cause (in this case, jazz promotion) runs high, it is clear that organizational issues become easily intertwined with personal relationship issues and must be dealt with together.

For veterans of such organizational crises, it will not be surprising that the loudest cries from various organizational leaders was for the removal or replacement of other leaders (of course, there were widely differing views on who that should be). The embattled Executive Director saw ridding the Board of "troublemakers" as his only route to salvation, while board members took turns taking pot shots at each other and the Executive Director (both in private and with hardly any veil in board

Relational Healing in Life-Cycle Transitions 271

meetings). After visiting our first board meeting, we were not at all sure that the organization would survive. But we were clear that severing one or another leadership "arm" of the organization would not in any way deal with the multiple levels of life-transition issues that were the real causes of organizational crisis. So we set out to uncover what those issues were, to help the system see itself in a more whole light, and to support it in making more informed organizational and personal choices based on an appreciation of the legitimately divergent points of view embodied in even the most threatening and polarized relationships.

In the course of our work, we came to realize how understudied the complexities of organizations in the midst of life transition are, how interwoven the personal and organizational issues become, and how crucial, therefore, is some dialogical process that engages the system in becoming more whole (what we call "relational healing") as a foundation for an enlivened future. We should say at the outset that we strived to carry no assumptions about what that "enlivened future" should look like, including even the possibility that the breakup of *JAZZ* as an organization could be a route to such a future. What we did hope was that whatever choices were made would be conscious ones rooted in a healthy appreciation of the real issues underlying the difficult organizational transition and of the people whose points of view differed on these.

We have structured this chapter as a rhythmical interchange between the story of *JAZZ* and our theorizing about it. Our intent is to seamlessly weave a fabric of grounded experience and intellectual understanding with a relational spirit that mirrors our topic. We begin in the next section by introducing our own relationship to *JAZZ* as action researchers. Then, we briefly introduce some of the theoretical constructs that will help contextualize the following two data sections: "A Life-Cycle History of *JAZZ*" and "Four Case Stories of *JAZZ*." We conclude by returning to a discussion of theoretical propositions and future research directions for the relational healing dimension of organizational development emerging from this in-depth case analysis.

OUR RESEARCH RELATIONSHIP WITH *JAZZ*

Conventionally, this is the section that would introduce the research "subject" and the research methodology. Our point in retitling it is to put primary emphasis on the relationship among us in this venture called research.[1] By avoiding the objectification of terms like "subject," we hope also to convey that we are not pretending to separate ourselves from what it is we research. Indeed, as will be seen, we became, over the course of time, quite involved in *JAZZ* in what continued after the initial study as a mutually rewarding relationship. The value of this approach can best be understood from an epistemology that sees knowledge as inherently relational, something that will be further discussed later in this section.

We began this research relationship as a result of a matching of *JAZZ*'s needs (as presented in their response to a grant-funded Request for Proposals from arts organizations desiring grant-funded field studies) and our interests and expertise in organizational development. Neither of us had previously been involved with *JAZZ*. Significantly, though we had been colleagues for one semester, we had not previously worked on a project together, so this was a relational beginning for us too, out of which evolved a close friendship.

Our dual mission in this field study was to assist *JAZZ* with its organizational challenges and conduct in-depth discovery research that could lead to propositions applicable to not-for-profit organizations facing the critical challenges of a life-transition point (or period). We began the inquiry without much in the way of preconceived ideas or theories of what we would find. It did not begin as an inquiry into relational healing—this is a concept that emerged later as we tried to make sense of our experience and to push into new arenas of theory building. What we did start with was a sense of an organization in life crisis and an inquiry approach that was inherently relational[2] and appreciative (Cooperrider & Srivastva, 1987). We resisted efforts by some organizational members to cast us in the role of problem solvers or experts in organizational design who could simply judge what should be done differently or determine who should leave the organization. Instead, we approached each member with curiosity, a genuine desire to appreciate them and their contribution, perspective, and stake in *JAZZ* (Cooperrider, 1996).

The major steps in our plan of action inquiry, and the primary avenues for gathering data, were as follows.

- *Formation of a Working Group:* Soon after starting, we recommended the creation of a working group to meet with us periodically and develop/discuss/ approve the ground rules of our work, the interview protocols and questionnaires we would use, the agenda for a proposed retreat, and other research-related issues. Significantly, we advocated that the composition of the group reflect some of the polarization in the organization, so that even in our interaction with the working group, we would be dealing with an organizational microcosm.
- *Attendance at Board Meetings:* As a regular forum for observing the entire leadership of *JAZZ*, the monthly board meetings provided us with a consistent window from which to assess changes in the interaction among members and the status of major organizational issues. Our practice in these meetings was for both of us to attend, listen, take extensive field notes, and occasionally (between 0 and 2 times per meeting) offer comments to the Board. As mentioned above, we actively resisted being treated as "experts" and refused to offer advice in general but especially any that could be interpreted to favor one member's point of view over another. Neither did we pretend not to be fully present—we sat in the same circle of chairs as the Board, and we participated as it seemed appropriate.

- *Personal Interviews of JAZZ Leadership:* We (sometimes together, but mostly separately) interviewed each of the 18 board members, 3 former board members, 8 Committee Chairs not on the Board, the Executive Director, and his part-time paid staff assistant. Our structured interview protocol allowed us to gather background information, to appreciate each person's perspective and concerns about *JAZZ*, its mission and challenges, and to begin sensing the themes that would help us shape the issues to be addressed as a group.
- *Questionnaire of JAZZ Membership:* In collaboration with the Working Group, we developed a membership questionnaire to be distributed together with a monthly newsletter mailing to all members. The 151 responses (22%) provided us with quantitative and qualitative information concerning the members' interests and concerns.
- *Board Retreat:* Using data from the interviews and membership questionnaire, we (together with the Working Group) designed a retreat (September 1992) attended by 80 percent of the Board and the Executive Director. Because we were fully occupied in conducting the retreat, we audio-taped the plenary and subgroup proceedings to assure preservation of our data.
- *Follow-up:* For one year following the retreat, we continued to attend monthly board meetings, the annual meeting of members, and periodic meetings with the Board President and/or the Executive Director. In addition, we have facilitated six additional board retreats during the three years since the original retreat, all of which have offered longitudinal vantage points from which to observe changes in *JAZZ*.

We bring to our inquiry an ontology and epistemology that are essentially relational (Alcoff & Potter, 1993; Wheatley, 1992). That is, we assert that organizations are, inherently, systems of relationships, and that to inquire into them is, unavoidably, to create a relationship with that system (Susman & Evered, 1978; Wheatley, 1992). The foundation of our relational epistemology is rooted in the social constructionist perspective (Gergen, 1985; Gergen & Gergen, 1991), which acknowledges that we cannot know apart from our relationship, that it is the nature of knowledge to be socially constructed and therefore, that knowledge is inherently relational. As such, one of the researcher's obligations is to be "choiceful" about his mode of inquiry and the questions he asks, knowing that the act of inquiring is itself helping to socially (re-)create the system of relationships we call the organization. We followed here an approach grounded in appreciative inquiry (AI) (Cooperrider, 1990; Cooperrider & Pratt, 1995; Cooperrider & Srivastava, 1987) as the theory underlying our choice of interview questions in the early stage of this research. Furthermore, we adopted throughout our interactions with the people in *JAZZ* the epistemological stance embodied in AI, inquiring into and holding up to the light that which gives life to the organization and the possibilities toward which it wishes to move. Finally, as will be seen, we contend that as systems of relationships, organizations are structured and restructured as the result of relational acts, and that far from existing in a relational vacuum, organizational structure follows from relationships. The model of relational healing evolved as a way to explain the process by which the polarized relationships characteristic of an organizational crisis period can be reintegrated into a new whole that is the relational foundation for the subsequent stage in the organization's development (including a likely organizational restructuring).

THEORETICAL INTRODUCTION TO RELATIONAL HEALING

It's not so much that we're afraid of change, or so in love with the old ways, but it's that place in between that we fear.... It's like being between trapezes.... There's nothing to hold onto.
—Marilyn Ferguson, American futurist (quoted in Bridges, 1991, p. 34)

This chapter offers a fresh perspective on the process of organizational development by introducing the concept and practice of "relational healing." A model of relational healing is explicated in the context of an arts organization that found itself at a crisis point, that mythical spot between two stages of development, when the very survival of the organization depends upon its successful navigation of a "healing crisis." For organizations, just as for individuals, the crisis occurring between life stages is hardly a point but rather a period of time, often fraught with troubling and disruptive forces that paradoxically are also the generative seeds of a more mature stage of development. This chapter takes an in-depth look at this period, proposing that it is in this "space between," specifically in the relational space between people and polarities during this period, that the power of healing manifests.

"Relational" refers to what lies BETWEEN in several senses: Between the typically recognized life stages of an organization, between the forces that polarize organizational groups, between the people who comprise an organization, and between the judgments that separate life into dualistic categories. What lies between are *relationships*. Although less visible than the polarized groupings or the differentiated life stages, *the relational space between*, it is argued, is the locus of healing energy for transforming both the people and the organization.

As people and their organization grapple with the polarized forces that seem to be tearing them apart, it is only by appreciating the relationships among them, bringing them together into a space that can hold a more diverse whole, that what appears as life threatening has the potential to be "re-cognized"[3] as life giving. In this sense, transformation is the outcome of healing, where healing is understood not as the elimination of "disease," but as a process of "wholing" consistent with a holistic model of health.

It is also important to understand that, from a holistic perspective, "healing" and "healthy systems" encompass both life and death, and that to be whole is not tantamount to experiencing no pain or loss. Rather, we posit that in times of crisis, organizations and their people have a choice—to deal humanely with one another through openness and dialogue or to succumb to rising accusations of doing wrong, being wronged, retaliating for all the wrongs, and so forth, thereby robbing the system of its life energy. It is like the difference between a caterpillar losing its legs to become a butterfly (or a snake losing its skin to reveal a new one), versus squishing the caterpillar (or severing the snake in two)—something is lost in both scenarios, but there is more life energy in the former result.

Methodologically, this research is grounded in a collection of stories, tales if you will, of the journey through a major life transition of one arts organization and its members. It is not a history nor a case study, per se, but a set of "case stories"[4] that depict the journey at various levels of analysis. Themselves embodying something between science and art, between fact and fiction, between objective and subjective, case stories are useful and artful means of peering into the normally invisible world "between." They convey a sense of relationships among the characters and organizational forces (and their narratives) that makes the between-ness almost visible. It is also within the "between space" created by narrativizing relationships that our relational healing methodology "can make significant change possible":

> Engaging multiple narratives also permits the interpenetration of structures in contradictory ways and generates choices that can make significant change possible. (Tenkasi and Boland, 1993, p. 98)

The next section is a contextualizing narrative. Telling the story of *JAZZ* in the context of a life-cycle history not only locates it within the organizational literature on stages of development but offers an historical context for the four vignettes that follow.

A LIFE-CYCLE HISTORY OF *JAZZ*

Numerous models of organizational life cycles and/or stages of development have been offered by management scholars (see Quinn & Cameron, 1983 for an extensive summary and integration of these models) and some of these have more recently been developed for or specifically applied to nonprofit organizations (Bailey & Grochau, 1993; Hasenfeld & Schmid, 1989; Messal, 1980; Wood, 1992). But rarely do these models offer any insight into the process of transition from one phase or stage to another.[5] In fact, the transitions are usually referred to as points in time that divide two stages, rather than as periods of their own.[6] This chapter reframes the transition as a crisis period with a time dimension of its own. Inasmuch as transitions usually entail significant choice points not only for the evolution of the organization but also for all the individuals who must work through the felt experience of crisis, what happens *during* transition is an appropriate focus of more attention. As concluded by Quinn and Cameron (1983) in their review of the life-cycle literature, "Little is known about ... processes by which organizations progress from one stage to another" (p. 34).

In the following life-cycle history of *JAZZ*, we have used Bailey and Grochau's (1993) model to map against, because it is the most recent and well integrated of the models dealing with nonprofit organizations. As the description will show, at the time of our entry, *JAZZ* was in a crisis period between the second and third stage of their model.

After *JAZZ* was started in 1978 by what one founder described as "twelve jazz zealots," it had for many years the feeling of a "club." It primarily acted as a jazz concert presenter, bringing gigs to town for the benefit of its jazz enthusiast members. A working Board was the club's inner circle and they, supported by other volunteers, carried out all the organization's activities from the site of their own homes and workplaces for over ten years.

Bailey and Grochau (1993) refer to this period as "Entrepreneurial" in their model of nonprofits' organizational stages of development. The words they use to describe the socio-emotional tone of this stage well match descriptors we heard in interviews describing *JAZZ*'s "club" period—"high activity, excitement, hope, risk-taking, spontaneity, ... intense commitment, responsibility, ... protective, desire for/fear of first director" (p. 38). This latter quality, ambivalence toward having a paid director, was especially evident toward the end of this first stage, when the decision was made to search for the organization's first executive director.

The transition into the second stage began around 1987 with the far-reaching vision of a new board president (we'll call him Alan)[7] who wanted to seek grant funding, hire an executive director, establish an office, and ultimately transform *JAZZ* into a major arts organization engaged in jazz advocacy and education as well as concert presentation. Through his vision, enthusiasm, and commitment of time and energy, this president channeled the entrepreneurial spirit of the club into a new era of organizational sophistication. *JAZZ* emerged in 1989 with grants to support the office formation and the hiring of an executive director (Bob), who started in June of that year. As befits an organizational transition period, there were high stakes; in order to serve a much larger constituency with more ambitious activities, *JAZZ* risked both the stability of its finances and the commitment of its volunteers.

This simultaneity of high risks and opportunities seems to be one characteristic of transition points, when the organization risks its former life to become something new. The Chinese symbol for "crisis" is an holistic way of thinking of this type of crisis—it combines the images of "danger" and "hidden opportunity." Echoing this evolutionary interpretation of crisis, Torbert (1989) suggests that fundamental organizational change—"second-order transformation"—may not have to be alarming. Indeed, this first life stage transition for *JAZZ* was relatively painless, at least on the surface. But the embers of its unresolved tensions were reignited during the next transition period, which fully lived up to its dual billing as danger and opportunity. Then, as we shall see, only by engaging its wholeness in a dialectical dance was *JAZZ* able to embrace the opportunity inherent in the crisis while simultaneously working through the process of the emotional upheaval induced by the change.

Relational Healing in Life-Cycle Transitions 277

"Team-building" is the label given by Bailey and Grochau (1993) to the second stage in nonprofits' development. The primary organizational activities they cite are expanding personnel, developing new programs, and seeking additional resources, all of which apply to *JAZZ* during this period. Their words to describe the socio-emotional tone of this stage are equally applicable: "optimism, confusion [e.g., about roles], conflict, good intentions, collaboration, pride" (p. 38). Many board members who were active during this time referred to early events in this stage as times that they felt most excited about their involvement in *JAZZ*. Among the high points were winning both local and prestigious national grants, bringing in a half dozen nationally known acts in the same year, holding a gala tenth anniversary celebration, launching a jazz education program for children, reaching peak membership in the same year (1990) as a record number of events were held (32 concerts, 13 pub nights featuring local artists, and 12 jazz education events), and aspiring toward the status of a major arts organization with a special social calling—bringing racially and economically diverse people together in the spirit of jazz music.

But at the same time that the funding and hiring of *JAZZ*'s first executive director was making possible these peak experiences, there began a decline of volunteerism in partial reaction to having paid staff, and some "old" board members felt a reduced sense of intimacy and influence as *JAZZ* became more business-like and less "clubby." The seeds of the next organizational crisis were being sown during this stage by the very socio-emotional characteristics Bailey and Grochau (1993) describe. In this stage, the executive director is likely to feel "anxious, overburdened, uncertain, and ambivalent" (p. 38) (and he did), and the Board is likely to experience tension between new and founding members (and they did).

The impact of these tensions came to a head in 1991 when Alan, the man who led *JAZZ* into the second organizational stage, stepped down as president. Visions of the future seemed increasingly polarized: Would the larger-than-ever *JAZZ* begin to grow into the next stage of organizational development (labeled "Bureaucracy" by Bailey & Grochau, 1993), or would it (for lack of funds, lack of leadership, and the conflicting preferences of its board members) revert to the first stage or even fall apart completely? As Adizes (1988) shows, there are "traps" at each crisis point between stages that endanger the life of an organization, making it particularly vulnerable to collapse at those times. Adds Greiner (1972), "During such periods of crisis, a number of companies fail—those unable to abandon past practices and effect major organization changes are likely either to fold or to level off in their growth rates" (p. 40). These fears and questions were very present for *JAZZ* and its members when our study began.

In the proposal requesting a field study, Bob, the executive director, and Carl, the interim board president, cited a list of "growing pains" for which they sought help. Five items were listed: board factionalism, ongoing debates over the extent of board involvement in administrative detail, an executive director mired in routine tasks and multiple directives from board members, a reduction in voluntarism despite growing membership, and the loss in many of the feeling that active participation in *JAZZ* was fun. What the letter did not say, but became apparent in our first meetings, was that some board members felt the executive director was incompetent and needed to be replaced, and the executive director felt some board members were dead wood or worse, interfering with the smooth operation of *JAZZ*. Such conditions have been associated with the covert cultures of less effective boards (Holland, Leslie, & Holzhalb, 1993), but perhaps it is also a sign of a board facing a transition crisis. Consider what Bailey and Grochau (1993) say about the pivotal issues that characterize the end of the second stage:

> The growth that had been desired and has been achieved now threatens to overwhelm the organization.... There is a greater need for systems, policies, and procedures to coordinate the multiple activities which have been set in motion. (pp. 29-30) Directors at this stage [and, we might add, their Boards] often question their competency and their ability to manage [everything]. (p. 32) Toward the end of this period, the critical issue for the board is: "I still care about the agency, but I'm worn out. There's too much to do, and I don't have the needed expertise for all issues or know all the answers." (p. 34)

In addition to the functional issues of control, the end of the second stage of *JAZZ*'s development reignited the smoldering fires of dissent encountered during the transition from stage one to stage two. There was great ambivalence on the board about moving the organization to yet another plane of organizational sophistication that seemed to further distance *JAZZ* from its grassroots membership and from the type and size of organization they felt comfortable directing. Financial issues were often a lighting rod for these tensions. The original granting agencies were planning to cease their administrative grants and they were demanding greater in-house fund-raising and financial self-sufficiency. But most board members felt quite inexperienced in such matters and wary of the "moneyed" people who might bring in such expertise but also change forever the character of the organization.

So the pivotal control issue in this transition period was not just a question of control systems and procedures, but *whose* control would *JAZZ* be in. Also, while control might be seen as the critical *functional* issue of this period, the range of issues seen from a relational perspective are far more complex. Questions of control are well known for triggering all sorts of debilitating interpersonal dynamics and personal defense mechanisms that make it difficult to be objective about what structure is in the best interests of the organization. Furthermore, the damage to relationships which accompanies the frustration and finger-pointing of this period drain and block the energy of the organization in a way that an imposition of policies and procedures by itself cannot cure.

In a study of nonprofit boards and their relationships with their executive directors, Fletcher (1992) notes that a host of interpersonal issues, pressing circumstances, and overly passive or overly intrusive board members can make it difficult to manage relations between them, concluding that both research studies and anecdotal literature are needed to address how these barriers can be overcome. Such barriers are rampant during an organizational crisis period. In our quick fix culture, the most popular solution is some kind of surgery, axing the executive director, changing the board, and/or redoing the organization chart. But as Adizes (1988) comments, surgery is both the fastest and the most dangerous way to produce change, and does not address the underlying dynamics of the pain; it only makes people fear to complain lest they be faced with another surgery.

As we considered what approach to take in our action inquiry, we looked for a way to work from the inside out, not assuming that a third stage of "development" was necessary or even desirable, not privileging some voices (including our own) over others, and attempting to honor all the people comprising the organization in the process of moving toward questions of the organization's desired future. Our working premise (and this was a case for testing it) was that functional issues can best be addressed in a context of well-attended relationships, such that revisions to structure emerge organically from a revitalization of the organization as a system of relations and from a fresh clarity about what constituent members want the organization to be and how each can contribute to that.

FOUR CASE STORIES OF JAZZ

The following case stories are vignettes offering insight into distinct levels of the system of relationships known as *JAZZ*. They are fieldwork "tales" in Van Maanen's (1988) sense, being accounts grounded in the data of our interactions and documented in our field notes, but also interpretations, given the unavoidable choice among stories and the lens through which we see and tell the tales. From the standpoint of a constructionist epistemology, these accounts are more than a representation of reality, they are, inevitably, a "construction" (Gergen, 1985) of it and also, grounded in our "constructs" (Kelly, 1955). The primary construct informing our choice of tales and the story line weaving through them is that of relational healing.

The stories are told from "the inside-out" (Hunt, 1987) as shown in Figure 1. Working through four levels of analysis, the stories begin with the *interpersonal* relations among organizational members, move next to the *intergroup* level, then to a level we call the *interfunctional*, and finally to the *interorganizational*. In each case, we look at how the relations were transformed during the life transition of the organization.

As explained earlier, the way we began our inquiry was to start with personal interviews of each board and staff member, committee chair, and several past board members. Entering with a lot of open curiosity and without prejudgment enabled us to hear (and freed board and staff members to tell) stories filled with soul that were deep and sometimes paradoxical, inspiring and sometimes depressing, filled with pride and sometimes disappointment. As McLean et al. (1982) illumine, OD consultants operating from an organic and spiritual vantage point, "work intuitively

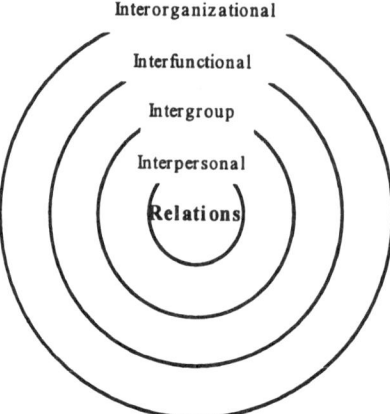

Figure 1.

with the intangible aspects of organizational life" and "have a high level of self-awareness and self-acceptance" (cited in Egri & Frost, 1991, p. 205). This OD approach is ideally suited for exploring the subtle, spiritual, and finely interconnected realm of between-ness, the metaphysical locus of our research.

Just as we got to know *JAZZ*'s people through individual interviews before the organizational stories emerged, so too would we like to begin by introducing readers to some of them, mirroring how we met them. We call them the STARS of *JAZZ*!

The Stars of *JAZZ*

As we begin to write about the people in *JAZZ* and their relationships with each other, we realize we are wary of "characterizing" them. Our hope is to honor them, for their uniqueness as individuals and their contribution to the soulful life of *JAZZ*. The only way we can fairly tell about them is to relate to you some of our experience in being with them, what we felt, what we noticed, some of what was said, all of which has given us a strong impression of each of them as a being. That is different from a character in a story, even though we are telling a story, because for us, these are real people, with real lives. It is an important distinction, as Omri realized in the book Gurudev just read to his six-year old, *The Indian in the Cupboard* (Banks, 1981)—the characters Omri's cupboard brought to life from plastic were not toys he could just play pretend with but real people, if tiny, who had lives of their own. In a similar vein, the people we write about here are not "characters" in a story, and not representations of reality, but soulful beings with rich and complex lives we wish to honor.

Bob Davis is the Executive Director of *JAZZ*. A long-time jazz historian, part-time musician, and now administrator, his passion for jazz is palpable—getting the chance to work for his passion full time could be seen as the opportunity of a lifetime. But when we first met him, he was looking and feeling beaten down by his work. He had been the executive director for two-and-a-half years then, the first and only one to hold that post. As the first paid staff in a formerly all volunteer organization, he had to shoulder a lot of board expectations, some articulated, many merely assumed, of what he should be doing. Also, the music he was hearing in the words and behaviors of several board members was discouraging and frightening—they obviously did not think he was doing his job well enough, he felt near the breaking point for overworking himself, and he was convinced some board members wanted to get rid of him and return the organization to its former all-volunteer status.

In 1992, **David Argus** was the new president of the jazz society, eager, respected, and anxious to exercise good leadership in service of his devotion to jazz. He had become friends with Bob during his stint as the organization's program chair, was sympathetic to Bob's feelings of frustration but uncertain how to deal with his often cantankerous Board. Like Bob, he had judgments about some of the more obstreperous board members, but was not quite as threatened or emotionally convinced that these "bad" directors needed to be removed from the Board. David appeared to have everyone's respect at this point, and didn't want to side unduly with Bob lest he lose that respect.

Philip Everly was one of the board members characterized by Bob as part of the "old guard," members he felt wanted to return *JAZZ* to its former feeling of a club and be done with him. Philip, a writer and self-styled jazz critic, had once been given the accolade of "the consummate curmudgeon" for his seeming readiness to acridly criticize most jazz, most jazz artists, and most people in general. One of his fellow board members, Paul Kraft (whose expressed views also lean toward the censorious), once said of Philip: "He has a whole wallfull of jazz recordings of people he doesn't like." As we got to know Philip, we discovered a man full of intelligence and wit as well as critical taste, giving rise to his very focused agenda for *JAZZ* to present only "high quality" (HQ) jazz. Some called him the chair of the Jazz Police for his outspokenness on HQ judgments that were far from universally held. But Philip is quite sincere, and his pursuit of organizational excellence is equally well-intentioned (although largely unappreciated because of the manner in which he goes about it).

The three people above were the ones we suggested become a working group tasked to work with us as we jointly planned our action inquiry. In various ways, these three represented a microcosm of the organization as a whole, and we wanted to ensure from the beginning that the polarities in the organization were in some way embodied in this committee. The formation of the working group was one of our first signals to the organization as a whole that we would listen to everyone, and be guided by their divergent inputs. It was also a chance to work directly with some of the most polarized personalities in the organization in the context of a small group where we could ensure all had voice.

Alan Ludlow, the former board president (an eminent veterinarian and African-American socialite), was the man responsible for catapulting *JAZZ* into community prominence and initiating its stage two development. Under his leadership, *JAZZ* received its first administrative grant funding, occupied its first office space, and hired its first professional staff. It also greatly expanded its programming and educational outreach efforts and became a much higher profile arts organization that eclipsed its former identity as a club. He is absolutely eloquent on the subject of the place of jazz in serving the community. After giving his all for four years as president, though, he was burnt out, and ready for someone else to take over. His biggest piece of unfinished business was the creation (without significant follow-up) of a big name "Advisory Board," the hope for which was to be a money-raising arm of the organization. But it never achieved that aim (indeed, it was probably ill-conceived), languished, and yet stuck in the minds of board members as the supposed locus of *JAZZ*'s fund-raising ability and responsibility—not a good attribution.

Paul Kraft, a tax attorney by profession, made a great contribution to *JAZZ* in his role as treasurer for four years. Alan Ludlow was responsible for recruiting him, and he brought a new sense of order to the previously disheveled accounting of the organization. Some might argue he brought too much order, for his treasurer's reports were often excessively detailed, feeding the predisposition of some board members to nit-pick operations, rather than focusing on issues of larger scope. In addition, Paul brought in a decidedly skeptical worldview (often wrapped in his distinctively caustic humor), which was a useful counterweight to Alan's facile optimism. But after Alan's departure, Paul became an increasingly dominant voice of financial fear for *JAZZ*'s future. Indeed, given the lack of attention paid to fund-raising at the board level, and the lack of skills and sophistication for it, it is clear he often felt alone in demanding attention for the organization's monetary needs. But his wolf cry was so frequent and so noisy (even in the face of a strong cash position), his warnings began to be seen by some as merely masking an agenda to get Bob out of the executive directorship. As with Philip, Paul's conversations with us showed a depth and sensitivity to his character that those who merely reacted to his facade missed out on. When another board member confronted him in a financial committee meeting about his "adversarial relations" with the executive director, he was genuinely taken aback. Later he commented reflectively to us, "I'm not afraid to disagree. Maybe it's my way. I've been told I intimidate people."

Oscar Count, the membership chair, was justifiably proud of the fact that he was elected by the grass roots of *JAZZ*. An African-American, Oscar is a jazz musician as well as an enthusiast, a telemarketer by profession, and as hard working a volunteer as they come. But like a lot of the board members, we doubt he felt fully appreciated for his contributions, and exhibited a curious combination of talking up the importance of his calling program but being unwilling to share the load of it with anyone else. It was a tremendous breakthrough when, mid-way through our years with *JAZZ*, he and **Katherine Feingold**, a white female board member and

Relational Healing in Life-Cycle Transitions 283

its volunteer coordinator, became a team to cochair the membership committee. What a team they make, and what a creative set of ideas they have developed and implemented, going even beyond membership generation into corporate fund-raising! To look at them, they are an unlikely pair—Oscar, a fast-talking, very large black man and Katherine, a slight, somewhat hesitant white-haired woman who speaks rather haltingly. Neither is the epitome of what one might imagine of a corporate fund-raiser, yet here they are, after many long years of being on a Board that shied away from the supposed sophistication of high-level fund-raising, digging in with enthusiasm.

* * *

So those are the STARS. Next come the stories corresponding to the four different relational levels depicted earlier in Figure 1. Each vignette is divided into two sections: fragmentation and healing. The first describes the fragmented situation as we observed it upon our entry into the organization. Then we describe the process of healing, what we did, and what happened.

Level 1. Interpersonal:
The Embattled Relations of the Director and Treasurer

Interpersonal Fragmentation

Of all the many embattled interpersonal relations at the time we began working with *JAZZ*, among the most problematic was the one between Bob, the executive director, and Paul, the treasurer. It had ramifications throughout the organization because of the obvious need for these two positions to work closely together and the combined high sensitivity and low sophistication among board members regarding money. Bob and Paul essentially kept two separate sets of books and presented their own reports to the Board in forms that could not be reconciled. Whenever possible, it seemed, their reports suggested errors in the others' reporting, making for embarrassment. Their embattled relationship was a clear impediment to the smooth functioning of *JAZZ*, both functionally and emotionally.

Bob believed Paul to be the greatest of his enemies, the most feared member of the Board, and the one controlling his destiny. In our initial interview, he said "Paul has been able to dominate the Board meetings.... [He determines] the terms under which I was hired ... and the conditions under which I work." He was also convinced that Paul wanted him to be fired in order to return the organization to its former all-volunteer status (the good old days). Between this and Paul's tendency constantly to nit-pick, Bob felt utterly victimized by Paul and saved his strongest criticism for him (which, not surprisingly recommended figuring out a way to get him off the Board).

In meetings we attended, there was certain evidence to support Bob's fears. For example, at a finance committee meeting in May of 1992, with Bob's contract up for renewal, things came to a head. The plan of the meeting, which Paul chaired and Bob did not attend, was obviously geared to persuade those attending toward a predetermined conclusion. Paul had compiled a list of fund-raising ideas, which he dismissed one-by-one as being too little too late, arguing for the need to cut back on expenses, justifying Bob's departure, to be replaced by an unnamed part-time executive director. He even had a severance package for Bob pre-considered. Here are a few quotes from Paul that give a sense for his demeanor and judgments:

"Bob has bad mouthed just about everybody.... I can't understand what takes so much of his [Bob's] time. Everything that Bob's doing used to be done by volunteers.... I don't see any money coming in in the next three months, and nothing guaranteed in the next six months.... Many an arts organization has gone the route of a part-time executive director. We should find someone retired or a part-time woman emerging from childbirth.... [Facing resistance from the Board president]: I will take this to the whole Board if necessary over your head."

What was never clear was what the driving issue was for Paul: Returning *JAZZ* to its voluntary "club" days, which he has said he preferred; a personal grudge against Bob, which the extremity of his comments sometimes suggest; a need to control, which his behavior indicates; or a genuine fear of the organization going down in financial flames. Particularly during the stressful environment of an organizational healing crisis, all of these personal/organizational issues tend to get highly muddled, and one of the challenges to moving forward is the difficulty of separating them. But rather than analyze, separate, and "problem solve" these issues, our approach was very simple—to appreciate, as fully as possible, each of them, and to support all the organizational members in greater appreciation of each other and their contributions to a common passion—jazz and *JAZZ*.

Interpersonal Healing

As a first step in this process, the appreciative interview revealed a very different side of Paul, a more open individual, reflective on his shortcomings, who is sincere in making a much needed contribution to *JAZZ* in the unique way that his talents allow. As he said, "I see it as my mission to see the glass as 1/2 empty," and while others may find his dark warnings overblown and unnecessarily critical, there is more than a grain of truth in his comment, "I sometimes feel like nobody's worrying about money except me." He acknowledged the need for someone who knows accounting better than he, to implement a fund accounting system—"I just hope the phantom will appear." In contrast with his regrets over the lost volunteerism and club atmosphere, he argued for a radical reorganization and a wholesale emptying of the Board's old guard that would leave *JAZZ* operating more like a traditional arts organization. He revealed his pride in helping to make *JAZZ* more financially secure and his concern that if the organization were forced to return to being a club, it would at least be no worse off than when he started.

As we worked with Bob, we supported him in valuing both himself and his board members, no matter how difficult he found them. Arguing that both he and Paul suffered from underappreciation, we asked him to experiment with finding ways to genuinely compliment (and complement) Paul's contributions. The culmination of this process came in February of 1993 when we facilitated a meeting between Bob, Paul, and David to confront and reconcile their long-standing accounting differences. Whereas Bob seemed eager for the meeting, Paul was still recalcitrant, and in an hour-long telephone call the morning of the meeting, we listened to what proved to be the last of his defensive diatribes in this situation, ending with his feeling the support necessary to come to the meeting prepared to reconcile with Bob through the symbolic vehicle of the *JAZZ* books.

The special meeting that night, and the two board meetings that followed, were watersheds for *JAZZ*. For the first time in two years, Bob and Paul were "in concert" regarding the finances and accounting, and making public statements of pride on how they had managed to reconcile their differences. For the first time in memory, Paul characterized the *JAZZ* cash position as robust and offered a set of projections that held enthusiasm for the future. Most significantly, Paul had at last relented on allocating a percentage of non-administrative grants to cover overhead (including Bob's salary) and presented to the Board a proposal (which they adopted) that henceforward, the books reflect this 20 percent allocation. It was an incredible relief, for this was the hardest-fought territory, swinging the battle of perceptions of whether the organization could afford to keep an executive director. At the next board meeting, Paul presented a thorough write-up of treasury policy, incorporating the newly agreed upon elements, remarking "I can't believe I wrote a page and a half and nobody quibbled with a word." Telling the Board he would not be standing for reelection, he took the opportunity to say gracious good-byes and received in return a unanimously approved motion of commendation followed by thunderous applause. After witnessing a year of irascible backbiting and gloom and doom reporting, we observed in ourselves and the board members' reactions enormous relief from the negative energy that had long held captive the board meetings. While it is tempting to cast Paul as someone who "lost," given that he decided to leave the Board, this was not at all the emotional tone of it nor the content behind it. In our initial interview with him, Paul had expressed the desire not to rerun—because "it's healthy for organizations to have new blood." Once he felt his contribution to the Board was properly appreciated, and the possibility present that others could adequately oversee finances, he was able to transcend the battle stance he had previously taken, and let go of his animosity as well as his board seat.

Evidence of the watershed continued during the meeting. Ironically, it was around a proposal for instituting term limits. Because Paul's action was uncharacteristic of board members, there was a felt need to confront the issue structurally (which again proved premature, without the relational foundations in place). The breakthrough at this stage was that after the controversial proposal was presented, every single board member spoke (this had never before happened in our memory), and it was done with a sense of openness and making space for others' viewpoints. In the past, this kind of issue would have been cause for critical diatribes and domineering attempts to cut off debate. Shaming would have been the norm of the day, and people would have left feeling hurt. (It was little wonder that it had been hard to attract and retain "good" people on the Board—a point that one board member thoughtfully reflected on during the meeting). The meeting ended with congratulatory accolades in all directions, including Paul and Bob both offering appreciation to a board member for his newly published book on jazz.

So in terms of the relation between different levels of the system, not only did the relational healing of Bob and Paul make it possible for them to feel good about their contributions, for Paul to exit gracefully and with appreciation for his efforts, but it set a whole new tone of interaction among board members in discussing difficult issues. We move now to the next system level, that of "intergroup," to explore the next incidence of relational healing.

Level 2. Intergroup:
Inside and Outside Groups—Dynamics of Including the Whole

All organizations are at risk for perceptions of inside and outside groups (Smith & Berg, 1990), and passionately driven nonprofits perhaps all the more so. In *JAZZ*, there were several possible dimensions of inside/outside group perceptions, which partially overlap, leading to a high likelihood for certain members to feel excluded or marginalized. This case story covers the journey from a fragmented sense of intergroup relations (and the accompanying feelings of alienation and exclusion) toward more integrated intergroup relations (and feelings of connection and inclusion).

Intergroup Fragmentation

JAZZ has always been a grassroots organization. Its board is a working board, and proud of it. Prior to the advent of its paid staff, all of its activities were conducted by volunteers. The volunteers and board members come from a variety of backgrounds, but most have traditionally been working class. Founded by a group of white jazz lovers, the organization grew into a substantial African-American membership and board participation, but little Hispanic participation. Among the "old-time" board members, there is a certain nostalgia for the days when the organization functioned exclusively as a "club," relying entirely on volunteers.

So the possible "in" groups can be framed in a variety of ways: All board members, white board members, professional board members, or old-guard board members. Here are some of the ways in which these groups have manifest fragmentation in *JAZZ*.

Board membership is itself an in-group, even more so because this is the only Board most of its members serve on. It is not just a volunteer job that the members have passion for, board membership offers an acknowledgment of value, a symbol of political power, and a sense of belonging—often in a way that non-board members feel excluded by. Especially in the "club" days, there appears to have been a sense that the Board's purpose was to attract to town jazz performances that it (the Board's) members wanted to hear. As one board member put it in response to our question, "Why are you on the Board?": "[To] keep the organization going and to sponsor concerts that I myself would like to go to."

One subset of the Board is the so-called "old-guard." Because the organization has no term limits (and terms themselves are three years), it is not surprising that one-half of the Board had been serving for more than three years in 1992. One of the most problematic manifestations of this "in-group" was the cultural norm that evolved of highly conflictual board meetings that seemed, especially to outsiders, intolerant and uncivil. It was a very effective, if unconscious, means of ensuring that newly appointed board members would not last. We heard stories of many board members who left because of it, heard the Executive Director bemoan the unlikelihood of attracting good board members, and heard board members in their interviews mention it as reasons for their irregular attendance. Said one member: "I'm a volunteer. Why do I need this?"

We were witness to this phenomenon during the very first board meeting we attended. Here is the account one of us wrote some time later:

> I remember my first board meeting. I'm sure Ron Bailey does as well. It was also his first board meeting as the new chair of the Education Committee. Ron was presenting a proposal for a program called Jazz Week that brings jazz artists and education into the public schools. A few old-guard board members put Ron on the veritable cross-examination hot seat. If one didn't know otherwise, one might think Ron was up on a criminal charge. No one had expressed any excitement over a program that is so good it's since become a cornerstone of the *JAZZ*'s education efforts, nor had anyone given Ron positive feedback on a plan he had clearly worked hard upon. What a welcome to the Board! To his credit, he said, after 15 minutes of this haranguing, that he was frustrated by all the nit-picking and that he hadn't heard anything but negatives. A discussion ensued in which no one engaged each other, but several people made barbed comments. David Argus, the board president, said "Certain people don't want to take anything on faith." Alan Ludlow, the former board president, opined that "most boards trust their board members" and clarified that there is an ongoing issue of this board being a working board that wants to micromanage everything. Philip Everly reframed it as a question of whether *JAZZ* should have a rubber stamp board or a suspect board, the sarcastic tone of his voice making clear he heavily favored the latter.
>
> There were indeed some substantive issues in the questions, but who could get at them with all the rancor in the room. These people weren't having a serious conversation about board norms, they were having a dart throwing match. To my ears, the undertone of the meeting's "music" was violent, powerfully present behind the melody of rational discourse of committee reports and board practice. And playing that music were a group of jazz aficionados who weren't listening to their fellow players, didn't seem to appreciate them or their contributions, and expressed their frustration with acrid solos played under the guise of dutifully carrying out the business of the organization. Imagine what a jazz concert would sound like under those

conditions!

During the rest of the business of the meeting, (which went on for another hour and a half) Ron didn't say one word. He closed his eyes later in the meeting, sat slumped in his chair, a little pushed back from the circle; he was about as checked out as he could be without actually leaving. He never came back to another board meeting, and officially resigned a couple of months later.

One of the clarion calls of the old-guard board members was the desire to maintain the grassroots integrity of *JAZZ*, preserving it from being taken over by "elites," even as they recognized the need for their money and expertise. In this light, any board members perceived as "moneyed" seemed to be viewed with suspicion. While the Board probably contained no members who could truly be said to be wealthy, elitism was easily ascribed to those who were professionals, or white collar, or sometimes simply white. Alan Ludlow's solution to this love/hate relationship with elitism was to form a separate Advisory Board comprising well-known names in the community holding some combination of money, prestige, and expertise. By virtue of the set-up, the Advisory Board would not interact with the working Board of Trustees, and each would be "protected" from the other. It was a bandaid solution at best, and in many ways, only served to exacerbate the divide between these groups. More will be said about this Advisory Board in the third case story.

Intergroup Healing

As was the case with the interpersonal healing story, the increasing polarization between groups in this case seemed anchored in an atmosphere devoid of appreciation. At least superficially, the in-group of the old guard held firm against the forces for change, the "club" advocates disdaining involvement and action by those seeking to make *JAZZ* a more professional arts organization. But upon deeper examination, it was not such a facile matter to categorize people into these groups. Our interviews revealed a much greater complexity of views than the polarized perceptions would suggest. People had become stuck inside their stereotypes of each other and in the battle against their own fears, may have projected their shadows onto those perceived to be in the "other group."

The appreciative interviews began to unleash admissions and statements that would not have found their way into the public space of the board meeting, as well as giving rise to hopes for *JAZZ* that were not all that different from one another. When all the interviews were completed, it would be difficult to support that any of the so-called "old guard" was truly against the growth and development of *JAZZ*. That some members articulated a wish for more of the "club" atmosphere and greater voluntarism could be seen not as an impediment to development but a valuable consideration in how *JAZZ* might optimally evolve. Here are a few excerpts from statements made by four board members who would most typically be stereotyped as old guard, the critical naysayers, in favor of *JAZZ* returning to its club days, and resistant to almost any change:

REFLECTING ON PAST CHANGES:

"We've undergone a lot of change. Some people have dealt with it very well. For others, they've been very threatened, just as change has been threatening to me. I like to think I've been *for* most of [the changes]."

"When we went from all volunteer to having Bob there was a shakedown period.... Initially, some [Board members] abdicated responsibility to Bob. Bob did a lot of stuff on his own with little help.... We are slowly returning to a balance of voluntarism and hired staff."

"[*JAZZ*] is bigger and better than what it was.... On the surface of it, it's where I wanted it to be." [Pause, then reflectively], "So why am I unhappy?"

REFLECTING ON COOPERATION:

"Considering the diversity of backgrounds, [we've] generally [cooperated] pretty well. One person's priority may not be known to others. Personality clashes come about because people love the music and disagree over how to do it. But the fights over the years have been sincere fights."

"There is cooperation, but it could be better.... Everybody gets a little defensive at times.... I've probably been guilty of that.... Apparently, there's a feeling we need more civility.... [At the retreat], perhaps we should get to the bottom of some feelings. Get them out on the surface in a civil way."

REFLECTING ON DESIRED FUTURES:

"We must go forward march. It is a death knell not to."

"[*JAZZ* needs to] explore sources of money to continue to have an office and a staff."

"If I could reconstitute [*JAZZ*], I would put the two Boards together and get rid of the old guard rank and file." [Remember, this from an "old guard" Board member].

"[We should] operate more like a traditional arts organization, ... but I would be impeached [for suggesting it]."

"[The *JAZZ* Board] needs more rotation.... I don't want to re-run because it's healthy for organizations to have new blood."

"[*JAZZ* serves] an evolving public. We don't need to serve the original group."

"Jazz is really the great combiner—blacks and whites together, working, playing and enjoying. This has great social potential, especially now in this day and age. I would like to see our mission turn a little more in the direction of education and proselytizing."

By creating space for reflections such as these to be uttered, the interview process itself already began the unfreezing of polarized group stereotypes, especially those held about ones' own. In this manner, fragments of perspective disallowed by the organizational culture's stereotyping could be reclaimed; and the ground for listening was being laid by being listened to. In addition to these reflective outputs, there was an opportunity, amply availed, to put out to willing ears (and in some metaphorical way, extinguish) the untold stories of frustration. What became apparent was the noble intentions and passionate zeal underlying much of the rancorous antics in board meetings. For example, even Philip Everly's relentless "critiquing attacks" seemed clearly rooted in a dedication to "quality" that, in itself, was admirable. Of course, not everyone agrees on what is high quality jazz, let alone high quality administration. Also, unbalanced by any compassion, either for his own shortcomings or those of others, that quality dedication ends up sounding more like neurotic raving about incompetence. Philip was especially intrigued by our process: claiming he had considered resigning before out of despair; now he was excited to hang around in order to see what came out of this effort.

Another important contribution to the recovery of lost fragments of the whole system came in the form of the membership survey. In what apparently was the first membership-wide survey done by *JAZZ*, the Board had an opportunity to discover the voices of a significant, often voiceless, "outgroup" in the membership. Cocreated by the working group and ourselves, the survey covered questions such as how members saw the purpose of *JAZZ*, why they are a member of *JAZZ*, their preferences in jazz music styles, types of programs *JAZZ* should support, suggestions for increasing volunteer support, and so on. Symbolically as well as substantively, the survey gave board members a stronger connection to the membership they serve, and enabled the possibility of moving toward genuine trusteeship as an alternative to "in-group" self-service.

In retrospect, the September 1992 retreat, titled "Orchestrating Our Future Together," was a space for healing, conceived in the sense of "wholing." Summaries of the interview data and membership survey data were provided to the attendees, giving them the breadth of perspective from which it becomes possible to "own" the whole of *JAZZ*. Starting with the peak experiences of the interviewees in their involvement with *JAZZ*, we began our collective reconnection to the heart and life force of *JAZZ*. The centerpiece used to reunite the polarized perspectives allegedly held by different groups was to present them using an adaptation of Senge's (1990) model of creative tensions (more will be said about the adaptation later). The atomic-like energy inherent in the creative tensions was depicted as in Figure 2.

As small groups discussed how to dynamically balance these creative tensions in their vision of the ideal *JAZZ*, it became possible for the Board as a whole to own the whole of this life energy, rather than relegating its polarities to stereotyped (and perhaps mythical) groups. Later, in the session devoted to addressing how the Board wants to work together and organize themselves, the wholing/healing energy carried forward to permit an acknowledgment of the challenges posed by the Board's diversity and tendency to fall into attacking each other. What emerged from the day was a passionate articulation of a *JAZZ* mission that allowed all the powerful elements of *JAZZ*'s life force to express themselves, and a set of consensual recommendations for improving how the Board is organized and works together.

Relational Healing in Life-Cycle Transitions

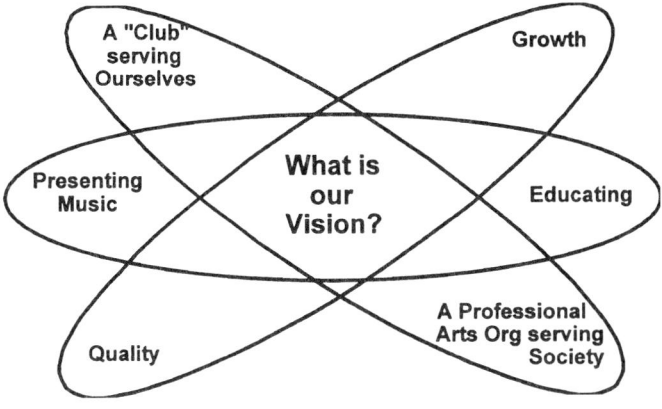

Figure 2.

While the intense work during 1992 was pivotal (including the appreciative interview process, the membership survey, and the board retreat), the process and outcomes of intergroup healing continued to manifest over the next couple of years. Board meetings became much more fun and appreciative as well as productive, often followed by social time where people stuck around to listen to jazz and share a beer. (In the earlier meetings we attended, people usually scattered immediately after a meeting, drained by both its length and acrimonious tone.) In this sense, the social club atmosphere was being revived. At the same time, the Board became less involved in picayune matters, more focused on policy concerns, and the meetings became more efficient. In this sense, it was a more sophisticated arts organization. Instead of becoming EITHER more "clubby" OR more professional, the Board became BOTH.

Another breakthrough in relation to perceived in- and out-groups occurred around an election issue with racial overtones. As mentioned in "The Stars of Jazz," Oscar Count had set a precedent in *JAZZ* of being elected by the grassroots, rather than the conventional nomination process. When another African-American candidate was nominated in 1993, and subsequently given access to board membership records, there were charges of unauthorized electioneering. When he failed to win, there were countercharges of unfair voting practices and possible racism. At the next board meeting, an African-American board member brought in a group of non-board *JAZZ* members to protest the election practices and sensitize the board members to some of the underlying issues. What could have been a highly volatile and fractious meeting instead was the occasion for an excellent dialogue that had two important results: (1) a committee was assembled to draft new election procedures that were more equitable and clearly publicized; and (2) the leader of the disgruntled group of *JAZZ* members was invited to join the Board, assuming the vacant role of the Education Committee chair. So this proved to be an exemplary healing at two group levels—between races represented on the Board and between the Board and the grassroots membership.

The final example of intergroup healing is so significant it deserves its own case story. What follows is the special case of the two *JAZZ* Boards and how the healing of the split between them also healed the split-off financial function.

Level 3. Interfunctional:
Reclaiming the Split-off Functions—The Case of the Advisory Board

Interfunctional Fragmentation

One of the most curious and relationally challenging features of the *JAZZ* organizational structure was the so-called "Advisory Board." Created ostensibly as a pathway for involvement and financial support from more elite jazz supporters, it functioned as a way of keeping them separate from the "working board" and uninvolved in the workings of *JAZZ*. Not long after the creation of the Advisory Board, it became an impressive set of names displayed on *JAZZ* stationery and a board in name only, never meeting and never actualizing its intended function of being a fund-raising arm of *JAZZ*. Interestingly, members of the Board of Trustees long maintained the fantasy that one day, the Advisory Board would be successfully engaged toward its purpose, without any evidence of past action to support this. Given that fund-raising was that aspect of the organizational self toward which some board members had the greatest ambivalence and negativity, it seems hardly accidental that this function became split off literally as well as psychologically.

At the center of this disconnected board structure was Alan Ludlow, the *JAZZ* president who had been so instrumental in leading *JAZZ* into its second stage of development. Alan was the sole member in common of the Board of Trustees and the Advisory Board (which he chaired). Unfortunately, Alan himself had become disconnected not long after creating the Advisory Board. Burning himself out after four years of dedicated service, he was rarely present at board meetings and hardly in a position to act as liaison to the Advisory Board. The Advisory Board reportedly had one meeting, which was curiously lacking in one fundamental aspect, any mention of its intended focus on fund-raising or even an explicit appeal for donations. Not surprisingly, the Advisory Board members we interviewed were at a loss to describe the nature of the Board they were on, and some were not even sure they were on it or that it still existed (which of course, in a way, it did not).

Given the lack of follow-through on implementing the Advisory Board strategy for fund-raising, it is tempting to believe (as it was for the Board of Trustees) that its failure was simply an implementation problem that could still be remedied, and not a strategic error. But over the course of our involvement, it became clear that the Advisory Board (and Alan Ludlow) were simply a convenient target for board members to place blame, given their own avoidance and discomfort with fund-rais-

Relational Healing in Life-Cycle Transitions 293

ing. Board members were stuck in a pattern of de-energizing complaints that seemed easier to maintain than reengaging the issue of fund-raising in new ways. The inability of the working board fully to "own" the fund-raising responsibility associated with *JAZZ*'s rising status as an arts organization was the most obvious artifact of its tendency to split off unwanted functions, severely criticize that which lay in its own shadow, and thereby block the process of development by reverting to behavior from an earlier stage of organizational life. One might speculate that members of the Board who felt most threatened by the possible addition of influential "moneyed" board members were themselves most interested in wielding the power inherent in their status as board members. But by keeping all of this hidden in the shadow, both personally and institutionally, there was no chance for healing and a high risk of institutional failure.

Interestingly, the same type of splitting-off behavior occurred in relation to the Executive Director's role in fund-raising. By some accounts, Bob was the most effective fund-raiser in the organization, spending countless hours writing and managing *JAZZ*'s substantial grants. But for other board members, he was a target of blame for the lack of a more versatile fund-raising strategy, this despite the fact that the Board had at one time proscribed him from leading other fund-raising efforts.

What was most striking about these dysfunctional relations around money was the absence of any real dialogue about it—hence, the institutional shadow. There were plenty of sideways comments, an abundance of blaming somebody else, a general lack of financial sophistication, and very little expressed appreciation for the substantial financial success of *JAZZ*.

Interfunctional Healing

Re-owning and integrating the financial shadow-side of *JAZZ* has been a long process of healing. While it will undoubtedly be the arena for further healing in the future, the accomplishments have been amazing.

As the interpersonal and intergroup healing described in the previous case stories began to take root, there began a natural process of board turnover that brought in new board members, including ones with greater financial acumen. Among these, the current treasurer, Sam Aston, stands out as a godsend, helping *JAZZ* to institute its first rigorous annual budgeting system. He has also been willing to act as a patient educator of Committee chairs and other board members on the requirements of budgeting, involving them all in the process.

In March 1994, a retreat was held that marked the first time members of the Board of Trustees met together with members of the Advisory Board. It was another watershed event. In the context of talking about the *JAZZ* organizational structure, members of both Boards had a chance to express their feelings about the present structure and articulate the history of their (largely unfulfilled) expectations of the Advisory Board. In the dialogue, board members also heard first-hand how the role of fund-raising really did not match the interests of the Advisory Board members and were able to broaden their notion of how these members might contribute. With the whole (governing) system in the room, at least in a representative sense, this was the meeting where a plan of reorganization for *JAZZ* was born. Its core idea was the elimination of the Advisory Board and the creation of an expanded Board of Trustees that would meet only quarterly and encompass the wider range of expertise of both Boards. An Executive Committee would also be formed that would continue to meet monthly and conduct the affairs of *JAZZ* in much the same way that the present (working) Board of Trustees now did.

So, from out of the shadow, there emerged relationships that at last embraced the split-off Advisory Board members and a structure that would encompass them all. Most importantly, the fantasy of relegating fund-raising to some "other body" was finally laid to rest.

It was not long after that the two board members in charge of Membership (Oscar Count and Katherine Feingold) together with the Treasurer (Sam Aston) took on a new corporate-focused fund-raising initiative (see their mini-story under "The Stars of *JAZZ*"). In 1995, after years of false starts and "talking about" fund-raising parties, *JAZZ* held its first ever major community fund-raising event. Also in 1995, a new board member was elected to become the first director of finance and development under the new board structure. Given the previous reliance exclusively on grant funding, membership dues and event fees, and the enacted belief that fund-raising was outside the "working" Board's purview, these developments were milestones in *JAZZ*'s maturation into a self-sustaining professional arts organization.

Level 4. Interorganizational:
Toward Community Collaborations for Jazz Appreciation

Interorganizational Fragmentation

In this final case story, we extend the perspective outside the organization to the interorganizational level of analysis. Not surprisingly, given the internal turmoil of *JAZZ* during the period we entered, *JAZZ* found it difficult to play on the same field as some of the major arts organizations in its region. Although some of its leaders aspired to such a status (notably starting with Alan Ludlow), the same internal issues that kept new members off the Board also kept potential collaborators at some distance. Its biggest collaborations for many years had been with the local community college, which sponsors an annual jazz festival and which had its director on the *JAZZ* board, and with a suburban theater in the park, where summer jazz concerts were held.

Interorganizationally, it could be said that *JAZZ* had fragmented relations with the arts community it wanted to be more a part of. As an outgrowth of the conflicted ideas about *JAZZ*'s constituency, the ambivalence on the Board toward major players in the local arts scene sometimes reached love/hate proportions. One meeting in early 1993 was especially illustrative. Some board members were discussing one of their favorite external whipping boys, the local radio station that has the most jazz programming. The tone was highly critical—of the talent of the jazz announcers, of the quality of the jazz they played (of which they had done an "audit"), of the management of the station for not giving jazz more playtime and programming leadership. While we cannot comment on the validity of their concerns, we can say that the tone of their complaints were not constructive, and of themselves would hardly lead to improvements of the sort they said they wanted. While the letter that they eventually wrote to the radio station management was somewhat toned down, it was hardly a letter of collaborative intent. Yet, *JAZZ* would have loved to see its programs better publicized on this station, and would benefit greatly from their joint sponsorship of events.

In the same vein, but less dramatically, *JAZZ* members dreamed of being able to hold regular concerts in some of the best local concert halls, attract series ticket holders like the symphony does, and be a major force for jazz education in the schools. But in its ambivalence, *JAZZ* did not quite see itself on a par with the likes of the local (and nationally prestigious) symphony or art museum and did not want to alienate its diverse and grassroots membership. To put it simply, *JAZZ* had a chip on its shoulder, and could not realize its visions until it changed internally.

Interorganizational Healing

Two events in 1992 catalyzed *JAZZ*'s venture into the world of collaborations with other major arts organizations. With funding from the largest private grant in the history of jazz, *JAZZ* commissioned a famous jazz musician to create a piece honoring Pablo Picasso, to be presented at the local art museum's concert hall during its major special exhibition of Picasso's work. The concert was a huge success, was broadcast live by the local radio station, and gained national attention for *JAZZ* as the recording was released in 1994. Later, *JAZZ* initiated a Thursday night jazz series in cooperation with the local Center for Contemporary Art.

These collaborative ventures together with the inside-out process of healing *JAZZ*'s internal relations, appreciating their history and energizing a new sense of mission led *JAZZ* into a greater sense of self-esteem and a new way of operating in the local arts community. No longer was it just the voice of one man (Alan Ludlow boldly held this vision while he was president, but did not have a Board that could effectively join him in it). Now more and more ambitious jazz programs began to seem viable to the Board. The long-standing hope of some of the local funding agencies for greater arts organization cooperation began to be realized.

In 1993, two major concert series featuring artists from different regions were held at the art museum, and in 1994 presentations there included the Lincoln Jazz Orchestra, which later released selections from the concert on CD. But the biggest local jazz collaboration in the history of this region was responsible for the jazz series that started in 1995. It brought together, in joint planning and sponsorship, the Symphony, the Art Museum, the Community College Jazz Festival, and the Radio Station. That concert series, presented at the prestigious symphony hall as well as the art museum, has now become a regular feature of the annual jazz scene. Furthermore, a member of the Art Museum and the music director of the Radio Station with which *JAZZ* had had such conflict now sit on the *JAZZ* Board. A cooperative program with local school districts to provide jazz education in the schools has been expanded to encompass a long-dreamed of used instrument donation program for needy kids.

In 1996, *JAZZ* and its hometown had the privilege of hosting the Lila Wallace–Reader's Digest National Jazz Network's annual meeting, where executive directors and programming managers from over twenty jazz service organizations like *JAZZ* gather to share ideas and plan cooperative booking, touring, and fund-raising. *JAZZ* is held in high regard by this network, a testimony to its cooperative stance in assisting newer and smaller jazz service organizations around the country. As it happens, the executive director of the National Jazz Network is also the founding president of *JAZZ*, nearly 20 years ago. Following his visit in 1996, his written comments to *JAZZ* Executive Director, Bob Davis, provide a fitting tribute that echo our sentiments in acknowledging the transformational journey of *JAZZ*: "It is truly gratifying to see how far [*JAZZ*] has come and how strong it has become."

THEORY BUILDING: TOWARD A RELATIONAL HEALING MODEL OF LIFE- CYCLE TRANSITIONS

This section uses the data of the preceding case stories to develop a preliminary "relational healing" model to explain the process by which this arts organization traversed a "crisis point" in its life cycle of development to reach a new stage of maturity. Following the description of the model and its grounding in the data of this study, a set of propositions are offered to expand upon the implications of the model and invite further research.

Creating the Relational Healing Model

As was observed in the beginning of this chapter, life-cycle theories typically treat the passage between stages as a crisis point, a one-dimensional boundary that might be depicted as in Figure 3. In our relational healing model, we reframe the transition as a crisis with a time dimension of its own. Indeed, the period encompassed by the four stories in this chapter span three years, a time when *JAZZ* can be said to be between stages of development, and working to achieve the transformative potential of the crisis period. A relational healing view of the relationship between stages of development, with a focus on the transition period as "the space between," might look like Figure 4.

By focusing our attention on the process by which one organization successfully navigated the transition between stages, we can learn more about what may be

**Stage Transitions in
Traditional Life Cycle Models**

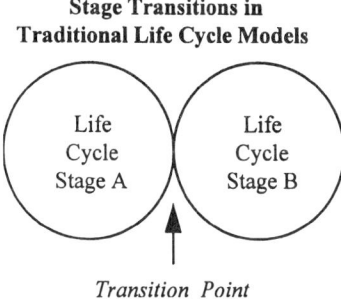

Transition Point

Figure 3.

desirable for arts organizations faced with the challenge of climbing to the next developmental stage. We have suggested that the "space between" is pivotal in this analysis in more ways than just the time dimension between stages. Indeed, we believe a crisis period is characterized by a set of tension-filled gaps between polarities that collectively provide the creative energy for transformation. These tensions may show up in the form of polarized interpersonal relations, polarized groups, polarized functions, and even extend beyond the organization during this period to manifest as polarized organizations. It can be said of all four of the case stories in this chapter (dealing with these four levels of analysis), that polarized forces at each level manifest in a fragmentation of the fabric of the organization. The attention in our relational healing model is on the space between those polarities, specifically the creative tension[8] that exists in that space tying the fragments together. The energy of that creative tension is the energy of relationship, it is that which holds parts together into a whole.

Peter Senge (1990) also refers to a space between polarities as "creative tension," in his case the gap between current reality and vision. He argues that this "gap is *the* source of creative energy" (p. 150) to move from the current reality toward the

**Stage Transitions in
Relational Healing Model**

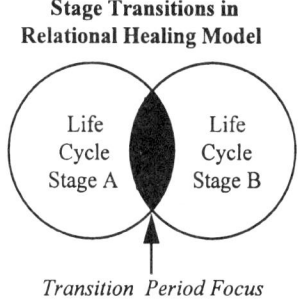

Transition Period Focus

Figure 4.

vision. But what about if an organization does not have a clear vision of its next stage of development? What about if there are conflicting visions of what it should look like? Building on Senge's model, we might create a tripartheid version that acknowledges another set of polarities in the current reality as being the source of the creative energy that fuels the future possibility (see Figure 5).

Given that the polarizing energy of an organization in transition creates the kind of discord and fragmentation we saw evidence of in each of the four case stories, one might ask what is it that holds the poles (and the organization) together. Why do they not simply split and go their own ways? We would argue it is because the poles are in fact, part of a larger whole, and that what holds them together is their relationship.

> To say that opposites are polar is to say much more than that they are far apart: It is to say that they are related and joined.... Polar opposites are therefore inseparable, like the poles of the earth or of a magnet, or the ends of a stick or the faces of a coin. (Alan Watts, quoted in Johnston, 1991, p. 13)

If poles are inseparable, does that mean it was just an illusion that the polarizing forces seemed to be tearing *JAZZ* apart? Not at all. For while the polarities around an issue can never truly be separated, there is always a question of whether a particular container, be that a person, an organization, or even a society, is able to hold the creative tension of polarities consciously together. Imagine a balloon as the gas inside is being heated; as the gas molecules move faster (symbolizing the

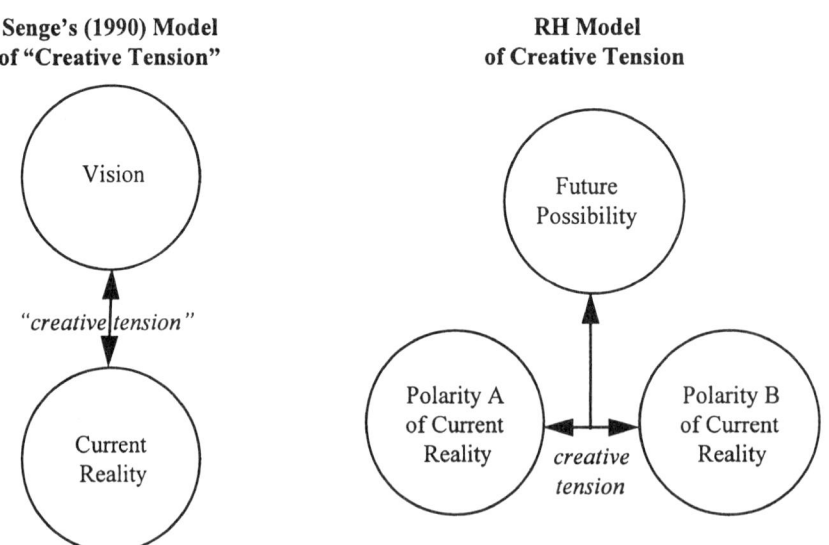

Figure 5.

Relational Healing in Life-Cycle Transitions

increasing creative tension), the balloon expands. But if the balloon is not strong (and flexible) enough to contain the expanding gas, it bursts. So too might an organization that cannot contain the creative tensions heated up during a crisis period. While the gases do not disappear when a balloon bursts anymore than the polarities of the issues that were in creative tension go away, they may no longer be contained within the organization, having burst forth into the surrounding environment instead. Even if the pressure is not sufficient to break the balloon, the pressure is certain to find the weak spot in its material, creating a balloon with a distended area. Now one pole of the creative tension is cut off from the mainstream of the organization and "hidden in the shadow" of its distended compartment.

So let us now suggest a model of what happened in the transition process documented in this chapter, the process we have labeled relational healing. As mentioned at the beginning of this chapter, healing is used in the context of "wholing," such that its aim is not the elimination of disease but the re-collection[9] of the whole within the organization. By attending to the "space between" the creative tensions, by honoring both poles and strengthening the entire organization, not just one force within it, one might say that the poles were pulled together into a tighter space (a space exemplified by the retreats, for example), from which their creative tensions had the possibility to grow into the zygote of the transformed organization, the organization at its next stage of development. Figure 6 depicts a sequence of five steps in the relational healing model of organizational transformation during the healing crisis of a life-cycle change.

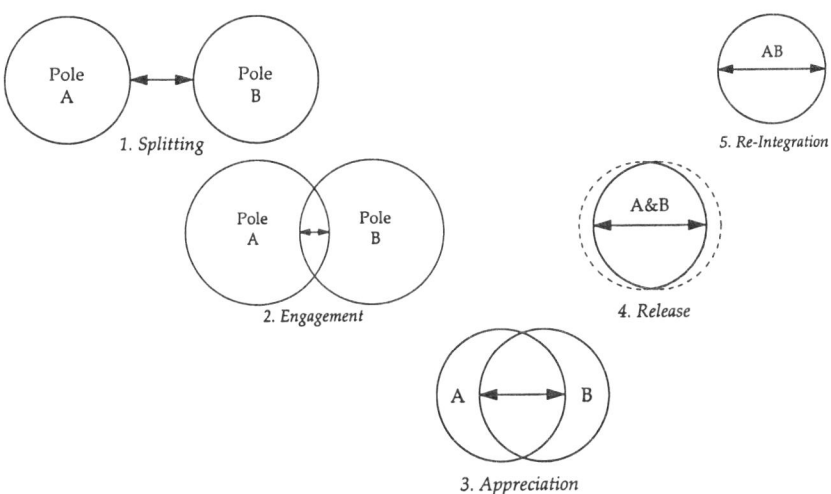

Figure 6.

1. *Splitting*—The creative tension of the crisis is experienced as splitting. The polarized forces are fragmenting the organization, people and groups are disengaging from each other, and the unconfronted tension threatens to tear the organization apart (or badly misshape it).
2. *Engagement*—The polarized forces are brought into contact with each other, initially in a small way, in the safety of a contained environment within the organization.
3. *Appreciation*—Both poles of the creative tension are appreciated, as are the people representing those poles, allowing them to expand (appreciate) the space in which the creative tension is held together.
4. *Release*—Once there is enough mutual appreciation between the polarities, and the space in which the creativity is held grows large enough, the tension in the relationship reduces allowing release of unneeded or outgrown elements.
5. *Reintegration*—Finally, a new whole is achieved wherein the creative tensions are still present but working inside the transformed organization, continuing to enlarge it rather than split it apart.

Applying the Relational Healing Model to *JAZZ*

Now let us see how this model fits the four case stories of *JAZZ*. For purposes of simplicity, Table 1 identifies, for each of the four levels of analysis, a single central issue in creative tension (although each involved several intertwined issues). Each issue is then taken in turn, mapping the case stories of healing against the relational healing model.

Table 1. Central OD Issues and Polarities in *JAZZ* Case Stories

Level of Analysis	Central OD Issue	Pole A	Pole B
1. Interpersonal	Staffing	All volunteer	Paid staff
2. Intergroup	Constituency served	Grassroots—Ourselves (Club)	Elite—Arts community (Professional arts organization)
3. Interfunctional	Fundraising function	Specialized responsibility (Advisory Board)	General responsibility (All members)
4. Interorganizational	Relationship with other community arts organizations	Being distinctive	Gaining acceptance

Staffing—All Volunteer vs. Paid Staff

When we entered *JAZZ*, staffing was a polarizing issue, with Bob (the Executive Director) at the center of it. Regardless of whether people truly lined up as definitively for or against having a paid staff, there was a widespread perception that some people in the organization would like to (or saw no viable alternative to) return to an all-volunteer organization. Thus, the polarities of this issue became translated into either Bob must leave or those ostensibly pushing for his departure must leave (personified in our case story by Paul). The tension was hardly creative at this point, as the two polarities were powerfully *split* (step 1), and could well lead to the split up of the organization. Meanwhile, in open forums, it was difficult for any one member to own both polarities of the tension. They were either perceived as for or against Bob.

By refusing to buy into this polarization and valuing both Paul and Bob, we began through the appreciative interviews to create a relational space for engaging the much more complex issues that underlie the staffing juggernaut. Once into this *engagement* (step 2), the larger issues of fund-raising, financial security, declining volunteerism, and so forth, could begin to be deciphered. What Paul was standing up for was much more than whether or not *JAZZ* should have an executive director; he was standing for financial responsibility and encouragement of more volunteerism. Even his resistance to allocating a portion of grant monies to cover administration could be seen as an overbalanced attempt to avoid the financial complacency that he might fear resulting from a bookkeeping maneuver. Because Paul's perspective was split off from the organization (partly due to his style, partly due to the lack of financial sophistication on the Board), the polarity he represented was not owned by the Board, and his increasingly ardent frustration-fed pleas were misconstrued by many as purely personal attacks and ill will.

During the *appreciation phase* (step 3), which included the first *JAZZ* retreat in September 1992, these larger issues began to be surfaced, engaged, and appreciated by the Board as a whole. The expansion by appreciation of the relational space holding the creative tensions was making room for the zygote of new possibilities to be born. During the retreat and at following meetings of the Board, expressions of appreciation across lines previously uncrossed were appearing. It was not as if the critiques disappeared, but they were less fervent, and the gifts brought by both Paul and Bob more often noted. Still, given the entrenchment around this sensitive issue, it was not until February 1993 that the expanding space of safety created by appreciation of both people and their ideas made possible a breakthrough on the question of Bob's future.

The symbolic agreement on accounting reached by Bob and Paul had no losers, for *JAZZ* retained its Executive Director, Bob kept his job, and Paul was able to declare a financial victory—feeling enough support for *JAZZ*'s financial responsibility, he offered his first ever positive projection for its future and acted upon his long-stated desire to step down from the Board. It was a *release* (step 4) in many senses of the word.

Bob released his animosity toward Paul, making possible a reconciliation; Paul released his fear that *JAZZ* would collapse without him acting the part of financial alarm bell; Paul also released his position as treasurer, making way for the entry of a new treasurer (who at the time no one knew about) who has led *JAZZ* into a new era of financial responsibility (without all the fear-based energy).

With the releasing of entrenched positions, and the making space for something (or someone) new, the creative tension of staffing fulfilled its transformative promise in a *reintegration of the polarity* (step 5). Not only was the paid staff retained, but the half-time position assisting the Executive Director was significantly upgraded, and volunteerism showed a dramatic upsurge. In other words, both elements of the polarity grew appropriately stronger in their no longer split-off relationship to each other. This is the significance of relational healing—the healing focus on the relationship created an enlarged space for the whole, for both polarities of high volunteerism and paid staffing to be present, honored, and grow in the reintegrated form of *JAZZ*.

To be sure, not all of this is a *direct* result of the events described in this case story, but the freedom created by the resolution of this most challenging interpersonal tension supported related breakthroughs. Bob, in his new-found security, was able to be much more supportive and appreciative of volunteers, the volunteer coordinator was admitted to the Board and quickly gained respect, and the new treasurer, with the help of the budgets he instituted, was able to demonstrate the viability of paying more in staff costs.

Constituency Served—Grass Roots vs. Elite

Prior to our entry, the *JAZZ* board had reportedly had for many years a style of operation that made it very difficult for new entrants to become part of the group. There was no formal orientation process, and the de facto introduction of the board member's first board meeting was a trial-by-fire awakening to the rifts and acrimony among some elements of the Board. The standoffishness was perhaps most evident in the old guard, whose attachment to their board positions made even the consideration of term limits a risky business. New members, especially any with wealth or other power, were perhaps both wanted and feared, anticipating that they would change *JAZZ* from the grassroots organization it started out as. So the pole of self- and past-preservation (symbolized by the "club" vision) was at odds with (and *split* off from) the expressed aim to serve the community of the region (symbolized by the "professional arts organization" vision).

As the complexity of the issue was *engaged* in the safe space of the appreciative interviews and later shared during the retreat, it became apparent that not even the old guard was against development; but they, like all of us, were afraid of change. Through the *appreciative* phase of the relational healing model, the stereotypes began to unfreeze as the consensus around moving the organization forward became more and more explicit. Presenting the data from the membership survey and the appreciative interviews during the retreat surfaced the atomic-like energy of several creative tensions (see Figure 2); then, in a process of dialogue that confronted each tension consciously, this energy was transformed into a powerful new consensus and statement of *JAZZ* vision (see Figures 7 and 8).

The *releasing* phase for this issue also hinged significantly around the same February 1993 board meeting mentioned earlier. The extraordinary discussion of term limits in that meeting was a transformative moment in the history of *JAZZ*, for it enabled the articulation of reasons for preserving the continuity of the past (embodied by board members and their memories) even while the logic of term limits was persuasively presented. Although the "hearing" did not reach a consensus on term limits, it clearly enlarged the space holding the creative tension to the point that over the next couple of years, board members voluntarily resigned by not standing for reelection (though sometimes not without some further play out of resistance).

Once again, the *reintegration* achieved is not a simple victory of one tension over the other. As documented in the case story, *JAZZ* became a more sophisticated arts organization AND regained more of its club flavor, as demonstrated by the growing sense of fun and socialization both during and after board meetings. What is more, the diversity of board members now participating on the Board, and the sense that the club now extends beyond the bounds of the Board (to the dedicated corps of volunteers who staff *JAZZ* functions and enjoy social gatherings as well as jazz events), have enlarged the membership in "the club," making it a richer and more common experience. Ultimately, the strengthened relations between the poles revealed that both contained the ideals of serving community; what was bridged were the split-off relations of two definitions of community—the grassroots and

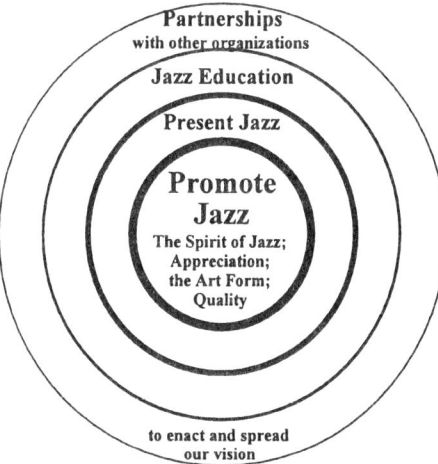

Figure 7.

The purpose of JAZZ is to be an effective advocate for jazz as an art form by furthering community awareness and appreciation through performance and education.

Figure 8.

the arts community. Now *JAZZ* serves both, with resources drawn from the widest, most integrated sense of community.

Fund-raising Function: Specialized Responsibility vs. General Responsibility

It is difficult to imagine the function of fund-raising being more *split off* than it was in the collective psyche of the *JAZZ* Board with its fantasy of the phantom Advisory Board. Because the process of appreciative inquiry encompassed the Executive Director and Advisory Board members (as well as the Board of Trustees), the healing of this fragmentation began to be *engaged* early in the relational healing process. But unlike the previous two case stories, the most dramatic phase of *appreciation* really did not happen until the pivotal first meeting of the Board and Advisory Board members in March 1994. The new first hand appreciation of long-shadowed opinions about one another's roles, and so on, was enlightening. When the new structure of *JAZZ* was born at this meeting (wherein fund-raising became a specialized responsibility *within* the Board), it was built on the foundation of a new set of relationships—among board members, Advisory Board members, and the Executive Director. While we (or any others who might have examined the structure of *JAZZ*) might have recommended such a restructuring two years earlier, it would have had no relational underpinnings. Without healing (wholing) first at the interpersonal and intergroup levels, and a collective revisioning of the mission and programs of *JAZZ*, such a reorganization would have been unworkable and potentially disastrous.[10]

As in the other stories, a phase of *releasing* remained to be confronted. In this case, it took the form of two key board members: One was another old-guard member of the Board who was not at the pivotal retreat and still feared the reorganization as a threat to the grassroots nature of *JAZZ*, and the other was the man who founded the Advisory Board, Alan Ludlow. No one knew how Alan would feel about the elimination of his creation. As it turned out, he was ready (perhaps even relieved) to release it, and when he came to the board retreat in September 1994 that presented the detail of the reorganization, he was eloquent in his praise

Relational Healing in Life-Cycle Transitions 305

for the personal and structural integration he had not thought possible only two years earlier. On the other hand, the old-guard board member was not ready to accept the change, and fought it pretty much on his own through the vehicle of the newsletter (which he edited) and letters and phone calls to *JAZZ* members. Despite the best attempts to work with him and understand his concerns, he never relented from his position. Because of his influence and the incredible job he had done as the newsletter's editor, fears abounded as to whether he might thwart the reorganization or resign from the Board and his editorship (as he threatened). As the story played itself out, all we could do was counsel the board chair and the Executive Director that it would be unwise to let one man hold the entire Board and membership hostage. He should be allowed to voice his opinion and respondents should do their best to reply, in respectful disagreement with his fears. In the end, it was the Board that had to be ready to release this much respected elder and his editorship of the award-winning newsletter, if he chose to resign over this issue. Indeed, when the entire *JAZZ* membership voted on the reorganization and bylaw changes in April 1995, it was overwhelmingly approved, and the dissenting board member did resign. It took a while and a few shaky issues of the newsletter to recover completely, but recover they did.

So, through a process of natural evolution, *JAZZ* had achieved a *reintegration* of its governance functions, especially fund-raising. In integrating the split-off polarities of special and general responsibility, *JAZZ* gained the best of both—fund-raising is now something the whole Board and the Executive Director participate in (and co-own), and a standing committee of the Board, Finance and Development, takes a specialized leadership role in it. What is more, the reintegration went beyond the re-owning of the fund-raising responsibilities of the Board to encompass a revitalized organizational structure that now was in complete alignment with the strategic plan of the organization. During the course of three board retreats (in September 1993, January 1994, and March 1994), all the elements of the strategic plan were thoroughly worked through. The reorganization plan approved by the Board at its September 1994 retreat created an Executive Committee (implemented following the Membership bylaw approval in April 1995) that included the Board Committee chairs responsible for each of the six major thrusts of the strategic plan: jazz performances, jazz education, finance and development, marketing and public relations, membership, and organizational structure (this last being represented by the president).

Arts Organization Relationships: Being Distinctive vs. Gaining Acceptance

The identity crisis of *JAZZ* had its most outward manifestation in its relations with other arts organizations in the community. On the one hand, the desire to be a grassroots organization, dedicated to jazz, and wary of the requirements of being a major arts organization, kept *JAZZ* in a competitive, critical relation to some arts groups (most notably the Radio Station) and in a dependency relation to others (the major local arts funders). But as strong as was the force for being distinct, so too was the desire to be accepted. To be sure, collaboration with other arts organizations, one at a time, and largely as a source of venues for jazz programming, was already underway when we first were engaged by *JAZZ*. But the predominant feeling of being a sort of second cousin, unwanted in the primary family circle, took time to dispel. The question was, how much did *JAZZ* want (or not want) to emulate and collaborate with the mainstream arts organizations of this city? While they wanted access to the funding and the venues, they did not want to be elitist in membership or presentation.

As some of the other issues were dealt with, the openness to (and readiness for) greater collaboration happened quite naturally. The affirmation, during the first retreat, of the outer circle of *JAZZ*'s vision (see Figure 7), "Partnerships with other organizations to enact and spread our vision," began gradually to become more and more prominent in *JAZZ*'s mode of operation.

The *engagement* and *appreciation* phases of the relational healing with other arts organizations occurred throughout the period of this study, but in a much less focused way, so that by the time *JAZZ* was ready to consider larger collaborative projects, the relational groundwork had been laid to make it possible. In the early going, this was largely due to the continued relationship building on the part of the president and executive director. It might be that what *JAZZ* had to *release* in relation to this issue was an image of itself as "less-than," the chip on its shoulder, if you will. In this regard, we cannot help thinking about Philip Everly, the board member in the original "Working Group" we formed for this project. More than any other board member, he embodied this "chip." He was critical of just about everyone and everything, and yet, he appeared truly to want *JAZZ* to outgrow its adolescence. What he could not seem to realize is how much his own participation—ringleading the fight against the Radio Station or writing editorials in the newsletter with racial slurs—was a part of that adolescence. In releasing its adolescence, it may be that *JAZZ* outgrew board members like Philip—like the other old-guard members, he left the Board when his time for reelection came up.

Once again, it can be said that the resulting form of *JAZZ*, following its transformational journey to the next life stage, *reintegrated* both the wish to be like (and work with) the other arts organizations in town (gaining acceptance) and to retain its grassroots flavor (being distinctive). It had to grow beyond its "angry young man" adolescence, but not without a lot of continued appreciation for the wide variety of society segments served by *JAZZ*. Both in its education efforts, with its instrument program, and its performances, with its program taking jazz into the inner city, *JAZZ* has not lost touch with the grass roots. The board's membership may be the best testimony—it is now occupied by people of more colors (Hispanics are now represented when before they weren't), people of a wider range of wealth (the range has increased substantially on the high side, without losing members of very modest means) and talents (finding a marketing/PR chair and a finance &

development chair were key), and people from a variety of community organizations with which *JAZZ* now collaborates.

Propositions of the Relational Healing Model

What follows in this section are a series of propositions that are suggested by the relational healing model and the data of these case stories. It is not considered that these propositions are "proven" by the cases, but are offered here as provisional possibilities warranting further research. Relational healing, as described in this chapter, is a model of the organizational change process occurring during the critical transformation from one stage of an organization's life cycle to the succeeding stage. It is an organic theory in that it proposes that the movement toward greater wholeness described by its steps is a natural process; the aim of facilitators such as ourselves is simply to encourage that natural process with minimal active intervention aimed largely at holding the space for the creative tensions to be safely and consciously explored.

Proposition 1. Relational fragmentation signals the beginning of a life-transition crisis.

At multiple levels of a system, the initiation of a life-transition crisis is observable by the fragmentation of relationships—between people, between groups, between functions within the organization, and between the organization and other organizations. This fragmentation is a natural result of the build-up of energy in creative tension between polarities of issues that must be confronted in order for the organization to grow to the next level of maturity.

Proposition 2. The creative tension between polarized fragments fuels the transformation in a life-cycle change.

Paradoxically, it is inherent in the very process that appears to be tearing the organization apart, that the seeds of the organization's next life stage lie. It is the engagement of these creative tensions that leads to a resolution and reintegration of their creative energy in the form of a transformed organization at the next life stage.

Proposition 2a. The transformed organization emerging from a successful life-cycle change reintegrates into a new whole the elements in creative tension.

In contrast to Senge's (1990) model of creative tension between current reality and vision, the relational healing model describes the tension between two polarities of the organization's critical development issues. While one pole may be more identified with continuity (the past) and one more with change (the future), the natural result of engaging their creative tension is not a victory by one over the other but the emergence of a newly integrated organization that now contains and creatively engages both forces.

Proposition 3. Relational healing (the process of growing a new whole from the polarized fragments) results from attending to the relationship between the poles, the "space between" that links them.

Relational healing, as its name implies, puts primacy on the *relationship*—of people, of groups, of ideas in creative tension—and supports the recognition of their polarities as part of a single whole. In this way, attending to the space between the poles means building conscious awareness of the relationship between them. Attention to the space between honors both poles (without which the relationship wouldn't exist), and encourages each pole to value the other.

Proposition 3a. Relational healing is initiated by bringing polarities into engagement in a safe space.

A relational healing practitioner can support this first stage by creating an environment (like the working group or the first retreat in our case stories) where all polarities among the relational fragments feel valued and safe. Appreciative inquiry (either by the practitioners or the members of the organization) is an excellent means of having everyone feel valued. Presentations in such spaces should bring the whole into focus, so that all the polarized fragments can see their perspective valued in the context of the whole.

Proposition 3b. Appreciation of the relationship between the polarized fragments strengthens the relationship and enlarges the safe space that contains them.

This appreciation can take many forms. It is best modeled by the relational healing practitioner and reinforced whenever possible among the organizational members. Among the forms:

- *Unconditional presence* (Welwood, 1992). We were often astounded how often our mere presence, without comments of any kind but with a conscious intent to observe, seemed to impact the consciousness in the meetings.
- *Caring (vs. a curing) attitude* (Moore, 1992). Minimizing defensiveness and maximizing healing potential, a stance of caring is powerfully neutral appreciation.
- *Spirit of inquiry* (Cooperrider, 1996). Modeling a spirit of inquiry, ideally an appreciative one, supports the enlargement of the safe space.
- *Unconditional positive regard* (Rogers, 1961). Applying this famous Rogerian principle whenever possible in interviews, observations and other encounters in the system supports appreciation.
- *Dialogue (vs. discussion)* (Bohm & Nichol, 1996). Dialogue creates a listening space where the exercise is not in debating the point but understanding the issue thoroughly.

- *Meditative stance toward critique.* Neither agreeing nor disagreeing with critiques, but hearing them out and meditatively letting them pass without reacting both honors the person and enervates the critique.
- *Seeking common ground* (Weisbord, 1992). Finding common ground (even where it is not expected) can energize passions that unify.

Proposition 3c. Releasing outgrown or unneeded elements completes the "death" of the old organizational stage, making room for the transformed organization.

Like the old skin of a snake, there are parts of an organization growing into its next stage that must be sloughed off before the new form (new skin) can take shape. It is a natural process, not a surgical one, and requires that the organization and the relational healing practitioner allow the old skin to fall off rather than severing it. In the realm of healing, where the aim is wholing and not the elimination of disease, cutting out unwanted parts eliminates something needed by the whole to complete the relational healing, and in one form or another, the unwanted will either return or hide in the organizational shadow, wreaking havoc from there.[11]

Proposition 4. The organization in relational healing is best regarded as a conscious organism, not a machine (see Morgan, 1986).

In line with the last point above, in relational healing, natural processes of evolution are being supported, rather than interventions being made. In this sense relational healing is both a natural theory and a spiritually rooted one. On the one hand, "thou shalt not kill" is a cardinal rule of relational healing. Simultaneously, death is accepted as a natural process. To put it in spiritual terms, there is not only a tolerance, but an appreciation for death (as the final timely releasing stage of the process) and a recognition that any attempt to defy death or to take life into one's own hands has karmic consequences.

Proposition 5. In relational healing, changes in the organization occur organically from the inside-out.

Organizational transformation in this model begins at the interpersonal level and migrates outward toward functions (organizational structure) and finally interorganizational relations (see Figure 1). The sequence in which the case stories were told is more than just a convenience of clarity. Although the processes described in each overlapped, the concluding step of each relational healing process (reintegration) occurred in the same time sequence as the stories. Not so simple as a linear process, it was definitely true of *JAZZ* that the interpersonal and intergroup relationships reached resolution before the interfunctional (organizational structure) issue was even possible to deal with, and that the interorganizational collaborations really blossomed after the personal and organizational issues were largely dealt with. This is supported by research that shows that boards that demonstrate the most extensive positive changes did so in the context of facilitated retreats where members had a chance to reflect together on their underlying values and patterns in their work as a group, thus dealing with interpersonal and intergroup issues as a foundation for change (Holland et al., 1993).

Proposition 5a. Relationships create structure.

Relational healing sees structural/functional issues, in their essence, as relational issues. As relationships among organizational members and their roles shift, a sensible organizational structure can be defined that is a natural outgrowth of the relations rather than an imposition on the organization. The organization in transition does not benefit from mechanistic restructurings; rather, changes in form result from changes in relationship. This is supported by research that shows that the effectiveness of boards and their organizations' performance are positively correlated with processes (like strategic planning) that have the Board engaged with each other, and hardly impacted by structural variables (Bradshaw, Murray, & Wolpin, 1992). The relational healing practitioner is like a midwife to these new structures, patiently awaiting and facilitating their birth more so than engineering their form and timing.

DISCUSSION AND CONCLUSION: INTELLECTUAL ROOTS AND FUTURE BRANCHES

The concept of relational healing has a number of intellectual roots. Appropriate to the theme of "wholing," these roots extend in several directions, pulling in sustenance from a variety of literatures, including "new science," psychological, spiritual and philosophical writings as well as organizational development and behavior.

To begin, it may be helpful to both honor the organizational development (OD) roots of relational healing and also to distinguish the branch we believe it represents from more conventional OD approaches, using analogies from the ongoing revolution toward new science (with its roots in quantum physics). Clearly, relational healing is informed by and has roots in a traditional team-building approach (Katzenbach & Smith, 1993). The primary ontological unit of analysis in team building is the team, a reified construct intended to encompass the wholeness value of the team, its members, and relational dynamics. Efforts at team building aim to strengthen the overall fluidity, performance, role clarity, and conflict-resolution dimensions of the team. Primacy is given to the team as a living, knowable *thing*. This construction of team is markedly Newtonian in nature because of its attention to discrete parts and laws of human behavior governing the proper functioning (or dysfunctioning) of the team (see Wheatley, 1992; Youngblood, 1997 for the distinction between classical, Newtonian and "new" quantum science/new science in organizations).

While a Newtonian analysis and team-building approach are certainly useful, by its grounding in new science, relational healing may offer an even more powerful concept and method to understand and transform. Relational healing animates and develops the objectified *team as thing into team as living process*; in that sense, relational healing "avoids making processes into things" (Reason, 1988, p. 200). Instead of focusing on the gross-level functioning of the team (material perspective) as an obdurate *thing* and set of mechanically interacting parts, the quantum focus of relational healing looks at the space between the various facets of the team, in this case the *JAZZ* Board. Because of its quantum roots, relational healing recognizes the fact that *things* are mostly comprised of empty space (Capra, 1975, 1982). Moreover, the observable aspects of the *thing* are very tenuous indeed, dependent on the participating observer (see Heisenberg's Uncertainty Principle in Talbot, 1981, p. 15; Wheatley, 1992, p. 35), and interdependent because of a lively interchange of energy. At this subtle, quantum level of analysis, relationships between, for example our *JAZZ* board members, are merely "probabilities to exist" (Wheatley, 1992), rather than definite, ossified *things*.

The practical importance of this paradigmatic distinction is that it offers the practitioner and client system greater flexibility to change, self-organize, cocreate, and transform almost instantaneously. For instance, had we intervened into the *JAZZ* project with a team-building approach, we would have privileged the extant team structure as immutable. The catalytic renergizing of the board team, ironically through the loss of one of its core members, is a demonstrable data point for how, when functioning from open, unbiased "spaces between," radical possibilities for evolution can emerge. The evolution of the team composition emerged as a possibility, precisely because it was neither assumed that the team must stay intact nor that anyone should be "removed."[12] Under these circumstances, people (and their organizational systems) experience the freedom of choice that was previously blocked by the construct of dissociated parts at war. Letting go of reified, culturally bound expectations of reality begins to encompass all possibilities, even those that, at first and superficially, appear to be destructive. The loss and replacement of board members, like death and birth, are both natural parts of life and organizing.

Relational healing also makes a departure from "second-order" types of change which "represent discontinuous shifts in frameworks" (Bartunek & Reis Louis, 1988, p. 100). Relational healing offers us the opportunity to manifest "third-order" changes where the frameworks themselves are dispensed with and the system is led to "a holistic (unitary) worldview in which all subject/object dichotomies have disappeared" (McWhinney, 1992, p. 60; see also Lefkoe, 1997, ch. 10 for more on "third-order" change).

> The meta-reality of third-order work is not a comfortable place to work in, being without the intellectual protection afforded by classification systems and expectations, but experience of it is necessary for the facilitators of change and, in the cases of deep renaissance, for all those who would be a party to the founding of a new mission or organization. (McWhinney, 1992, p. 236)

After the unconventional rearrangement of the Board, a "deep renaissance" toward more functionality and harmony ensued. Freed from the stereotype roles of a more Newtonian approach, relationships can emerge and recreate themselves spontaneously, in a self-organizing dance unfettered by structural dictates or planned interventions. In this sense, relationship is the processual unit of analysis of relational healing. As the microcosmic part reflects the macroscopic whole, each "pair" of members on the Board holds the promise of integrity for the whole organization:

> The pairing assumption [of Bion (1961)] leads the group to focus on two of its members—a couple...to symbolize the group's hopeful expectation that the selected pair will "reproduce" itself. (Kernberg, 1980, pp. 213-214)

By holographically focusing on the relational pairs and polarities, relational healing stimulates the entire organizational system to "reproduce" healed, high functioning at the organizational level.

The power of transformation resides in the appreciative, nurturing, and healing energy injected or rediscovered in the space between people. Ultimately, it is the living relational bonds, and not the externalized, intellectual team conception, that determines the health (wholeness) of the team—spaces between engender unlimited possibilities for creative metamorphosis.

In the spiritual literature, seeing the oneness inherent in polarities is a worldview dear to many eastern faiths—from Zen Buddhism to Taoism to Hinduism:

> The notion that all opposites are polar—that light and dark, winning and losing, good and evil, are merely different aspects of the same phenomenon—is one of the basic principles of the Eastern way of life. Since all opposites are interdependent, their conflict can never result in the total victory of one side, but will always be a manifestation of the interplay between the two sides. (Capra, 1991, p. 146)

Drawing upon psychology as well as the wisdom traditions, Charles Johnston (1991) presents a powerful case for the need to embrace polarities in our cultural thinking and action. Arguing that we live in a time when such bridging of polarities is essential to our future, he demonstrates both the ubiquity of polar dynamics and the thesis that creative integration is the second half of the creative process of differentiation that created the polarities to begin with.

In its recognition of the connective tissue between polarities and the embracing of the whole they encompass, relational healing is rooted in this eastern spiritual/philosophical tradition cum emerging Western zeitgeist.

Turning next to Western philosophical traditions, the most obvious inspiration for relational healing comes from Hegel's dialectic. In his affirmation that the true is the whole (Gray, 1970), Hegel is in alignment with eastern philosophy and the process of relational healing described here. But the reintegration of polarities in relational healing has a slightly different emphasis than does the synthesis of Hegel.

Where Hegel speaks of thesis and antithesis canceling out or annulling the past, giving rise to a future that preserves essentials in a higher synthesis (Gray, 1970; Miller, 1977), relational healing takes an appreciative approach, valuing both poles and the relationship among them in order to create the space where reintegration (healing/wholing) can occur. Relational healing holds open the space between, and invites everyone into it; where relational healing puts the emphasis on engagement and mutual appreciation, Hegel characterizes the process of achieving synthesis as more oppositional/negative. The "releasing" phase of relational healing is the place where outgrown elements of the past are shed, not so much out of combat, as following the natural course of a life and death cycle brought to its organic conclusion. Making the analogy with approaches to mediation, the practitioner of relational healing takes a position not so much of being neutral (and certainly not of taking one side over the other) but of taking all sides. One is reminded of Gandhi's famous exhortations: "I am a Hindu, I am a Muslim, I am a Jew,...."

Regarding the relational healing concept of healing as becoming more whole (vs. fixing a disease), modern movements of psychological therapy have increasingly taken this point of view and developed processes in alignment with it. Starting with Rogers (1961), the process of honoring all that is present in the other (with unconditional positive regard) is seen as a simple and powerful approach to healing. Similarly, Gestalt therapy (especially as practiced by the Cleveland school) regards the goal of therapy as the enlargement of awareness and the process as one of supporting engagement throughout the contact cycle (Nevis, 1992; Wheeler, 1991). It is especially interesting that Wheeler (1991) has reframed what Perls and others referred to as resistances into "contact styles," casting these heretofore negative poles into a new light of wholeness. More recently, Josselson (1992) has affirmed the value of relationship as a therapeutic goal (feeding the interconnections) as opposed to the historic emphasis on "self-development." She advocates a therapy rich in tending (care). Similarly, Moore (1992), in his popular *Care of the Soul* has made care (vs. cure) a widespread reinterpretation of therapy's role in life. Finally, Heard (1993) has affirmed "the between" as the focus of his dialogical form of therapy. From all of these exemplars of a trend in psychology and therapy, our theory of relational healing draws sustenance.

Finally, returning to our own field of organizational behavior, two works stand out as pivotal in informing the spirit and process of relational healing. One is Smith and Berg's (1990) *Paradoxes of Group Life*, whose appreciation of the role of paradox in moving groups forward is instructive. The other is Bohm's (1996) *On Dialogue*. In addition to being informed by the spirit and process of dialogue as Bohm explains it, relational healing benefits from his understanding of fragmentation:

> Fragmentation ... originates in thought—it is thought which divides everything up. In actuality, the whole world is shades merging into one. But we select certain things and separate them from others—for convenience, at first. Later we give this separation great importance. (p. 9)

So it is that in relational healing, through a process that draws upon the dialogical perspective, the parts that have been separated by the mind (with one pole often being disowned) are brought back into a reintegration that honors the whole before it was split.

For the future, our hope is that this explication of relational healing can benefit both the realm of OD practitioners dealing with organizational life-cycle transitions and contribute to their theoretical understanding of the process by which organizations, especially nonprofits, traverse the transformational space between life stages. Egri and Frost (1991) express how ultimately the enlightened OD practitioner is knowledgeable about the external dynamics of life-cycle transitions, but just as aware, internally, of the spiritual dimension of shamanic organizational transformation processes such as relational healing:

> Instrumental to this process is the ability of the OD consultant to engage in systematic self-reflection in order to become knowledgeable about the dying-death-recreation cycle. Perhaps most importantly though, shamanic development require individual and personal introspection in the spiritual sense as to one's role in guiding organizational transformations. (p. 207)

Relational healing cultures the OD practitioner's "spiritual sense" necessary to facilitate effective life-cycle transitions. We hope this chapter aids all those in the often painful process of transitions to recognize (and re-cognize) the benefits of the polarized tensions they are experiencing, and to use them as the source of finding the relational bridges that lead to dynamic reintegration. Given that this model was developed in large part based on a single case, we look forward to future research that attempts to apply and test this model in other settings.

NOTES

1. Our choice to use first person descriptions is consistent with this relational construction of research.

2. Inquiry that is "relational" puts primary emphasis on relationships with and among the people of the organization, acknowledging its centrality to every aspect of our work. As a mode of inquiry, it is inspired by people like Moustakas (1990) and his heuristic inquiries, Gergen (1994) and his social constructionist inquiries, Reason and Rowan (1981) and the participatory approaches to human inquiry that they champion, and feminist inquirers such as Alcoff and Potter (1993), Ianello (1992), and Mauws (1995).

3. The hyphenated form, "re-cognized," is here used to emphasize the roots of the word "recognized," making clear that re-cognition is a process of coming to know again, reconstructing awareness in such a way that a different cognition, a different knowing results.

4. See Kaczmarski and Khalsa (1998) for a description of case stories, building upon the case study method (Stake, 1994; Yin, 1994) and the recent use of storytelling in organizational studies (Boje, 1991, 1995; Phillips, 1995).

5. William Bridges (1986) has a nice model of managing organizational transitions, but it is not connected to life-cycle change periods.

6. One exception is Greiner (1972), who refers to the period precipitating a life-cycle change as a "stage of revolution"; beyond describing its characteristics, however, his model does not give much

in-depth explanation of what happens during such a stage to successfully bridge two "stages of evolution."

7. The names of all *JAZZ* members referred to in this paper have been changed, as has the name of the organization, in order to preserve anonymity.

8. Various writers have addressed the positive qualities of "dynamic tension" in nonprofit boards, struggling with questions of role conflict, and so forth (Conrad & Glenn, 1976; Leduc & Block, 1985).

9. As with "re-cognized" used earlier, we hyphenate "re-collection" to suggest a process of collecting again, of regathering the disparate elements of the whole into a container, initially a container even smaller than the organization.

10. For a case study of what can happen when restructurings are pursued first, before organic changes in relationship and mission, see Khalsa (1991).

11. Gurudev is reminded of his wife's experience with a cervical polyp that conventional medicine would have surgically removed. Cut off, such polyps almost always return; the alternative treatment of using tea tree oil to naturally shrink it allows it to be removed without a knife, and then it almost never comes back.

12. As the cases of Alan and Bob Davis reflect, participants quite naturally cycle in and out of organizations. As was experienced in *JAZZ*, these departures were actually evolutionary events, catalyzing substantive positive changes to the organization—understanding the "inevitable departure" from organizations helps members become "more receptive to transformational changes" (Dyck, 1995, p. 176).

REFERENCES

Adizes, I. (1988). *Corporate lifecycles: How and why corporations grow and die and what to do about it*. Englewood Cliffs, NJ: Prentice Hall.

Alcoff, L., & Potter, E. (Ed.). (1993). *Feminist epistemologies*. New York: Routledge.

Bailey, D., & Grochau, K.E. (1993). Aligning leadership needs to the organizational stage of development: Applying management theory to nonprofit organizations. *Administration in Social Work, 17*, 23–45.

Banks, L.R. (1981). *The Indian in the cupboard*. Garden City, NY: Doubleday.

Bartunek, J.M., & Reis Louis, M. (1988). The interplay of organization development and organization transformation. In R.W. Woodman & W.A. Pasmore (Eds.), *Research in organizational change and development* (pp. 97–134). Greenwich, CT: JAI Press.

Bion, W.R. (1961). *Experiences in groups*. London: Tavistock.

Bohm, D., & Nichol, L. (1996). *On dialogue*. London: Routledge.

Boje, D.M. (1991). The storytelling organization: A study of story performance in an office supply firm. *Administrative Science Quarterly, 36*, 106–126.

Boje, D.M. (1995). Stories of the storytelling organization: A postmodern analysis of Disney as "Tamara-Land". *Academy of Management Journal 38*(4), 997–1035.

Bradshaw, P., Murray, V., & Wolpin, J. (1992). Do nonprofit boards make a difference? An exploration of the relationships among board structure, process, and effectiveness. *Nonprofit and Voluntary Sector Quarterly, 21*(3), 227–249.

Bridges, W. (1986). Managing organizational transitions. *Organizational Dynamics, 15*(1), 24–33.

Bridges, W. (1991). *Managing transitions: Making the most of change*. Reading, MA: Addison-Wesley.

Capra, F. (1975). *The Tao of physics*. Berkeley, CA: Shambhala.

Capra, F. (1982). *The turning point: Science, society, and the rising culture*. New York: Simon and Schuster.

Capra, F. (1991). *The Tao of physics* (3rd ed.). Boston, MA: Shambhala.

Conrad, W.R., & Glenn, W.R. (1976). *The effective voluntary board of directors*. Chicago: Swallow Press.

Cooperrider, D.L. (1990). Positive image, positive action: The affirmative basis of organizing. In S. Srivastva, D.L. Cooperrider, & Associates (Eds.), *Appreciative management and leadership: The power of positive thought and action in organizations* (pp. 91–125). San Francisco, CA: Jossey-Bass.

Cooperrider, D.L. (1996). The child as agent of inquiry. *OD Practitioner, 28*(1 & 2), 5–11.

Cooperrider, D.L., & Pratt, C.S. (1995). *Appreciative inquiry: Relational realities and constructionist approaches to organization development.* Workshop presented for the Organization Development Network 1995 National Conference, Case Western Reserve University, Cleveland.

Cooperrider, D.L., & Srivastva, S. (1987). Appreciative inquiry in organizational life. In R.W. Woodman & W.A. Pasmore (Eds.), *Research in organizational development and change* (pp. 129–169). Greenwich, CT: JAI Press.

Dyck, B. (1995). Transformational change and organizational half lives. In R.W. Woodman & W.A. Pasmore (Eds.), *Research in organizational change and development* (pp. 145–180). Greenwich, CT: JAI Press.

Egri, C., & Frost, P. (1991). Shamanism and change: Bringing back the magic in organizational transformation. In R.W. Woodman & W.A. Pasmore (Eds.), *Research in organizational change and development* (pp. 175–221). Greenwich, CT: JAI Press.

Fletcher, K.B. (1992). Effective boards: How executive directors define and develop them. *Nonprofit Management & Leadership, 2*(3), 283–293.

Gergen, K.J. (1985). The social constructionist movement in modern psychology. *American Psychologist, 40*, 266–275.

Gergen, K.J. (1994). *Realities and relationships: Soundings in social construction.* Cambridge, MA: Harvard University Press.

Gergen, K.J., & Gergen, M.M. (1991). Toward reflexive methodologies. In F. Steier (Eds.), *Research and reflexivity* (pp. 76–95). Newbury Park, CA: Sage.

Gray, J.G. (1970). Introduction: Hegel's understanding of absolute spirit. In J.G. Gray (Ed.), *G.W.F. Hegel: On art, religion, philosophy: Introductory lectures to the realm of absolute spirit* (pp. 1–21). New York: Harper & Row.

Greiner, L.E. (1972). Evolution and revolution as organizations grow. *Harvard Business Review, 50*(4), 37–46.

Hasenfeld, Y., & Schmid, H. (1989). The life cycle of human service organizations: An administrative perspective. *Administration in Social Work, 13*, 243–269.

Heard, W.G. (1993). *The healing between: A clinical guide to dialogical psychotherapy.* San Francisco, CA: Jossey-Bass.

Holland, T.P., Leslie, D., & Holzhalb, C. (1993). Culture and change in nonprofit boards. *Nonprofit Management and Leadership, 4*(2), 141–155.

Hunt, D.E. (1987). *Beginning with ourselves: In practice, theory, and human affairs.* Cambridge, MA: Brookline.

Ianello, K.P. (1992). *Decisions without hierarchy: Feminist interventions in organizational theory and practice.* New York: Routledge.

Johnston, C.M. (1991). *Necessary wisdom: Meeting the challenge of a new cultural maturity.* Seattle: ICD Press.

Josselson, R. (1992). *The space between us: Exploring the dimensions of human relationships.* San Francisco, CA: Jossey-Bass.

Kaczmarski, K.M., & Khalsa, G.S. (1998). Spiritual leadership beyond the bounds of religion: A case story of Bishop William E. Swing. *Journal of Management Systems, 10*(4).

Katzenbach, J., & Smith, D. (1993). *The wisdom of teams: Creating the high-performance organization.* New York: HarperBusiness.

Kelly, G.A. (1955). *The psychology of personal constructs. Vol. I: A theory of personality.* New York: W.W. Norton.

Kernberg, O. (1980). *Internal world and external reality.* New York: Jason Aronson.

Khalsa, G.S. (1991). *The rise and fall of Pacific Northwest Consulting*. Unpublished manuscript, Case Western Reserve University, Cleveland, Ohio.

Leduc, R.F., & Block, S.R. (1985). Conjoint directorship: Clarifying management roles between the board of directors and the executive director. *Journal of Voluntary Action Research, 14*(4), 67–76.

Lefkoe, M. (1997). *Re-create your life: Transforming yourself and your world with the decision maker process*. Kansas City, MO: Andrews and McMeel.

Mauws, M. (1995). *Relationality*. Paper presented at Administrative Sciences Association of Canada, Organization Theory Division, Windsor, Ontario.

McLean, A.J., Sims, D.B.P., Mangham, I.L., & Tuffield, D. (1982). *Organization development in transition: Evidence of an evolving profession*. New York: Wiley.

McWhinney, W. (1992). *Paths of change: Strategic choices for organizations and society*. Newbury Park, CA: Sage.

Messal, J.L. (1980). Organizational growth and change: The life cycle of a community mental health center. *Administration in Mental Health, 8*(1), 12–22.

Miller, A.V. (Trans.). (1977). *Hegel's phenomenology of spirit*. Oxford: Oxford University Press.

Moore, T. (1992). *Care of the soul: A guide for cultivating depth and sacredness in everyday life*. New York: HarperCollins.

Morgan, G. (1986). *Images of organization*. Newbury Park, CA: Sage.

Moustakas, C. (1990). *Heuristic research: Design, methodology, and applications*. Newbury Park, CA: Sage.

Nevis, E.C. (Ed.). (1992). *Gestalt therapy: Perspectives and applications*. New York: Gardner.

Phillips, N. (1995). Telling organizational tales: On the role of narrative fiction in the study of organizations. *Organization Studies, 16*, 625–649.

Quinn, R.E., & Cameron, K. (1983). Organizational life cycles and shifting criteria of effectiveness: Some preliminary evidence. *Management Science, 29*, 33–51.

Reason, P. (1988). Experience, action, and metaphor as dimensions of post-positivist inquiry. In R.W. Woodman and W.A. Pasmore (Eds.), *Research in organizational change and development* (pp. 195–233). Greenwich, CT: JAI Press.

Reason, P., & Rowan, J. (Ed.). (1981). *Human inquiry: A sourcebook of new paradigm research*. Chichester, UK: John Wiley.

Rogers, C.R. (1961). *On becoming a person*. Boston: Houghton Mifflin.

Senge, P.M. (1990). *The fifth discipline: The art and practice of the learning organization*. New York: Doubleday.

Smith, K.K., & Berg, D.N. (1990). *Paradoxes of group life: Understanding conflict, paralysis, and movement in group dynamics*. San Francisco, CA: Jossey-Bass.

Stake, R.E. (1994). Case studies. In N.K. Denzin & Y.S. Lincoln (Eds.), *Handbook of qualitative research* (pp. 236–247). Thousand Oaks, CA: Sage.

Susman, G., & Evered, R. (1978). An assessment of the scientific merits of action-research. *Administrative Science Quarterly, 23*, 582–603.

Talbot, M. (1981). *Mysticism and the new physics*. London: Arkana/Penguin Books Ltd.

Tenkasi, R.V., & Boland, R.J. (1993). Locating meaning making in organizational learning: The narrative basis of cognition. In R.W. Woodman & W.A. Pasmore (Eds.), *Research in organizational change and development* (pp. 77–103). Greenwich, CT: JAI Press.

Torbert, W.R. (1989). Leading organizational transformation. In R.W. Woodman & W.A. Pasmore (Eds.), *Research in organizational change and development* (pp. 83–116). Greenwich, CT: JAI Press.

Van Maanen, J. (1988). *Tales of the field: On writing ethnography*. Chicago: University of Chicago Press.

Weisbord, M. (1992). *Discovering common ground*. San Francisco, CA: Berrett-Koehler.

Welwood, J. (1992). The healing power of unconditional presence. In J. Welwood (Ed.), *Ordinary magic: Everyday life as spiritual path* (pp. 159–170). Boston, MA: Shambhala.

Wheatley, M. (1992). *Leadership and the new science*. San Francisco, CA: Berrett-Koehler.

Wheeler, G. (1991). *Gestalt reconsidered: A new approach to contact and resistance*. New York: Gardner.

Wood, M.M. (1992). Is governing board behavior cyclical? *Nonprofit Management & Leadership, 3*(2), 139–163.

Yin, R.K. (1994). *Case study research: Design and methods* (2nd ed.). Thousand Oaks, CA: Sage.

Youngblood, M.D. (1997). *Life at the edge of chaos: Creating the quantum organization*. Dallas, TX: Perceval Publishing.

THE PROFESSIONALIZATION OF ORGANIZATION DEVELOPMENT
A STATUS REPORT AND LOOK TO THE FUTURE

C. Ken Weidner, II and Orisha A. Kulick

ABSTRACT

The chapter asks whether organization development (OD) should be professionalized and submits for consideration several ideas relating to professionalism, professionalization, and organization development. First, we provide an overview of the issues related to professionalization and organization development. Next, we examine three meanings attributed to professionalism and the process of professionalization. We also explore the advantages of professionalism for individual practitioners, clients, organizations, and society, and the relationship between professions, the state, and capitalism. We use the medical profession as an illustration of each of the foregoing concepts. In the second portion of the chapter, we explore the changing focus of organization development practice toward strategic thinking and whole-system change, and discuss the current status of organization development in terms of its identity, professionalization, and professionalism. We examine potential action steps for the future of organization development, including clarification of its body of knowledge, and the use of professional liability insurance as a means to increase professionalism.

Both the emerging profession of organization development (OD) and the conditions inside and surrounding established professions such as medicine are changing simultaneously. Is this chapter, we explore questions surrounding the professionalization of organization development. First, we provide a brief overview of arguments supporting and opposing the professionalization of organization development. Next, we examine three meanings that have been attributed to professionalism. We also explore professionalization as the means by which an occupation may seek to become a profession, and we use the profession of medicine as an example of how interaction between the state, capitalism, and the profession plays a role in the loss of professional control. Finally, we examine the changing nature of organization development, the current state of organization development's progress in becoming a profession, and the professionalism of practitioners. We also identify possible choices that organization development faces regarding its future professional occupational status.

SHOULD ORGANIZATION DEVELOPMENT BE PROFESSIONALIZED?

One of the ongoing questions in organization development is whether organization development should be professionalized, that is, should it follow the formula that other professions have followed in the past to become a recognized licensed profession in the same way that medicine and law are recognized licensed professions? While we do not presume to be in a position to make decisions for OD as to what its membership should do in regard to professionalization, we do suggest discussion of the merits of arguments both for and against professionalization. This paper is intended to provide both information and a context to frame a discussion of professionalization, and to suggest potential future directions the field might take.

Arguments against professionalization are often based on concerns about retaining control, whether that control is over individual practice, practice standards, or entry into the profession. For example, another professional group, such as psychiatrists, might seek partnership with state regulators to exclude or limit organization development practitioners who are not physicians. In such a scenario, organization development practitioners might need psychiatric supervision in order to practice. This situation is less likely today than twenty years ago, given the current professional freedom that other mental health practitioners such as social workers enjoy in performing their clinical duties without psychiatric supervision. Nevertheless, an attempt by another profession to dominate organization development practice is still theoretically possible, and as such it is a factor in this debate.

The argument against professionalization might also include a caution against allowing state regulators to have authority over standards for fair entry into the field. Assurances that regulators would look to established organization development leaders for the content of examination questions might allay the fears expressed in this argument. On the other hand, this argument raises another question: How can state regulators accurately test for organization development core knowledge when organization development practitioners do not consistently agree among themselves as to what core knowledge is needed in the profession? It seems logical that state regulators would not be able to test for organization development knowledge unless organization development leaders agree on the content of the organization development body of knowledge.

Consideration of the "body of knowledge" approach to the development of OD leads to a third argument against the professionalization of organization development: that written testing for a body of knowledge cannot evaluate how well a practitioner might use oneself as a tool in the practitioner–client relationship. Therefore, written testing should not be used as the exclusive means to allow entry into the field of organization development because it is unable to test how well an individual is able to function as a tool with the client. On the other hand, other professions, such as social work, also deal with people, organizations, and change under difficult circumstances. The profession of social work has managed to effectively use written licensing exams in conjunction with the entire professional education experience to select candidates who are able to function as effective tools of change. Organization development might look to social work for guidance on how to select practitioners who can both use who they are personally with clients and pass a written licensing exam. Indeed, some personal characteristics that are not tested in a written exam, such as the capacity for empathy, are very important for clinical service in social work.

Independent external organization development practitioners might argue against professionalization with a narrower control-based objection. They may claim that they personally do not want anyone looking over their shoulder while they work because they are currently doing well without consultation or supervision. Unlike internal practitioners, who are more likely to have supervisors within the structure of their employing organization, independent external practitioners may view professionalization as imposing a restriction on their independence. In contrast to this concern, professionals like the second author, who have experienced supervision and peer consultation, can report about the positive aspects of these activities. The second author actively sought out supervision and consultation because those activities were opportunities to increase knowledge, sharpen skills, and most importantly, to better serve a particular client. She looked forward to regularly sharing knowledge and experience with professional colleagues in a supportive environment. In other words, supervision and consultation were experienced as positive performance enhancers with no fear of restriction on client service. It was her choice to follow what she understood to be the ethical and professional course of action that best served her clients. Seeking external input only made better client service more likely and it was her professional goal to give her clients the best possible service.

Ironically, one argument supporting the professionalization of organization development also involves control, in the form of a professional's autonomy in their work. Professionals tend to enjoy their relatively autonomous work environment and respect that accompanies their status. More importantly, peer support and long-term economic rewards reinforce professionals putting their clients' interests above their own immediate financial interests. Thus, professionals are supported in combining the body of professional knowledge with "who they are" as a person in order to serve clients with their professional judgment.

The "professional autonomy" argument for professionalization can be challenged because sometimes professionals abuse their professional position and make decisions that are not in the best interests of their clients. No one can guarantee that professionals will always place their clients' interests above their own economic interest. Although professionals are encouraged to seek supervision and consultation, there is no guarantee that the availability of supervision and consultation can always prevent professional abuses. Successful malpractice lawsuits are examples of cases in which professionals have served clients poorly. If professionalization cannot prevent client abuses, why should organization development engage in the expensive, arduous process of seeking professional status in fifty states? Because client abuses occur both with and without professional status, why seek professional status? Why not simply encourage professional behavior by individuals and allow market forces to reward those who behave professionally?

A second argument for professionalization notes a more limited but positive aspect to preventing professional abuses. The argument is that while professionalization cannot completely prevent ethical abuses, it can greatly extend the arm of ethical enforcement because of the state's role in licensure. For example, if a licensed professional has his license revoked by the state after a fair hearing, he cannot seek employment in that professional role anywhere in that state. Instead of removal from the profession, the errant professional may be required to experience corrective education or other measures to improve his future practice. In contrast, if an organization development practitioner abuses his relationship with a client and harms that client, he can escape from the scene without censure or corrective action merely by severing his relationship with that client or by seeking new employment in a firm down the street.

A third argument for the professionalization of organization development relates to quality control within the field. If all schools that taught organization development knew that their students would be sitting for a statewide exam for licensure, these schools would become keenly aware that it was in their institutional best interest to teach certain subject matter because the profession required it. Without such a requirement, it is possible that organization development programs may be tempted to bow to economic pressure and please prospective students with courses that students enjoyed in order to fill classroom seats as opposed to offering courses that students needed to best serve clients. An organization development licensure exam would thus help encourage quality education in all organization development academic or training programs and set a minimum standard of knowledge below which no applicant would be allowed into the profession.

In the following sections, we examine the companion concepts of professionalism and professionalization more closely. It is our hope that a fuller understanding of both concepts will help the organization development community define how to best move the field forward.

PROFESSIONALISM AND PROFESSIONALIZATION

What is Professionalism?

Professionalism is a concept that has multiple meanings. Freidson (1994), a leading scholar in the study of professions, has summed up the confusing circumstances around the issue of definition. He states that when one examines work on the professions closely (p. 107), "one finds sufficient confusion and contradiction in the use of the word 'profession,' and related words such as 'professionalism' and 'professionalization'." Freidson believes that people may be talking to each other in general and talking past each other in particular as a result of this confusion. Therefore, we have selected three meanings that have been ascribed to professionalism which we believe can inform the debate about a desired course of action for organization development in the future.

The first meaning of professionalism that we consider focuses on the position of the professions as one part of a four-part occupational system. Each state in the United States awards professional designation to only one of four occupational categories (free market, technical, scientific, professional) because of the complexity of knowledge and perceived criticality for societal welfare of the work done by these professional occupations (Torres, 1991). Research has shown that occupations acquire rather than inherently possess the attributes associated with the occupational category of "profession." Given this view of professionalism, it is possible for an occupation to follow a set of key criteria in attempts to achieve state-sanctioned professional status (Wilensky, 1964).

The second meaning of professionalism that we consider refers to an individual's attitude of caring combined with a pursuit of excellence (Maister, 1997). Under this definition, a professional need not be a member of an established profession (e.g., medicine, law), but may be found in any occupation (e.g., a secretary). Trust and respect are accorded to individuals earning respect and trust in the service of their client(s), rather than given to an occupational category or group.

In the third meaning of professionalism that we consider, professionalism is viewed as a changing historic concept that describes one of three methods in which the performance of work is organized. The first of these, professionalism, is different from both of the other methods—free market and bureaucracy—in that it revolves around "the central principle that the members of a specialized occupation control their own work" (Freidson, 1994, p. 173). At the foundation of professionalism is "specialized knowledge and skill thought to be of value to human life" (Freidson, 1994, p. 167). In this definition of professionalism, control means that the members of the occupation determine the content of the work they do. Absolute control means controlling the goals, terms, and conditions of work as well as the criteria by which it is evaluated. In contrast, in a free market, consumer demand and the free competition of workers for consumer choice determine what work will be done, who will do it, how they will do it, and how much pay they will receive. In a bureaucracy, the market for labor and its products is institutionalized by rational-legal methods, thereby giving executives of organizations the decision-making control over what product will be made or service offered, who shall make it, by what methods, and how it shall be offered to consumers (Freidson, 1994).

It might be helpful for the reader if we clarified the applications of the term "free market" as both Freidson (1994) and Torres (1991) use it. The free market way of organizing work, which is one part of the three-part paradigm that Freidson outlines (free market, bureaucracy, and professionalism), and the "free market" that Torres uses in his four-part description of occupational control (free market, technical, scientific, and professional) are both similar and different. They are similar in that they both refer to the process of commerce that people in capitalist societies experience in routine life in which consumers buy goods and services from suppliers. This process of free commerce provides the contextual backdrop for both Torres' and Freidson's models.

Torres' (1991) and Freidson's (1994) usage of the term "free market" is different in how each of them examine the issue of control. Torres uses the word control in its relationship to criticality of knowledge, complexity of knowledge, and licensure. He uses the term as one occupational category of a four-part occupational control theory in which there is no quality control by the state in the free market occupation category, in contrast to the professional occupation category in which licensure provides state quality control. Freidson's usage of the term "free market" focuses on the differing amounts of control that workers have available to them in the performance of their jobs within the free market, within bureaucracy (organizations), and within the professions. Under Freidson's paradigm, professional workers have the greatest amount of individual control over their work because they can choose to resist consumer demand and competition. In contrast, consumer demand and competition have the most control in the free market.

Professionalism as a Form of Occupational Control

Torres presents the position that professionalism is a form of occupational control. He states (1991, p. 49) "the formal border between professional and nonprofessional occupations is quite clear—state legislative acts granting occupa-

Professionalization of Organization Development

tions professional status." Under this definition, the state, with its licensing authority, has the ultimate power over professionalism, separating the professional occupation category from the three other forms of occupation, that is, the scientific, technical and free market occupational systems. It is also by means of licensure that the state transfers regulatory and policing powers to the occupation itself thereby giving that occupation control over entry and practice within the occupation. Such autonomy is allowed by the state because professionals are deemed to have mastery over complex knowledge that is beyond the understanding of the laity. Therefore, professionals are the only ones who can knowledgeably exert control over their practice. Thus, while free market businesses operate on the basis of caveat emptor ("let the buyer beware") professional occupations must of necessity operate on the basis of trust or credat emitter ("let the buyer believe in us").

Torres' (1991) paradigm of professionalism allows for movement into and out of the professional control system. For example, while horseshoers are currently members of a free market occupation, at one time they were licensed professionals in the state of Illinois. Licensure granting horseshoers' professional status was repealed after the importance of horses in society declined. A brief description of each part of the four-part occupational control system follows.

The free market occupational system: "Buyer beware." The free market form of the occupational system is illustrated by occupations such as janitors and auto mechanics. They have the lowest level of both perceived criticality of knowledge and complexity of knowledge.

Technical occupational system: "Higher criticality, lower complexity."
Information technologists could be considered as one occupation or they could be considered as several occupations (e.g., computer programmers, software designers, network administrators). Police officers provide another illustration of technical occupations. These occupations have higher perceived criticality of knowledge than janitors in the free market and less complexity of knowledge than professionals like physicians and lawyers.

Scientific occupational system: "Respect without being highly valued."
Torres (1991) uses sociologists as an example of a scientific occupation. While sociologists have a high degree of complexity of knowledge (equal to that of professionals), sociologists' perceived criticality of knowledge is relatively low. It is equal to the criticality of knowledge found in the free market.

Professional occupational system: "Believe in us—with licensure." T h e professional system of occupations clearly includes the well-established professions of medicine and law. These groups have both the highest level of perceived criticality of knowledge for societal welfare (as do the technical occupations) and the highest level of complexity of knowledge (as do the scientific occupations).

Torres (1991) suggests that there are two necessary environmental conditions for the creation and survival of professionalism. These conditions are (1) the need for specialized knowledge, the application of which is deemed to be necessary for the solution of complex problems; and (2) perceived criticality of some tasks or services to societal welfare. There is no objective standard by which we can assess either of these environmental conditions. On the other hand, a major hurdle for occupations seeking professional status from the state(s) in the past has been to convince the state that these two foregoing environmental conditions exist.

The gate-keeping function that Torres and other sociologists place upon licensure as a means of entry into professional status is not shared by all sociologists. According to Krause (1996, p. 21), "doctors, attorneys, engineers, and academics...constitute between half and three-quarters of those people that most sociologists call professional." While it is more conceptually clear that academics are not a licensed group (except for faculty in professional schools), the professional requirements for engineers are subject to some confusion because engineers are offered credential options such as state registration that allows them to use the term "professional engineer."

The existence of a registration relationship with the state or use of the term "professional engineer" may lead the reader to question whether engineers are a licensed group. In casual conversation, engineers may refer to themselves as being "licensed" when they are technically "registered" with the state. The term "certification" does not appear to be favored by the engineering community. Due to this technical legal distinction, engineers are not licensed as a result of registration or certification with the state, because neither registration nor certification reflect the same legal relationship as a license. Licensure is the strongest form of legal regulation because it not only verifies education, knowledge, and skill requirements, but it also uses the state's regulatory powers to ensure that the professional's behavior complies with standards. In addition to actively involving the state in enforcing legally binding professional standards, licensure also prevents people from outside the profession from using a professional title with consumers. Employers cannot confer a licensed professional title upon their employees— only the state can. State certification, another means of professional relationship with the state, furnishes certificates to people who document that they satisfy the qualifications for the certificate. States may sanction professionals who claim to be certified when they are not. Registration is the least powerful form of state regulation in which state registration provides a register of professionals and may discipline practitioners who claim to be registered when they are not (DuBois & Miley, 1999).

In anticipation of further questions about the engineering profession, we offer a brief general description of the American engineering profession. This comparison provides an opportunity for organization development practitioners to reflect upon the development of a profession that appears to be ambivalent about licensure as a means for structuring its legal relationship with the state. In their report, *Engineer-*

ing Licensure: Summary and Analysis, the National Society of Professional Engineers (NSPE, 1999) has expressed the opinion that the use of the term "registration" instead of "licensure" in many state engineering laws is a misnomer. This may have occurred because the distinction between these two frameworks was less pronounced when engineers began to fall under state regulation in the early 1900s. The NSPE uses the terms "registration" and "licensure" interchangeably.

According to Krause (1996, p. 62), "[e]ngineers continue to be a profession without community." Krause (1996, p. 60; Larson, 1977) continues, "It is a category of work but not, in most of its essentials, an occupation acting in its own interest." Krause (1996) included the engineering profession in his extensive study of professions for precisely this reason. In this respect, engineering "is similar to many other occupational groups that are basically employee categories rather than potent political entities" (p. 60). Records are kept on fifteen major American engineering subspecialties and there is practically no action as a group across work settings. The absence of any overall unitary shape to the profession prevents the development of any oppositional group consciousness (Krause, 1996).

Employers determine who is given the job title of engineer. Corporate sectors in manufacturing hire the majority of college engineering graduates. A 1985 study of a high-tech engineering firm and a traditional metalworking firm in the Northeast found that the percentage of engineers with no college degree was 25 percent at the former firm and 28 percent at the latter (Krause, 1996; Zussman, 1985). The percentage of employees who are high school and junior college graduates and who are moved into engineering roles after many years of experience approaches 25 percent in many companies. Krause further observes that there is no "engineering monopoly" because "neither the colleges of engineering, the corporations, nor most engineering specialty societies support a monopoly licensure system" (p. 62). Thus, engineers present a classic case of a bureaucratically situated profession. Success for engineers means promotion up the corporate ladder and education for them means education for both subordinate technical employment in and responsible management of corporate industry (Krause, 1996).

Regarding the state's role in engineering, Krause (1996, p. 65) observes that "[t]he state plays a minimal, consultative role in the American system, and there primarily in education, except for the state's involvement in the military-industrial complex." Krause (1996, p. 67) concludes, "engineering in the United States is a very poorly organized, middle-level employee group, with a series of scientific societies for each specialty."

The NSPE is working to strengthen the engineering licensing laws in all fifty states. Currently, the NSPE represents approximately 61,000 professional engineers out of the two million engineers in the United States. It is estimated that 15 to 20 percent of all engineers in the United States are either registered or licensed. Consulting engineers are most likely to be registered or licensed (A. Schwartz, general legal counsel of NSPE, personal communication, May 17, 1999). In addition, the NSPE report (1999) indicates that many state engineering registration and licensing laws exempt categories of individuals employed in industrial or manufacturing firms, corporations, public utilities, public transportation, government agencies, academic institutions, manufacturing, or scientific research. Thus, even the existence of a licensing law within a state does not mean that most engineers in that state will be required to be registered or licensed, because most engineers work in industry and as such are likely to be in an exempt category (A. Schwartz, personal communication, May 17, 1999). In addition, the NSPE report on licensing reveals that state engineering licensure laws also include exemptions that are connected to specific types of activities rather than to places of employment. These exemptions include practice of another licensed profession (including incidental practice), temporary practice, practice as an employee or subordinate, expert witness services, work on one's own property, and work on public or private works including buildings of specified costs, sizes, or types. Thus licensure (or registration), as it exists in the engineering profession, does not have the same acceptance and wide application as licensure in the more established professions of medicine or law which require all members to be licensed in order to perform any of their professional tasks.

Constant Struggle Between Professionalism and the Free Market Orientation to Occupations

Torres (1991) points out a struggle inherent in his four-part model of occupational systems. He notes (p. 53) "professionalism's social domain is continually being challenged and redefined. Especially evident is the struggle between free market and professional control which is at the core of many of the landmark court decisions." Torres refers to several court decisions in which prohibitions on advertising were successfully challenged by members of various professions (e.g., pharmacy, law).

In the 1980s, lawyers resisted efforts to allow corporate ownership of law firms on the grounds that the "bottom line" rather than the code of ethics would determine the course of action for legal services. A similar struggle occurred in the field of medicine during the same decade. Administrators in publicly owned health maintenance organizations (HMOs) began to monitor the costs of services provided by physicians, dismissing those who exceeded the national average for their specialties. This resulted in a screening process whereby primary care physicians determined the utilization of specialist physicians for patients covered by HMOs. We agree with Torres' (1991, p. 53) observation that, "[w]ith the imposition of standards from differing occupational systems, one emphasizing profitability and efficiency and the other emphasizing knowledge and safety, professionals may find themselves in a quandary as to which set of standards to follow."

Professionalism as Individual Responsibility vs. State-sanctioned Professionalism

Sociologists have observed that seven criteria are met by state-sanctioned (licensed) professions (Wilensky, 1964). These criteria are examined later in this chapter. We now ask: Is there an alternative way to bring professionalism to the workplace?

We believe the answer to this important question is affirmative. Individuals can choose to conduct themselves in a professional manner regardless of the occupation in which they work. For example, Maister (1997) offers his business manager (former secretary) as an example of someone who exhibits professionalism to a greater degree than many of the lawyers, consultants, accountants, engineers, and actuaries he meets. Maister asked his business manager for an explanation of why someone who was not sharing in the firm's profits would want to demonstrate a professional level of commitment. She responded (Maister, 1997, p. 17):

> Professionalism is not a label you give yourself—it's a description you hope others will apply to you. You do the best you can as a matter of self-respect. Having self-respect is the key to earning respect and trust from others.

Thus, an *individual* earns the kind of professionalism that Maister and his business manager describe. Maister observes that in the past, professionals were often given respect and trust automatically because of their position and that is no longer true. Now respect and trust must be deserved and earned.

At first glance, one might conclude that all that is needed for true professionalism to exist in our workplaces is for people to learn about the concepts in Maister's (1997) book; and then if they find the ideas desirable, to implement them. But Maister recommends realistic as well as principled action. Maister understands that good ideas need support in order to be implemented on a wide scale. Therefore, he suggests consideration of the following.

> It is hard to sustain the faith, and the commitment to excellence,...if one is the only person doing it. It is easier if you are part of a community of true believers, all committed to practicing the same values. In fact, one could argue that it is only by having a common purpose backed up with *rigidly enforced* shared values that a firm can define itself. Without such common purpose and values, a firm becomes nothing more than a convenience for practitioners wanting to share space, support services, and a name....A firm will have functioning values only to the extent that it has an effective *management* system that is intolerant about deviations from those values. Most firms have—implicitly or explicitly—a catechism or creed which says in effect that to be one of us, to be a member of this firm, there are certain ways of behaving to which you *must* subscribe, including and covering the usual categories of supervising people, dealing with clients, and exhibiting collaborative behavior. However, what firms also need (but mostly do not have) is a method to *enforce* these rules of membership in the "family." (emphasis in original, pp. 80–81)

Maister (1997) recommends that management in firms establish a method for identifying departures from excellence and schedule more management time devoted to closed-door counseling and coaching discussions with those straying from the firm's values. Maister (1997, p. 18) states: "Professionals must know that if they fail to strive for excellence, and settle instead for 'OK,' they can expect to hear the managing partner say, in private, something like 'Come on, old chap—bad show! Not the way we do things around here, don't you know?'"

The foregoing recommendation by Maister (1997) for management to help individual firm members sustain their faith and commitment to excellence (professionalism) sounds very collegial and similar to the socialization and policing functions that state-sanctioned professions offer their members. It appears that Maister recognizes the need for ongoing institutional guidance for individual professionals just as Freidson (1994, pp. 213–214) acknowledges that professionals need peer review:

> The method of regulating work that is most compatible with professionalism is collegiate rather than hierarchical, and is loosely denoted by the term, 'peer review.' As a method, peer review is interactive and, unlike bureaucratic methods, can employ qualitative judgments finely tuned to variable individual circumstances, problems and clients....To be true to the claims of professional virtue, peer review must be judgmental and demanding even while being supportive.

Freidson notes that professional peer review is a method that is facilitated by working in a large-scale organization (versus a solo practice) where all are familiar with the concrete and often idiosyncratic circumstances connected with their work. We would add that the proximity and the role of peers in a large-scale organization also make peer review possible. Solo practitioners can avoid peer review unless they receive unsolicited guidance and support after coming to the attention of their professional association because of a reported ethical violation. All professionals are also encouraged to voluntarily and regularly seek peer input to their practice. Problems arise when professionals avoid peer review.

The difference between the support and policing function provided by professional service firm management as recommended by Maister and that provided by the professions is that a firm employee who violates firm values and harms clients can quit that job and obtain employment with another firm without changing his behavior and without substantial detriment to his career. Such a person can then resume unprofessional behavior and continue to harm clients in the new firm. In contrast, a member of a profession cannot as easily escape censure by his professional peers. If a professional has harmed a client and that client complains to the professional association of which the professional is a member, or if that client complains to the state, the state can, after appropriate investigation, either order rehabilitative measures for the professional or it can remove his license to practice anywhere in the state (e.g., losing a medical license, being disbarred). Thus, the professions, in partnership with the states, have a broader reach of influence than individual firms in monitoring the professional behavior of their members.

Professionalism as a Changing Historic Concept Describing the Organization of Work

Although scholarship among sociologists in the emerging field that studies professions is in its formative stages, there are three common elements to the various descriptions of the professions: expertise, credentialism, and autonomy (Freidson, 1994). Expertise exists because professionals are full-time specialists who are committed to their work as a source of income. Their specialized knowledge and skill is called expertise. There is not enough time in a single person's life to learn, perform, or otherwise be engaged with each specialty often enough to sustain competence in order to be expert in the same way that a professional is expert.

Credentialing, the second generally accepted element of the professions, relates to the formal training thought to be prerequisite for competent performance or for providing the background necessary for training on the job. Expertise and credentialing are inseparable in a complex society such as ours because there is too much to know to be able to know everything directly. We have no alternative but to rely on indicators of expertise such as credentials due to the volume of knowledge and skill and the limitation of time (Freidson, 1994).

The third identifying common element of the professions, autonomy, is not as easily explained as the first two (expertise and credentialing). Professional autonomy has been the source of much hostility toward the professions based on the occasional abuse of autonomy by professionals in which economic self-interest appeared to outweigh client service. Freidson (1994) and Krause (1996) agree that professionals need autonomy in the functioning of their professional tasks so that people can receive better service. Freidson (1994) argues that although the totality of professional autonomy may be altered in the future (e.g., professionals need not determine their own rewards), the discretionary nature of technical professional functioning remains necessary to preserve the quality of our lives. He also observes that the connection of professional work with income is a constant stimulus to self-interested exploitation of the sheltered autonomy that professions provide. Yet, Freidson (1994, p. 9) concludes that while "we may hardly consider professionalism to be the optimal solution to the problem of organizing work; it is merely the better of the three alternatives [of free market, bureaucracy, and professionalism]."

Change in the Extent of Professional Autonomy

In the early twentieth century, individual practitioners within the established professions had greater autonomy over their work. Freidson (1994, p. 9) argues that today "[p]rofessionalism is being reborn in a hierarchical form in which everyday practitioners become subject to the control of professional elites who continue to exercise the considerable technical, administrative, and cultural authority that professions have had in the past." In addition, he suggests that "[o]utsiders, lay and professional" oversee the peer review process so that "the internal peer review actually does go on within the profession, that it is practiced honestly and effectively, and that more than a narrowly collegial point of view is taken into account" (p. 166).

Accepting the notion that professionals need to exercise professional discretion and judgment in the performance of their services means that they need autonomy. The extent of that professional autonomy is the fundamental question. Absolute autonomy means that each professional goes about their work without having to answer to anyone about their professional decision making. Historically, the professions have laid claim to providing peer review as a means of preventing abuses of absolute autonomy. Unfortunately, traditional collegial peer review has been unsuccessful in preventing abuse of autonomy in which personal economic gain was placed before the client's best interest (Freidson, 1994). On the other hand, while hostility to or suspicion of professionals' motives continues to exist in the United States, this sentiment is diffuse and far weaker than hostility or suspicion directed against others. The existing evidence shows that the public is much more dubious about the motives of politicians, "big business," trade unions, and others (Mitroff & Pauchant, 1990; Morin, 1990). Neither is there any evidence that the relative prestige of professionals has declined (Freidson, 1994).

Professionals characteristically justify their autonomy by asserting that their work is sufficiently complex and uncertain so as to require the exercise of discretion and judgment from case to case. A frequent claim is that neither professional products nor professional services can be standardized. Freidson (1994) persuasively suggests that this claim is not true. He explains that any service or product can be standardized and mechanized. He uses the example of the shoemaker who can claim correctly enough that every foot is different and thus that every shoe must be custom-made by him as he exercises judgment and discretion in adapting to each individual's feet. Yet, history has shown us the success of developing standard sizes and standardizing the production of shoes. Therefore, the shoemaker's claim for discretionary need is refuted by a history of success with standardization.

Freidson continues his analysis of professional autonomy with a look at medicine, law, and other professional enterprises. He states (pp. 164–165):

> As is the case with shoemaking, so also is it with the pursuits of medicine, law, and other ostensibly professional enterprises; they too can be broken down into small units and reduced to standard problems and standardized services. Just as the consumer's feet are made to accept standard shoe sizes, so consumers' problems can be made to conform to standard solutions. Of course, it is also true that when we reduce the producer's discretion in dealing with individual consumers, we risk proletarianizing the producer and forcing the consumer into inappropriate categories. Such a consequence may not seem serious for shoes, transistor radios, or uncontested divorces, but it might seem serious to us in the case of other goods and services. When at least some of the consumer's needs are reduced to standard categories, thus reducing the consumer to a standard object, this may seem oppressive and disabling. I think it can be argued that the

producers of some goods and services should be able to exercise discretion and judgment not only for the sake of their own humanity, but also for the sake of the humanity of the consumer.

We can envision many circumstances in which clients hope that their service professional exercises professional autonomy with only the clients' best interests in mind. Desire for self-preservation would logically lead clients to avoid service professionals who are distracted by concerns of profit for an employing organization or cost savings for an insurance company. For example, a man who contracts a rare debilitating physical illness would want not only excellent immediate care, but also the best follow-up treatment available. Such a patient would want his physicians to exercise ethically sound discretion and judgment for the benefit of his health—both short-term and long-term—and not for the benefit of the hospital's profit or to satisfy an insurance company's desire to save costs. Such a patient would hope that the physician would, as Freidson suggests, exercise discretion and judgment not only for the sake of their own humanity, but also for the sake of the humanity of the patient.

Freidson (1994, p. 166) provides what we believe to be a possible version of professional autonomy for the future in the United States:

> Insofar as professional work remains discretionary and adaptive, its technical judgments are not easily or fairly evaluated by anyone other than a colleague. Thus, the characteristic mode of supervision and evaluation of professional work should remain collegial, involving peer review. But this does not mean that all other sources of evaluation and pressure are properly excluded. Only part of any kind of work is technical, based on truly esoteric knowledge and skill; only part of what enters into discretionary action is technical rather than social and moral....It is in supervising those social and moral areas of work where technical peer review is overly narrow that evaluation and control by those outside the profession becomes essential. Outsiders, lay and professional, are the ones who are appropriately engaged in assuring that internal peer review actually does go on within the profession, that it is practiced honestly and effectively, and that more than a narrowly collegial point of view is taken into account.

"Outsider influence" on peer review already occurs in a large number of organizations. For example, currently, laypersons often serve as members of Institutional Review Boards, groups that oversee research protocols in academic medical centers, universities, and for-profit research centers.

In the interest of respect for the quality of human life, we support the inclusion of clients in Freidson's mix of "lay outsiders" who assess the quality of professional peer reviews so that clients' interests are fully protected. Krause (1996) points out the bleak result if clients are excluded from providing input into the determination of what is expected from professional service. For the purposes of his discussion, Krause defines capitalism as a political-economic system with organized corporations in production and finance. States are defined as bodies that possess a monopoly over the means of force, as well as most of the means of sustaining the society through education and professional training. Elites are described as professionals who are in supervisory positions over other professionals and who (Krause,

1996, p. 281) "have increasingly bought the capitalist model itself and have imposed capitalist rationalization upon those lower in the professional pecking order." Other scholars similarly refer to "elites" (both administrative and full-time research elites) who focus on the aggregate performance of rank-and-file professionals (Freidson, 1994). In describing the late 1970s and 1980s, Krause (1996, p. 285) observes:

> As capitalists, the states, and the elites within the professions worked to rationalize the services to attain efficiency in the service of profit, the consumer once again became irrelevant. Consumer organizations still existed, and sometimes were formally consulted before actions were taken. Yet in the long run, the strength of the consumer—a potential fourth major force joining the triad of profession, state, and capitalism—has remained minimal. On the other hand, if we count the capitalist firm as a consumer of services—as the corporations often do—the situation looks markedly different. Even nonprofit health plans try to tailor their services to please large corporate buyers; the corporation's desire to have cheaper legal services has led to price-cutting. But the ordinary consumer is little regarded, for his weight is not enough to influence the new and larger professional service bodies. Practically anything that this consumer is likely to want is viewed as an unnecessary cost by the profit-conscious provider of service, be it extra time in the hospital, more individual service in the law firm or a car that provides safety and low-cost maintenance.

Krause (1996, pp. 3, 20) introduces the concept of "guild power over workplaces" to mean the social influence of the professions to exercise power and control over their association, the workplace, the market, and their relation to the state. The alternatives to guild power are capitalist power, state power, or some combination of the two. Krause argues that when one of the three holders of power of the workplace (capitalist, state, or guild) loses control over the workplace, that power (control) shifts to either one or both of the other two power holders.

What has begun to change for professions since the late 1960s is that professions and the work that professionals do has increasingly become the focus for actions by states, working with sectors of capitalism. There has been a shift from guild power to state and capitalistic control. These new elements have led to an expanded economic focus beyond the professionals' personal economic gain. Krause (1996, p. ix) observes:

> For professional work can be profitable if it is organized in capitalistic forms, forms that no longer place the person who needs the service as the first priority. This trend seems to be leading to a redefinition of what professions are, from something special to just another way to make a living.

Although Krause chooses not to definitively answer the question, he asks us to consider (p. x) "whether the consumer of professional services benefits by the loss of guild power by major professional groups or whether the consumer is falling out of an uncomfortable frying pan directly into the fire."

The Fiduciary Character of Professional Work

Freidson (1994) submits that the character of professional work suggests two basic elements of professionalism. These elements are (1) a commitment to practicing a body of knowledge and skills of special value, and (2) a commitment to maintaining a fiduciary relationship with clients. Knowledge and skills are obtained through a demanding period of training. According to Freidson, this course of training tends to create commitment to knowledge and skill so that the professional's work becomes a central life-interest that provides its own intrinsic rewards. Professionals develop intellectual interest in their work and they are concerned with extending and refining it. They do not merely exercise a complex skill, but they identify themselves with it.

According to Freidson (1994), the second element of professionalism, the professional's commitment to maintaining a fiduciary relationship with clients, comes about because of the special value in the professional's knowledge and the complexity that clients are not able to evaluate accurately. Therefore, clients of professionals must place more trust in them than they do in other occupations. Freidson (1994, p. 201) states that "[a] fiduciary relationship must exist between professionals and their clients. Professionals are expected to honor the trust that clients have no alternative but to place in them." Thus, professionalism calls for a client's needs to take precedence over the professional's need to make a living.

The Implicit Contract: Privilege in the Marketplace in Exchange for Competent and Ethical Service

Freidson (1994, p. 202) suggests that an implicit contract exists between a profession and both the state and the public: "Protect my members from the unfettered competition of a free market, and you can trust them to put your interest before their own. I will select them carefully and train and organize them to provide competent and ethical service." It is this implicit contract between the profession and society that both allows and requires us to trust the individual professionals we consult.

Consistent with the foregoing implicit contract, professions in the United States developed institutions designed to control the selection, training, and credentialing of their members. In return, these professions gained privileges providing marked advantage in the marketplace. As a result, thousands of individuals can be identified as members of the same profession and their interaction with hundreds of thousands of clients are very similar in form and content. These institutions also produce the circumstances that encourage and reinforce professionalism in individuals (Freidson, 1994).

Freidson (1991) explains the reasons why professionals are given a privileged legal position in the marketplace, a position that handicaps their competitors from outside the profession even if it does not always completely exclude them. The argument supporting professional privilege is generally based on three assurances that the profession makes to society. First, the profession's body of knowledge and skill deals with problems of great importance to the public good, whereas that of other occupations does not. Second, the profession's body of knowledge and skill is so specialized and complex that lay people cannot act as rational consumers capable of protecting their own interests in the marketplace. For the public's own protection, therefore, only the profession's members should be allowed to offer such services. Third, the profession is worthy of public trust because it is a profession rather than an ordinary occupation. At first glance, this might appear as a tautology or circular logic. However the argument can be made that because members of the profession are properly selected and trained, they may be trusted to put the good of their clients ahead of their own material interest. While a very few may abuse that trust, the profession conducts itself in such a way as to discover abuses quickly and discipline violators effectively. In addition to these three assurances, higher education and middle-class status give professions "general public esteem and trust without which legal support alone would be inadequate" (Freidson, 1991, p. 197).

Professional Cohesion: Both a Strength and a Weakness of Professionalism

Freidson (1994) describes the professional "maintenance project" which is the undertaking by a profession of the maintenance of sufficient cohesion of the whole profession in order to be able to undertake common action both to sustain its status and privilege and to advance its own "cultural" projects. This professional cohesion gives a profession the ability to adapt to the changing political and economic environment so as to be able to continue to control its own affairs. An overview of Freidson's review of elements that contribute to the maintenance of professional cohesion follows.

First, members of all professions, unlike most other occupations, have a distinct public identity that provides a foundation for solidarity and mutual sympathy. Second, when training is attached to a university, lengthened by requiring a college education as a minimum prerequisite, and standardized, all members of a profession share a highly similar socialization experience. The relatively long period of training required to enter the professions also encourages their members to commit themselves to a life-long career; this training creates a "sunk cost" that encourages commitment to a career in the profession and creates another bond joining its members. A critical—but often ignored—method of sustaining the solidarity of the profession lies in norms governing relations among its members and between its members and lay people. These norms may be written as rules or practiced as unwritten custom. These norms are not designed to prevent competition among members as much as to control it so that there is enough income for all (Freidson, 1994).

A downside to professional solidarity is that it may be considered "inappropriate" to criticize the work of a colleague to a client. Even if one feels that a colleague has done poor work, it may not be perceived as "proper" or "professional" to inform the client of such an opinion or judgment. Thus, professional solidarity and cohesion may institute an unwillingness to confront errant colleagues, much less to take any punitive or corrective action in the interest of clients' well-being. Even when there is intervention, predominant norms of confidentiality act to protect both the reputation and career of the errant colleague and the public face of the profession itself. This etiquette expresses an important part of the ideal typical of professionalism—namely collegiality. On the other hand, this collegiality tends to prevent the use of adequate regulatory procedures that protect the public; thus, at its base collegiality may result in a violation of the profession's implicit contract with the state and the public. Freidson (1994) concludes that there may be an intrinsic conflict between the profession's efforts to maintain the solidarity of its members and its fiduciary relationship with society. Despite this conflict, professions are useful to society as a whole and also within organizations. In the following two sections, we examine the usefulness of professionalism in society and in organizations.

The Usefulness of Professionalism in Society

Freidson (1994) describes ways in which professionalism serves a useful function in society. First, professions create prestigious social identities to which an ideology of high purpose or ideals is attached so that young people may be led to aspire to them. Second, professions "provide the conditions under which it is reasonable to expect ordinary or a bit more than ordinary people to sustain something resembling the high purposes claimed for professionalism" (Freidson, 1994, p. 205). Freidson argues that professionalism provides the materialistic and institutional support that people need in order to avoid feeling a conflict of interest between their own financial interest and their client's interest. Thus, professionalism encourages otherwise ordinary people to exercise sound discretionary professional judgment.

Professional support comes in part in the form of fairly stable and firm jurisdictional boundaries that minimize competition from other occupations and rules that control competition between colleagues. In addition, professional devices, such as norms, aim at preserving the independence of judgment from the interference of colleagues, clients, and the lay world. These conditions provide for sufficient economic security—but not necessarily great wealth—to make long-term commitments to the profession feasible. Furthermore, these conditions deflect efforts by clients, employers, and others to exercise control over its members' work. Freidson (1994, p. 205, emphasis added) observes that:

> With such protection, professionals can afford to be devoted to the integrity of their craft and to use it for the benefit of others....While the extraordinary person may very well rise above discouraging circumstances to be an exemplar, if their very living is threatened it is *unlikely* that most professionals—which is to say the profession as a whole—will put the good of their clients and the public before their own.

If one reflects upon the above statement, its logic becomes apparent: professionals are not born into this world. Instead, professionals achieve professional status by education and socialization; and they are sustained in their professional functioning by their professional associations. As Freidson (1994) suggests, we would not expect professionals to behave differently than any other workers except for the entire panoply of benefit and support that their profession provides to them. This professional environment makes it reasonable for these otherwise ordinary people to exercise professional judgment and discretion in the interest of the client even when the client's interest conflicts with their own immediate financial interest.

The Usefulness of Professionalism in Organizations

Benveniste (1987) emphasizes the important role that professionals can play *within* organizations. He proposes that organizations recognize the value that professionals can add in increasing organizational effectiveness and contributing to organizational adaptability and flexibility. According to Benveniste, professionals acquire selected desirable and predictable patterns of behavior in their socialization for complex roles. Benveniste (1987, p. 256) states:

> Since rules and routines work well when tasks are predictable, unvaried, and well understood, they are used extensively. However, when tasks are varied and unpredictable, when learning is important in the task situation, and when adaptability is required, discretion is necessary. Discretion and trust have to replace routines. This is where the professional in the organization takes on new importance....[I]n an uncertain environment and task situation, the professional has the necessary knowledge and experience from practice, to act independently. The professional can search for solutions, determine which alternative to adopt, and implement new approaches.

Benveniste (1987, p. 263) also observes that "[t]he routinized bureaucracy does not depend as much on trust." He suggests that rather than staying with the less trusting and less flexible bureaucratic structure and functioning, an organization might benefit from the trust generated by professional socialization. Benveniste (pp. 268–269) concludes:

> The transformation of the bureaucracy into the profession-oriented organization is an alternative. Instead of relying on trust generated by family socialization, we rely on trust generated by professional socialization. The two approaches have commonalities. They also have important differences. Family ties are replaced by qualifications for the task. The incompetent uncle is replaced by the trained expert. More important still, the values and ethics of the profession can be oriented toward the public well-being.

Thus, the kind of professionalism found within the professions offers a measure of trust to an organization that is not available from other occupations. This trust makes it possible for professionals to exercise relatively speedy, informed, professional judgment within organizations that employ them so as to give their organizations a competitive advantage in a continually changing world.

Freidson (1994, p. 211) also raises important considerations about the nature of the relationship between the client and professional in an organizational setting: "Is the policy to provide whatever customers or clients desire, even if their capacity to evaluate the service or product is seriously limited and what they desire contradicts the better judgment of the professional? In short, is the policy to substitute the discretionary judgment of the consumer for that of the professional?" Freidson (1994, p. 211) answers these questions in the following manner:

> [I]f the customer (or the citizen, in the case of publicly owned organizations) is to be always right, then professionals are no longer able to exercise the authoritative discretion, guided by their independent perspective on what work is appropriate for their craft, that is supposed to distinguish them. They become mere servants in a cafeteria, doing whatever is demanded of them and seeking above all to please. Professionalism can flourish only when practitioners in organizations have firm but by no means absolute support from their employer for the consequential exercise of judgment that is independent of their clients.

Freidson's conclusion on the foregoing point has important implications for occupations seeking professional status if they are currently operating with the free market goal of selling services to maximize profit. While selling more services to generate profit without prior consideration of client need or to please the customer despite personal reservations may be one way to achieve free market success, it is not the professional way to achieve success. The professional way to be successful depends upon acting in line with professional judgment and discretion even when that judgment opposes what a client seeks to purchase. Freidson also notes that professional behavior operates within the professional's subtly balanced relationship with the client in which the professional asserts technical authority while respecting the client's right to self-determination. Within such a relationship, clients expect a "client best interest" focus from professionals. Clients trust that professionals will do what is best for the client, whether that means avoiding additional sales or advising the client to use additional services.

Maister (1997) offers organizations as well as individuals a practical observation about the benefits of selflessly giving clients honest advice, even when that advice may be counter to one's own interests. He states (p. 80): "The more you act selflessly and give clients honest advice, even when it may be counter to your own interests, the more trust you earn, and the more future business you get." Maister goes on to lament that, "What is fascinating is that this simple point seems to need stressing. Doing the right professional thing is not a *moral* point, it's just *good business*. If this is true, why don't more people get it?" (p. 80, emphasis in original). It appears that Maister's logic about good business parallels the accepted norm behind good professional service, that is, the client must receive honest advice/sound professional judgment even when it is counter to the professional's own financial interests. Following Maister's logic, it could be argued that professionals, as an occupational category, are good for business. Therefore, organizations that employ professionals should respect and encourage their professional behavior in providing clients what they need versus selling clients whatever they will buy.

What is Professionalization?

If professionalization of an occupation is desired, the question arises, how does that occupation become a profession? Wilensky (1964) proposes the following model of sequential events as descriptive of the process of *professionalization* that leads to state-sanctioned professional status:

1. becoming a full-time occupation;
2. establishing the first training school;
3. establishing the first university school;
4. establishing the first local professional association;
5. establishing the first national professional association;
6. establishing the first state license law; and
7. creating a formal code of ethics.

There is some disagreement among sociologists on the need for a specific sequence of the foregoing events for professionalization of an occupation to occur. However, there is a general consensus that the ultimate purpose of the behavior under observation is to construct the infrastructure necessary for generation and transmission of complex knowledge and for the regulation of services in the interest of societal welfare (Torres, 1991). The foregoing set of criteria seems straightforward and simple enough. In contrast to this linear simplicity, the implementation of each of the items on the list is a monumental task. Each item on the list is a substantial hurdle that our society requires each profession to complete in order to achieve the goal of recognition as a professional occupation.

A logical question about the role of occupational associations arises once the members of an occupation decide that they want to engage in the process of professionalization. All of the professionalization hurdles, except one on the Wilensky set of criteria, can be completed by an occupation without reaching outside its boundaries for approval. Licensure is the one hurdle that blocks entry into professional occupational status and that requires reaching outside occupational boundaries for approval. Each of the fifty states conducts their own licensing activities because licensing, with some exceptions such as aircraft pilots, is not a function of the federal government in the United States. As such, licensure is likely to be the most difficult and time-consuming hurdle for an occupation to overcome.

It is generally agreed among practitioners and academics that professional societies play an important role in convincing the state that a professionalizing occupation has the necessary criteria to be granted licensure. Krause (1996) outlines the history of the medical, legal, engineering, and academic professions in the United States, including the organized political action of the medical and legal professional societies. The professional societies in medicine and law worked for licensing laws that were enacted from the 1880s through the 1920s. These laws were enacted state by state with the accompanying requirements that those who had completed the university-based training programs obtain new credentials and undergo examination by the new licensing boards. Several generations of professionals who had trained under the old system (prior to licensure) continued to practice in the new era. Not until the 1920s–1930s was the modern university degree plus license exam system firmly established nationwide. Our observation is that this transition took over two generations to occur.

Freidson (1994, p. 68) provides further explanation of the power-based relationship between the professions and the state:

> Insofar as privilege is deliberately organized on a legal basis, it has a political foundation. It is the power of government which grants the profession the exclusive right to use or evaluate a certain body of knowledge and skill. Granted the exclusive right to use or evaluate a certain body of knowledge, the profession gains power. It is in this sense that the professions are intimately connected with formal political processes.

Freidson (1994) goes on to note that in addition to the development of a profession, the maintenance and improvement of the profession's position in the marketplace and in the division of labor surrounding it, "requires continuous political activity" (p. 68). Thus, it appears that lobbying efforts by professional associations are necessary activities as a means of communicating importance and urgency to state legislatures. Indeed, as Torres (1991, p. 49) observes, "agreement on which occupations warrant professional status and which do not may ultimately depend on the resources that the occupations can bring to relevant forums." Because licensure (the official grant of professional occupational status) is a state legislative activity, state legislatures are the "relevant forums" in which aspiring occupations can seek professional status. In light of this, a time-saving option for an aspiring occupation would be to coordinate its licensing efforts across all fifty states.

Licensure may also be forced on occupations as a result of public debate in order to protect the lay population. Medicine, law, and optometry were licensed either before they established a university connection or before they formed a national professional association (Wilensky, 1964). The structure of our system of government and the history of the established professions indicate that in the absence of public demand for licensure, organized political activity by an occupational association is necessary if that occupation seeks to gain and then later maintain professional occupational status.

A FALL FROM POWER: THE CASE OF THE MEDICAL PROFESSION

Weidner (1998) demonstrates the striking similarities between OD practitioner–client and physician–patient relationships. Based on this analysis, he also argues that study of the parallel relationships between organization development and the helping professions might inform the advancement of organization development. Therefore, we look at the medical profession to explore what might be learned to benefit the field of organization development and avoid decline.

The medical profession had a phenomenal rise in status in the early part of the twentieth century, ultimately becoming the preeminent profession of the century (Kimball, 1992). Burnham (1996), a medical historian, questions whether professionalism and profession formation ceased to be effective forces in the 1980s. Burnham (1996, p. 22) notes that "[i]t may be, therefore, that 'profession' is not, at present, as important to physicians as it once was." He ends his historical review with a hopeful reference to the centuries old lingering "special spirit of profession" (Burnham, 1996, p. 23). McArthur and Moore (1997, p. 985) discuss the "epic clash of cultures between commercial and professional traditions in the United States" and recommend a national agency to assist the entry of commercialism into medicine without compromising patient care. It appears that the medical profession is feeling pressure from outside forces and that it no longer has the professional power it once possessed.

In this section, we draw upon Krause's (1996) history and analysis of the formation and current status of various professions to explore what factors contributed to the decline in power of the medical profession. Krause (1996) suggests that we visualize a triangle, with the state, capitalism, and the professions at the corners in order to better understand the interaction between these factors. Each of the corners of the triangle act to influence the others. According to Krause (1996, pp. 1–2): "The state influences and shapes capitalism and professions, capitalism influences and shapes both the state and the professions, and the professions act to influence and confront the power of both capitalism and the state." Krause's analysis of the relationship between the state, capitalism, and the professions describes a continual shifting of power among them. Thus, he frames the issue as one of shifting control. Krause stresses that, as consumers, we have an interest in knowing which corner of the triangle will control the services that are critical to our lives.

In the United States, the state is currently acting with the majority of capitalist sectors against the profession of medicine and it is gradually restricting for-profit medicine. Physicians who were once thriving as owners of for-profit facilities are losing their advantage as federal regulation tightens. In addition, physicians are increasingly being sued for malpractice. In response to the increase in lawsuits, physicians order more tests and procedures to protect themselves against legal action. But as the state closes in, working with capitalism to cut costs, it begins to ration precisely the tests that the physicians seek in order to avoid lawsuits. Krause (1996, p. 49) concludes: "The crossfire between cost control and malpractice cases is profoundly demoralizing to the profession."

Medicine came to its current condition as a result of a reduction in the profession's ability to control its own professional association, the workplace, the market, and its relation to the state itself. These elements were all attacked by the federal government, by changes in the mode of providing health care, by court decisions, and by divisions within the ranks of physicians themselves. In addition, the federal government gained greater power over the profession, as the power of each individual state became less relevant in controlling the profession.

The medical profession's first experience with substantial federal control was with the Medicare-Medicaid legislation. The American Medical Association (AMA), whose membership was primarily composed of practicing physicians, opposed passage of this legislation on the grounds that it was "socialized medicine." When the legislation passed in 1967, the AMA suffered a clear defeat, and the AMA has not held the same commanding position in the profession since. Krause (1996, p. 45) reports that AMA membership rose from 51 percent of all physicians in 1912 to a high of 73 percent of all physicians in 1963. Membership in the AMA has continued to decline as more liberal and younger physicians who were repelled by the AMA's stance on services to the poor and old either quit or never joined the AMA at all. By 1970, AMA membership had fallen to 65 percent of all physicians. In 1990 membership was less than 50 percent, which was below the 1912 level. *Business Week* reports that current AMA membership is merely 35 percent of all physicians (Weimer, 1999). Any organization that speaks for such a low percentage of the total profession cannot have the same political influence as an organization that represents more professional solidarity.

The AMA was not able to stop the passage of the Medicare-Medicaid Act because academic physicians (who were generally not active AMA members) favored the 1967 Medicare-Medicaid legislation and a new and powerful lobby also supported it—older Americans. It was also at this time that community sentiment, which had previously been in favor of the medical profession, began to change. People still tended to trust their own physicians, but they began to view the profession as greedy and, sometimes, heartless.

At the same time that AMA membership was declining, membership in medical specialty associations was rising. Unfortunately for the physicians' bargaining power, each of these specialty groups had its own interests, partly because the federal government had fixed different rates of reimbursement for different specialty groups. By the late 1980s the specialties were often working against each other. In addition, after 1970 academic medicine took a much more active stance against the interests of the AMA by working with the federal government on a variety of measures to control costs that did not affect its own interests.

A second source of loss of power for the medical profession was loss of control over the number of physicians that were graduated from medical schools. Krause

(1996, p. 45) reports that through the mid-1960s, AMA policy kept the ratio of physicians to the population fairly constant from 126:100,000 in 1931 to 132:100,000 in 1962. This ratio jumped to 151:100,000 in 1970, 180:100,000 in 1975, 202:100,000 by 1980, 252:100,000 by 1986, and almost 300:100,000 by 1990. This was more than twice as many physicians per capita as in 1962. Medical schools were forced to increase enrollment because they were dependent on government research grants from Washington and they were also under pressure from state legislatures. Money was also made available to aid medical schools for expansion. The number of medical schools also expanded from 87 in 1959 to 126 in 1978. Both higher enrollments and new schools characterized the 1970s and 1980s. As the AMA lost power, it lost the ability to control or even bargain with the academic wing of the profession (whose members were increasingly alienated from the AMA).

The Medicare-Medicaid legislation of 1967 had an additional ripple effect on medicine. Physicians in practice (non-academicians) were won over to participate in Medicare and Medicaid through generous fees that were called "usual and customary charges." Overnight, charity medicine disappeared in the United States, as physicians and hospitals not only began to charge for their services to all patients, but they charged the maximum allowed by law. The result was an immediate giant increase in the cost of serving the old and the poor.

The rise in costs as a result of huge Medicare-Medicaid physician participation began to mobilize American capitalism on the side of state cost controllers (with the support of the salaried physicians in medical schools and public health). The national Washington Business Roundtable established the Washington Business Group on Health in order to represent the 200 largest corporations in the United States. This group worked with government and academic cost controllers and researchers on the problem of rising medical costs and recommended solutions. All of the solutions that were recommended by this group involved the end of fee-for-service medicine, which was blamed as a major incentive to increase patient care past the point of need.

By the late 1960s, American capitalism was experiencing stiffer competition from capitalism in western Europe and Japan. Profits were being eaten up by health care. As a result of these factors as well as others, physicians were confronted with a powerful and well-organized opponent in Washington at the same time that their own lobbying strength through the AMA was seriously declining. Responding to rising health care costs, Congress called for cost control within a year of the passage of the Medicare-Medicaid Act. The AMA, which had been responsible for including generous provisions into this legislation in the first place, was powerless to stop the cuts. The new era of professional regulation for the medical profession and more direct forms of cost control had begun.

Recent developments in health care also reflect the loss of control by the medical profession. It was recently reported that Columbia Advanced Practice Nurse Associates, a group of advanced practice nurses (APNs—registered nurses with a master's degree in a primary-care specialty such as pediatrics or family medicine), have been given hospital admitting privileges by New York Presbyterian Hospital. Some insurers now include APNs in their list of primary-care providers and reimburse for their services. This change occurred in part because "the swelling ranks of APNs—140,000 in the United States have pushed for it" (Foltz-Gray, 1999). APNs have traditionally practiced in rural areas and in inner cities. APNs now seek to offer to commercially insured patients the same option that the medically underserved have had for thirty years. APNs emphasize educating patients about how best to manage an existing or potential medical problem. It is not unusual for APNs to offer one-hour appointments to discuss preventive care with patients. Their focus on education and prevention appeals to insurers who consider the savings potential of their services; APNs have been described as "the up-and-comers in the healthcare field" (Foltz-Gray, 1999).

The nation's 230,000 licensed pharmacists are another group of medical providers who are moving in on physicians' markets. Berner (1999) reports that the Health Care Financing Administration (HCFA), which oversees the federal and state insurance program for the poor, made Mississippi the first state to reimburse pharmacists under Medicaid for their services in advising patients with diabetes, asthma, and high cholesterol, and those in need of anti-clotting drugs. These pharmacists are now able to offer medical consultation and alter prescriptions. Pharmacists call these new services "disease management." Twenty-one states have given pharmacists the authority to initiate or modify drug treatment as long as they have "collaborative agreements" with physicians. The doctor provides a blueprint of care, and grants the pharmacist the authority to fulfill it. Twenty-two other states are weighing giving pharmacists these powers (Berner, 1999). The entire pharmacy profession is seeking additional authority to not only counsel patients and modify prescriptions, but to also initiate prescriptions. The AMA is reported to support pharmacist efforts to offer disease management. Although many of the nation's 620,000 practicing physicians are unhappy about this trend, a member of the AMA board of trustees is quoted as saying, "As long as the captain of that team is a physician, we all think that would provide better care" (Berner, 1999). We cannot but wonder what the role of physician/team captain will be 10, 20, or 50 years into the future.

Gorman (1999) sums up the current professional environment for the medical profession in the title of her article, "Bleak Days for Doctors." According to Gorman, managed care has slashed revenue as overhead continues to climb. Physicians feel pressure to see more patients in less time. In addition to the nurses and pharmacists encroaching on their market, physicians have bureaucrats second-guessing their decisions.

In response to these conditions, some physicians have joined unions (Franklin & Japsen, 1999). Other physicians have started moonlighting outside of their profession selling things like nutritional supplements to their patients. Plastic surgeons and dermatologists have long sold facial creams to their patients. The AMA ethics committee recommended a no-sales rule, but the AMA agreed to reconsider the rule in response to protests from physicians. Some physicians have become so discouraged that they are leaving the profession entirely. A personal friend of the second author is a physician who has recently completed the training and testing needed to become a licensed esthetician. In Illinois, estheticians provide skin care and facial wax and make-up services in beauty salons and spas; this physician/esthetician plans to open her own spa. Other physicians are learning how to live without managed care by dropping their managed care patients and living with a substantially reduced income (e.g., two-thirds less income). Gorman (1999) concludes that it is almost as if the medical profession is suffering an identity crisis.

The experience of the profession of medicine provides occupations seeking professional status with enough questions for many extended discussions. Was it the loss of power by the AMA that allowed the powers of capitalism and the state to take control of the practice of medicine? Did physicians lose their professionalism when they lost their previously higher level of autonomy over their own work? Was the increase in the power of capitalism and the state the best remedy for meeting the medical needs of the elderly and the poor? Will physicians continue to hold a position of professional status in our society or will that position slip with time as more people encounter disillusioned physicians who are just trying to make a living? Will medical schools continue to attract the best and the brightest under the current conditions?

Implications for action by occupations seeking professional status emerge from the physician experience. First, it seems clear that occupational solidarity in a professional association is necessary in order to assert the interests of the profession for market protection and to protect professional discretion and judgment on the job. Second, occupational solidarity requires consistently strong support for one main professional association from a large percentage of the total number of professionals in a given profession. Third, occupational solidarity makes it possible for the main professional association to push back against other professions that are encroaching on market share; against capitalist interests seeking profit from service delivery; and against the state in its desire to implement social policy by regulating service delivery. Thus, professionalization entails more than achieving professional status. It involves *retaining* control *within* the profession.

PROFESSIONALIZATION OF AND PROFESSIONALISM IN ORGANIZATION DEVELOPMENT

Changing Focus of Practice

Concurrent with the evolution of notions of professionalism over the past thirty years, organization development has been evolving as well (Beckhard, 1969; Beckhard & Harris, 1977; French & Bell, 1995; Woodman, 1989). Organization development practitioners have been encouraged to become more strategic (Jelinek

& Litterer, 1988), and are said to need to possess a better base of business literacy. Jelinek and Litterer describe a physician who works from ill patient to ill patient and then realizes that changing living conditions might have more impact on the health of the community. This will call for different skills, for preventing rather than curing disease. Using this metaphor, Jelinek and Litterer call for a change to strategic organization development, to relate organization development skills to the changing competitive context of organizations and members.

One might interpret Jelinek and Litterer's perspective as: "If it's not strategic, if it's not long term, if it's not working at the levels of system, then it's not organization development." This interest in whole-system change can been seen in developments in the arenas of management technologies (e.g., Dannemiller & Jacobs, 1992), practices for managing change (e.g., Worley, Hitchin, & Ross, 1996), and the emergence of academic programs with a systems-perspective emphasis. It is against this increasing systemic practice background that we examine the progress of organization development's emergence from an occupation to a profession.

At the same time that organization development has been moving toward more systemic approaches, the practice of organization development has been increasingly practitioner-dependent, and increasingly fragmented. Not just are there new entrants to the field, but also the backgrounds and preparations of practitioners are becoming more varied.

Another element of the contemporary organization development milieu that cannot be ignored is a heightened and popular cynicism regarding consultants in general (Adams, 1996; Stewart, 1997). It is possible that by way of easy access to consulting (and to organization development), that the consulting "label" has been debased almost as much as some traditional organization development people feel that organization development has been debased.

Is Organization Development a Profession Today?

Many students of organization development argue that the next step in professionalizing the field is professional licensure. Before suggesting alternative courses of action, we believe a brief assessment of the current state of the professionalization of organization development is in order. We make this assessment first by using the three perspectives of professionalism presented earlier. First we use Torres' (1991) framework, then we compare the state of organization development to Maister's (1997) notion of professionalism, and finally we assess contemporary organization development relative to Freidson's (1994) model of professionalism.

Organization Development: To License or Not to License?

Torres (1991, p. 49) notes that "the formal border between professional and nonprofessional occupations is quite clear—state legislative acts granting occupations professional status." According to this one requirement, organization development cannot be a profession until it is granted professional status by the state.

Several questions arise from this "requirement." First, is pursuit of licensure the only way in which organization development may become a profession? Second, are there *other* means by which organization development can become a profession? Third, is the professionalization of organization development prohibited by an historical "checklist" of requirements? All of these questions, which have been asked early and often by students of organization development, seem to focus on establishing a professional identity and an identity and recognition of organization development as a profession *in the eyes of others*. Such a "cry for respect" may not only be counterproductive, but also may reinforce whatever negative or unfocused stereotypes might exist about organization development. As an example, witness the contemporary position by human resources for a desired role as a "strategic partner"—perhaps a desire that betrays some level of insecurity stemming from not being at the executive decision "table."

The Identity of Organization Development

Beyond a concern for the emergence for organization development as a profession is the question of the identity of organization development (Church, Hurley, & Burke, 1992; Miller, 1993). Organization development faces an identity crisis on several fronts: (a) from practitioners self-identified as working in "performance improvement," "human performance," or "human resource development;" and (b) from the burgeoning array of practitioners who identify themselves as working in the areas of "organization effectiveness," "change management," or "change enablement." What might characterize the former is the formation of a professional membership association with those identifications (in the case of the above identities, represented as the International Society for Performance Improvement, American Society for Training and Development, and the Academy of Human Resource Development). Practitioners who self-identify as the latter might be members of or attend learning events sponsored by the same associations, but they are not engaged or "wrapped up" in the need for a professional identity as conferred by a professional membership association.

Perhaps more telling, nearly twenty years ago, Burke and Goodstein (1980, p. 9) observed, "[l]ike the young adult, OD is searching for its identity, its uniqueness. Some of the blush of youth is gone. We are not everything that we had hoped or thought; reality and experience have tempered dreams and enthusiasm." This lack of clear identity is likely at the heart of the results reported by Church, Burke, and Van Eynde (1994), in which a duality of values for humanistic concerns and the "bottom line" were expressed by organization development practitioners in both their current work as well as in an "ideal" state. Complicating the identity question further may be the relatively less well-explored world of the internal organization development practice, which McMahan and Woodman (1992) found to be significantly different than the external world of practice. For example, one of the practitioners in McMahan and Woodman's study said his firm "no longer does OD," and that total quality management (TQM) was the preferred label within his organization. However, this respondent described their TQM activities in a way highly consistent with that of general organization development activities.

It is possible that the use of "organization development" as a label may have become too laden with baggage of old to retain its relevancy—note that ours is an observation about *perceptions* of the *label*, not the field's people or their practices. This may or may not be true. A more important question may be whether the proliferation of labels described above are counterproductive, or merely transitory. Will "organization development" as a label stand the test of time? Such questions suggest that now is a critical time in the history—and development—of organization development.

Such lines of inquiry (or even simply worry) may serve to obfuscate more important issues. Perhaps just as important as focusing on organization development becoming a profession (or on being recognized as a profession) is the question of bringing professionalism to the practice of organization development. Despite theoretical and empirical advances in research in organization behavior, psychology, organization development and change, and related areas (e.g., Adizes, 1988, 1997; Bunker & Alban, 1997; cf. Porras & Robertson, 1987), it is possible that many practitioners: (1) are inconsistent across the body of their collective practice, (2) such practice demonstrates wide variance because it is not guided by a (a) widely used body of theory and/or (b) centrally codified body of knowledge, and (3) enter the field of organization development that has made possible (if not encouraged) acceptance of many entrants over low thresholds of entry to practice. Thus, if organization development has a "credibility gap," such a gap may have to do with an uneven level of its public's trust in the field; this may be a result of our public's lower level of trust in the opinions and ideas suggested by its practitioners (Anderson & Dedrick, 1990).

Organization development could seek a legislative grant of professional occupational status based on the specialized knowledge needed to resolve complex problems found in the practice of organization development and on the criticality of tasks for social welfare that occur in organization development. The field of OD, with its specialized knowledge that is critical for the welfare of the public, can provide substantial benefit to employees, volunteers, and management in organizations in many different ways. For example, organization development can be used to help organizations in ways that are critical for social welfare through strategic planning, helping an organization become more clear about its shared values and future direction (Schein, 1992). This might enhance the organization's viability, and increase access in the organization's community. Similarly, team building could remove destructive tension between coworkers and help them work together more effectively. Thus, by focusing on organization development's use of specialized knowledge in its helping role with people within organizations (versus focusing on organization development's role in increasing the profit potential for organizations that engage in high performance work practices) organization development can establish its claim to legitimate criticality to social welfare. This claim can be used as evidence of a grant of professional status from the state.

Organization Development as a Profession: An Absence of Control Over Entry to the Field

Organization development has done a relatively poor job of establishing occupational control over entry to the field and the arena of practice. As a result, a large variance exists in terms of what is organization practice, both as it might be described and as it is performed. In this section, we briefly compare the current state of organization development to Torres' four kinds of occupational control described earlier.

Free market OD: "Buyer beware." Some of the "change management" training occurring in recent years certainly reflects the idea that change management can be taught by people without an in-depth understanding of organization development. This is a concern among many OD practitioners for a variety of reasons. Usually the first concern is that organization development without guidance by OD values is dangerous, or at the least, manipulative. To the extent that organization development may have an image problem related to the competence of individual practitioners, some of this may be related to practitioners doing "change management" or reengineering or other work without an organization development background, while some may be related to clients' difficulty in assessing practitioners before deciding to proceed with a given project or specific consultant.

Technical OD: "Only your OD practitioner knows for sure." When organization development skills are viewed as technical skills, the skills are "booked" for an appointment or other short period of time (e.g., "seminar" "workshop"). Organization development practitioners encounter confusion when asked to explain what they do to prospective clients or even friends. Some of this confusion may stem from confusion over what organization development is, and some may stem from a lack of name recognition for organization development within organizations. That is, organization development may not have reached the mainstream of business vocabulary, in part because of past attributions of poor credibility or trendiness (e.g., failed T-groups, result-less adventure courses, uneventful team building). Thus, a discussion about the profession of organization development is necessarily a discussion about the positioning and marketing of organization development. This is somewhat awkward for many in organization development, in part because a segment of the field tends to eschew "business" terms or metaphors, in our observation, sometimes at the field's own peril.

Scientific OD: "Respected but not highly valued." Organization development faces challenges when skills are perceived as scientific skills in that the skills are respected, but those skills are not valued in the competitive marketplace compared to other sets of skills. In response to this perception, managers or executives may appear critical of OD efforts, for instance, saying that organization development is "too theoretical," "takes too long," or indicate that they do not want "touchy-feely stuff," or say "let's not waste time here." In other cases this nonsupport may be far more passive-aggressive, and manifest itself in either nonparticipation or withdrawal.

Professional occupational system: "The best of criticality and complexity—with licensure." Like physicians and lawyers, some OD practitioners are respected and perceived by their sponsors/clients as critical to their organization's success. These practitioners (or practice groups) demonstrate both the highest level of perceived criticality of knowledge for public welfare as well as the highest range of complexity of knowledge. However, this value has not been recognized by state legislatures in any of the United States, and this lack of recognition of *value* is a limiting factor on organization development's recognition as a *profession*. This lack of recognition may be due in part to a relative minority of practitioners achieving high levels of perceived criticality and complexity. Yet another contributing factor could be perceptions of organization development (and its practitioners) being perceived as being in the lower three levels of occupational control (free market, technical, scientific). Thus, a "catch-22" arises in that perceptions of professionalism have not coalesced at the high end of the spectrum. However, for professionalism to be institutionalized (and mandate change), recognition is necessary to create a demand (and a market) for professional licensure. If practitioners can make a good living and enjoy a (professionally and personally) profitable career without licensure, why would they need it or advocate for it? Why would new entrants to the field want licensure? We believe the answers to these questions lie not only in creating demand for licensure (in itself a worthy goal for organization development), but in *creating client-driven demand* for professional behaviors.

Organization Development and Professionalization

Where is organization development on the path that leads to state-sanctioned occupational professional status? In other words, where is organization development vis-à-vis Wilensky's set of criteria for a profession? In this section, we provide a critical analysis of the status of the profession vis-à-vis each of Wilensky's criteria for a profession. Summarized in Table 1, we indicate which criteria have been met (Yes), which have not been met (No), and which merit further examination, exploration, and/or development (Closer Look).

Full-time occupation. At first glance, it may appear that organization development is a full-time occupation. Many OD practitioners—indeed, entire firms—are able to develop a professionally and personally rewarding practice. However, a noticeable segment of the OD practitioner population may engage in being a "part-time practitioner"—not that there is anything wrong with such a notion—but

Table 1. Assessment of Organization Development vis-à-vis Wilensky's (1964) Criteria for a Profession

	Yes	No	Closer Look
1. Full-time occupation	☐	☐	☒
2. First training school	☒	☐	☐
3. First university school	☐	☐	☒
4. First local professional association	☒	☐	☐
5. First national professional association	☐	☐	☒
6. First state license law	☐	☒	☐
7. Formal code of ethics	☐	☒	☐

Key:
Assessment Description
Yes   Criteria have been met.
No   Criteria have not been met.
Closer Look   Criteria merit further examination, exploration, and/or development.

this may detract from the desired "professional" image as a "full-time" occupation. For example, an academic who does some organization development consulting would fit this description as well. Because of professors' primary roles in teaching and research, their consulting is by definition "part-time." Such a perception of organization development professionals might also be influenced by the contemporary notion that many white-collar workers now call themselves "consultants" when they are between jobs or are searching for jobs. Thus, even though organization development may be a full-time occupation for some, whether organization development is perceived as a full-time occupation by society at large may be a far more important criterion.

Training school. At this writing, National Training Laboratories (NTL) has offered training programs in organization development for over 50 years. During that time, many other training programs in organization development have emerged and provided training for new entrants to the field as well as advanced training in specific techniques or methods. This criterion appears fully met.

University school. As with the occupational criteria (#1 above), the series of graduate academic programs profiled in each issue of *Organization Development Practitioner* and surveys regarding academic programs in the field (e.g., Varney & Darrow, 1996) would appear to suggest there is a robust and perhaps even growing market for OD graduate education. This is an important form of legitimacy for an emerging profession (Abbott, 1988). However, a closer look reveals that many universities identified in Varney and Darrow's study do not identify themselves as "organization development" programs *per se*; instead, some programs are identified as "Management and Organizational Behavior," "Human Systems Development," and "Whole Systems Design." This reveals at best a mixed picture regarding consensus about what organization development is among "organization development" programs themselves, much less what organization development should be called in academia. Perhaps more insidiously, this lack of convergence of language suggests that the language of organization development is not yet settled; without such convergence within organization development, it is difficult to conceive of how individuals outside of the field will understand what it is. Although some academic programs identify themselves as an "organization development" program, a consensus on which the "first" program is, and whether or not that first program is called "organization development" leaves this criterion at best ambiguously met.

Furthermore, texts that might be used for teaching organization development at the graduate level have done a poor job of coming to terms with the issues discussed in this chapter. For example, the index of Smith, Houston, and McIntire's (1996) text contains an entry for "Professionals, OD," that refers readers to "OD Practitioners," an entry that has five subentries (including one called "ethical issues," and a listing of fifteen individual practitioners named in the text. The index of Cummings and Worley's (1997) revised text has no entry for professionalism or the profession of organization development.

This problem suggests that graduate education programs or some other preparatory mechanisms will play a pivotal (if not central) role in developing organization development practitioners whose skills and results "raise the bar" of organization development practice while raising the credibility of the field. This problem has been recognized simultaneously in the fields of law and medicine; Wilkins (1995) describes an attempt to create a multidisciplinary program to teach professional ethics.

Local professional association(s). The Organization Development Network (ODN) operates as a true network in that regional ODN groups are completely autonomous and, by extension, self-sustaining. The health of these networks is at an all-time high, as evidenced by the 45 local/regional ODNs at this writing. This level of sustained, dispersed, focused professional activity suggests there are local "communities of practice" that are as vibrant as they are vital. This criterion appears fully met.

National professional association. There is not one national association that speaks for or to organization development as a field. There are several organizations that identify themselves as being primarily engaged in "organization development," namely the Organization Development Institute (ODI) and the ODN. There is a modest level of collaboration between these two groups, including linking of web resources, and cross-organizational collaboration on developing an OD credo (ODI,

1996). However, while the ODI has pursued being a registrar of professional designations (e.g., Registered OD Professional, Registered OD Consultant), the ODN has focused on being an organization that provides individual members with professional support (e.g., books, conferences, professional liability insurance, long-distance telephone discounts, meta-networks, and listservs).

The universe of organizations that organization development practitioners call "home" is not limited to ODI and ODN. Other organizations competitively vie for the attention of organization development practitioners. The American Society for Training and Development (ASTD), International Society for Performance Improvement, and the Association for Quality and Participation all serve various cross-sections of the organization development practitioner population. The Academy of Management Organization Development and Change Division is home to many academics and practitioners in organization development, as is the Academy's Managerial Consultation Division. Finally, the Academy of Human Resource Development, which split from ASTD in 1993 (AHRD, 1998) has become a professional "home" to research-informed practice in Human Resource Development. AHRD reflects a wide range of diverse disciplines and approaches around this common interest, including both practitioner and academics from HRD, organization development, adult education, and vocational education.

In addition, the existence of an organization is not sufficient for this criterion to be met. Such an organization must be credible both within and outside of the field. Internal credibility alone is self-conferred without external validation. External validation may be easily eroded if an organization becomes an instrument of aims other than advancing the profession as a whole. This is the situation that gained national attention in the United States when the American Medical Association fired the editor of its flagship journal, raising questions about the editorial independence (and professionalism) of the association, the journal, and the profession of medicine (Jensen, 1999).

Thus, instead of these organizations converging and establishing a national professional association as a base or basis for professionalism, the various interest groups within organization development seem to be defining themselves in more fragmented rather than in pluralistic ways. This suggests that this criterion for becoming a profession has not yet been met. Perhaps more importantly, these events may signal a regression from past progress toward this criterion.

State license law. The ODI has been energetic in its efforts to exhort practitioners to get involved with licensure efforts in a substantial way and to encourage voluntary registration by practitioners (the value of such registration in raising standards of practice and improving the practice of organization development has not been established). However, the notion of licensure for organization development practitioners has not progressed significantly in the last three decades. There is little to suggest that this will change soon. According to both the structural/functionalists and process theorists (Torres, 1991), organization development would not be considered a profession because it is not regulated by state license and it does not have a formal code of ethics.

Our perspective is that licensure is desirable but looks like it will be a long time in coming. Certainly, OD practitioners work regularly and thoughtfully with issues of power to understand the high complexity and plurality of political processes (in fifty states) and understand the long lead time necessary to effect such a change. While the desirability of licensure continues to be the subject of heated debate and dialogue within the field, we see no sufficient commitment or critical mass of organized resources within the organization development community to reasonably suggest that such an effort will result in licensure in the near future. This does not suggest that all is lost, or that we need despair for the field. It does suggest that we consider other steps that might improve the practice of organization development within our lifetimes.

As noted earlier, Torres (1991) noted the inherent struggle between free market and professional control of emerging occupations. We believe that licensure (in the form, of state monitoring and control) can serve as a means of controlling professional behavior to protect clients from irresponsible OD practice arising from commercial pressures on individual or firms' behavior. We further take the view that although many organization development practitioners may choose to behave in a professional manner despite "bottom line" pressures (as described by Maister, 1997), other practitioners may be unable to withstand these forces as free market pressures cloud their professional judgment. Maister might respond that "bottom line" concerns are consistent with professional/ethical behavior in that clients are sophisticated enough to know when they receive good services, and that such services will benefit the ethical practitioner with more business, reinforcing ethical/professional behavior.

Do clients really know when they receive good service or not? An intervention might initially look like a smashing success saving the client large sums of money (e.g., reengineering) and then after a year or two that same intervention might have longer term negative consequences for that client (e.g., deteriorating employee morale). We believe that clients may or may not know when they receive good service at the time that service is provided. At some later point, the consultant has gone on to other clients, having received initial rave reviews for his past intervention. The client may or may not associate current or emerging problems with an intervention that occurred a year or longer in the past. This line of thought would seem to provide a solid argument for more rigorous evaluation of change initiatives.

Two other considerations support our suggestion that licensure is needed to improve the likelihood that most OD practitioners will behave in a professional manner. The first consideration is the logical parallel between Maister's argument on values in action and the monitoring and enforcement function that licensure provides. Maister (1997, pp. 81, 82) states that:

A firm will have functioning values only to the extent that it has an effective management system that is intolerant about deviations from those values....However, what firms also need (but mostly do not have) is a method to enforce these rules of membership in the "family." Enforcement would require two things. First, management must have a method of identifying departures from excellence....The second thing needed to enforce values is yet more management time devoted to closed-door counseling and coaching discussions with those straying from the firm's values....It's about what is and is not acceptable behavior in this "club." Unfortunately, in most firms today there are very few nonfinancial things that would cause the managing partner to drop by. The range of tolerated behaviors is very broad....where the value system is weak or nonexistent, significant management time and effort are required to build agreement among the professionals that strict accountability systems should be introduced to keep everyone honest regarding company values.

Such thinking need not be confined to a large firm environment. Prospective clients should ask the same questions about small firms as they would about large firms; however, many clients may be more sophisticated about expressing their own needs than they are in assessing a consultant's or consulting firm's competence and capabilities.

Formal code of ethics. The ODI (1996) reports that it was formed over a disagreement within the ODN about the creation of a Code of Ethics for organization development. Interestingly enough, the ODN posts "A Statement of Values and Ethics by Professionals in Organization and Human Systems Development, along with an "Annotated Statement" and a summary CREDO. An "Organization and Human Systems Development Credo" is available on the ODN's web site (ODN, 1999), while the ODI's "International OD Code of Ethics" (ODI, 1999) links one directly to the ODN CREDO site cited above. However, the ODI includes a copy of the "Organization Development Code of Ethics" in its publication (ODI, 1996); this code is also included in Gellerman, Frankel, and Ladenson (1990).

Confusion about ethics, credos, and values does not a profession make. Weidner (1998) reviewed the code of ethics of several helping professions and found striking similarities between the content of the ODN's values statement of responsibilities and those of other helping professions. Our observation in what is called for now is parsimony rather than fragmentation, solidarity rather than splintering. We further believe it would be helpful for organization development practitioners to consider the implications of this struggle for their own practice and for future actions by the entire field of organization development as it considers professionalization.

Torres (1991, p. 53) raises a challenging consideration for organization development practitioners: "With the imposition of standards from differing occupational systems, one emphasizing profitability and efficiency and the other emphasizing knowledge and safety, professionals may find themselves in a quandary as to which set of standards to follow." If we assume for the sake of argument that organization development is currently in Torres' free market occupation category, it is subject to the controls of the free market. How well can an individual organization development practitioner's desire to live according to unclear and unenforceable OD professional ethical standards withstand the pressure for short-term profitability exerted by market forces?

Upon reflection, we think that organization development sometimes exhibits characteristics of all four occupational systems, that is, free market, technical, scientific, and professional. This lack of clarity about the occupational status of organization development might help to explain some of the continuing confusion regarding the conceptualization and articulation to those outside the field of organization development of what organization development is and what knowledge and behavior can be reasonably (and consistently) expected of an OD practitioner (Church, 1999). Much like the story of the blind man and the elephant, varying perspectives on what knowledge and behavior is expected of an organization development practitioner may be accurate—but unfortunately, all too incomplete.

In summary, organization development is very incomplete as a profession when one critically examines the field's progress against Wilensky's criteria. The field has suffered from a lack of unity of direction and voice, from fragmentation, and from what could be critically viewed by some as a self-perpetuating identity crisis. None of these symptoms are in and of themselves abnormal or even fatal, but they do present not only vital signs of organization development, but also a sense of how well (or incompletely) organization development has responded to emerging notions of what it means to become a profession.

Organization Development and Professionalism

How well does the field of organization development live up to Maister's (1997) notion of professionalism? It is difficult to assess organization development as a field vis-à-vis Maister's behavioral professionalism without an empirical study of how practitioners are perceived in the field. That said, there is evidence of professionalism characterizing high performing consultants (Olthuis & Hiebert, 1995). Additionally, both of the present authors have personally met enough talented, capable people in the field to feel confident that a large number of practitioners meet Maister's standards. But such impressions mistake who we (as persons) meet for how we (organization development as a field) behave. Rightfully, there is concern in the field about both (a) the low extremes (Weidner, 1995) and (b) the "average" of organization development practice. We believe exemplary organization development practitioners exhibit Maister's described behaviors; we also believe that organization development as a field does not exhibit these behaviors consistently enough to receive external recognition as a profession. We would suggest research be conducted on these questions because the question of licensure, from the above discussion, requires a combination of a managing perceptions of organization development and the building a critical mass, a coalition of solid support for change. Olthuis and Hiebert's (1995) research on highly effective internal consultants may provide a point of departure for such a stream of study.

Maister's notion of professionalism often comes down to a discussion of "the firm's management." One of the inherent challenges for organization development in dealing with professionalism of the field is the number of independent practitioners whose individual practice is inseparable from that of their firm. In other words, the challenges of institutionalizing these practices, no matter how desirable, are inherently different for an individual (or a small practice group) than a mega-professional services firm in excess of 50,000 employees. Future research should explore notions of professionalism and professionalization across the spectrum of practice models and across a range of sizes and types of practices.

To return briefly to the example of medicine discussed earlier, we note that physicians have maintained their identity as a profession and standards or practice despite many physicians working in individual practices. The market pressures, combinations of knowledge, and scalable deployment provided by today's large professional service firms are the same pressures that drive consultancies into firms of ever-increasing size. Organization development would do well to study (a) the professionalism of medicine in the days when single practitioners were more prominent and (b) the prominence and relative success (or failure) of effective peer review during that period.

The question of values and professionalism within a firm is ultimately a question of enforceability and the quality of the firm's management. What are these values? Who will decide what those values are? Where will the leaders of consultancies learn how to tackle professionalism? To begin to approach these questions, we turn to an analysis of organization development vis-à-vis Freidson's framework of professionalism.

Professionalism Reborn

Torres' and Wilensky's view of professionalism appear to be highly consistent with the predominant "mental map" of organization development practitioners concerned with professionalization of the field. While Maister's perspective provides important behavioral guidelines for practitioners and practice leaders, a review of Freidson's approach to professionalism reveals how past efforts at professionalization within organization development have failed to (a) operate with adequate solidarity, (b) establish a codified body of knowledge, (c) establish a strong, positive identity in the minds of those outside of the field, (d) provide an agreed-upon and recognized credential that identifies professional practice, and (e) police and enforce standards of professional behavior. In summary, organization development has become somewhat market-driven and technique-driven rather than maintaining its own professional autonomy (Burke, 1997). Organization development is not alone among fields in this regard (e.g., medicine, engineering), but we recognize this realization may be difficult for some within organization development to accept.

Freidson's (1994) basic elements of a profession, a body of knowledge and fiduciary responsibility to clients, shed some light on how far organization development must mature as an occupation before it becomes a profession. Organization development has thus far failed to codify adequately its body of knowledge, as described above. Equally as important, organization development has failed to establish its unique value-added contribution to the world of organizations, that is, it has not *systematically* demonstrated its values in excess of (a) activity, (b) technique, or (c) resources consumed.

Organization Development: Some Choices for its Future

Different Road Maps

Perhaps unaware of Freidson's (1994) review of the evolution of professionalism (as well as others' work in this area), it appears that the field of organization development may have been seeking to establish itself in a "old world order" of professions rather than attuning itself to emerging notions of professions and professionalism. At this point, prospects for the professionalization of organization development take the form of options for the future. Our sense from the above discussion is that there exists a few "certain next steps" followed by a variety of paths to becoming a profession, rather than a limited range of mutually exclusive options. Bringing greater professionalism in organization development might include elements of two new approaches: (1) a "body of knowledge approach" and (2) a structural approach to "marshaling market forces" on the side of professionalizing the practice of organization development. We explain each in the next section, and briefly discuss their relative strengths and weaknesses.

Body of Knowledge Approach

One approach that holds some promise for the professionalization of organization development is through development, codification, and refinement of the field's body of knowledge (Curry, Wergin, & Associates, 1993). Several other professional organizations have done an admirable job of this, one model to emulate is the Project Management Institute (PMI Standards Committee, 1996), which has not only published its body of knowledge, but has provided skills development and training based on this body of knowledge.

Another example of the body of knowledge approach is the efforts of the Society for Human Resource Management (SHRM). SHRM offers certification exams in human resources; in addition to learning the body of knowledge, one obtains credentials that identify the level of demonstrated competence through successful completion of the exams.

The body of knowledge approach requires assembling a panel of experts in the subject, making the body of knowledge widely available, and creating, contracting, or partnering with an entity to run training, testing, and other programs. This has been particularly problematic to resolve in organization development, as a unified position on "the basic nature of OD" (Church & Burke, 1995, p. 3) has not been developed. Not surprisingly, one-half the respondents in Church and Burke's (1995, p. 28) study of organization development practitioners agreed that "new entrants in the field of OD are lacking in the appropriate background, training, values, and appreciation for theory."

Strengths. The body of knowledge approach has several strengths that make it appealing for advancing the professionalization of organization development. One is a function of the convergence necessary to create a recognized (or at least codified) body of knowledge—greater clarity about what organization development is and is not, and how different aspects of organization development fit together. Dissemination of that body of knowledge may expose organization development to a large audience and attract a pool of increasingly talented people; this has certainly occurred for the Project Management Institute and SHRM. Pursuing such an approach would precipitate development of (or partnership with) an entity or entities as the source(s) of training programs in this knowledge area.

Weaknesses. At least two primary weaknesses characterize the body of knowledge approach. The first is related to the obstacles to be overcome when creating and implementing such an approach; the second is related to perceptions of organization development among people not in the field.

Several substantial obstacles would need to be overcome to create a common body of knowledge, as evidenced by the Academy of Management's current efforts to do so championed by Glenn Varney and Arthur Darrow of Bowling Green State University. Aside from these concerns, getting consensus about the content and form of the body of knowledge of organization development would be—by necessity as well as due to the nature of the organization development community—a pluralistic process. It is highly likely that organization development practitioners will exhibit the same degree of resistance to this idea as has been shown when discussions of licensure or certification arise on the contemporary ODNET listserv sponsored by the Organization Development Network. The problems of organizing independent organization development practitioners is not at all dissimilar from the difficulties faced by the National Farmers Organization when attempting to organize fiercely independent American farmers. Walters (1971) provides a compelling portrayal of this story; his analysis is that the decline of independent farmers represented an end of a way of life and of valuing and caring for the land. Woodman (1993) suggests that in organization development his problem is in part due to a schism between the academic and practitioner subworlds of the field. This fragmentation compounds the fragmentation borne of different professional identities (e.g., OD vs. HRD) or roles (e.g., internal vs. external). Deaner and Miller (1988, p. 16) earlier posited that "OD as a field has been around too long to continue in this lack of clarity."

The second area of risk is that a body of knowledge may serve to erode rather than strengthen the posture of organization development. As an illustration, despite all the work that went into the design, marketing, and implementation of SHRM's certification, the SHRM certification was characterized by Thomas Stewart of *Fortune* (1998, p. 79) as a *"quasi-professional* credential" (emphasis added). An open question is whether such a perception would develop regarding an organization development body of knowledge, or whether such a credential would enhance or detract from current perceptions of organization development held by others in the field.

Related to this notion of popular marginalization (French & Bell, 1995) is that organization development may become perceived to be more a knowledge area than a profession. This is likely the worst fears of organization development practitioners who lament what they view as the "co-option" of organization development by such areas as change management.

Marshaled Market Forces/Structural Approach

Almost all of the models of large-scale change suggest that it is inadequate to depend purely on internal motivation to effect lasting change. As organization development professionals, we know this from our work with performance management interventions. Is the usefulness of licensure as a stepping stone to professionalism outdated in light of Maister's free market description of professionalism, or is licensure a necessary but insufficient step toward professionalization of organization development? In this section we describe what we call a "marshaled market forces" approach to professionalization of organization development.

Market forces could conceivably be marshaled to significantly advance the professionalization of organization development. The roots of this notion come from the first author's first-hand experience. A large professional services firm was called to explore an engagement in January of 1998. The firm asked the first author for information that had not been sought by the firm on the earlier engagement. The firm asked whether the first author possessed current professional liability insurance and whether or not the author would indemnify the prospective client for risks associated with the engagement. What had transpired between the first engagement and the inquiry for an additional engagement?

The answer is a court decision against Microsoft (Bernstein, 1998), in which the court found that software developers that Microsoft had regarded as contractors were, in the view of the IRS, to be employees of the company. Microsoft had to pay huge settlement costs, including not only back pay, but also back benefits, including pension, profit sharing, and stock options totaling hundreds of millions of dollars (Bernstein, 1998). Needless to say, this sent a chill through not only companies that use subcontractors, but also professional services firms who use outside consultants.

We were struck by the notion that professional liability insurance was a way of not only defining the contractual, arms-length nature of the relationship, but also was a way of defining professional responsibilities of both the client (contractor) and the consultant. Further examination of the professional liability insurance application (offered by a private carrier but by special arrangement made available to ODN members) revealed that professional liability insurance might hold significant promise for not only improving an individual's practice but also the overall practice of organization development.

Maister (1997, p. 125) provides a clear description of today's emerging consultative marketplace: "There was a time when clients said 'I don't understand or want to understand the details of what you do—just take care of me and my problems, and I'll pay your bill!' This is not an accurate description of today's marketplace. Today's clients are sophisticated buyers." Maister offers two additional observations on client trends. First, he sees that clients are buying fewer services as if their problems were totally unique. Instead, they want to tap into a firm's accumulated experience and methodologies in order to benefit from the efficiencies that come from dealing with providers who have done it before.

"Accumulated expertise and methodologies" need not be the exclusive domain of large consulting firms. Several small firms have established themselves firmly within a defined niche by being methodical about methodology and transferring their knowledge to their clients (e.g., Dannemiller-Tyson Associates, Scanlon Plan Associates). That said, one implication of the above shift is that small firms will be especially challenged by large, systemic work, especially to the extent that their practice is practitioner skill-specific rather than backed by specific and replicable methodologies.

Second, clients increasingly desire to be involved in the process or, at a minimum, they want to be kept informed of their options and up-to-date on progress. They also want to be assisted in understanding what is going on and why. This expectation calls for practitioners to be able to explain why they are doing what they doing (i.e., not by saying "this is the way our firm does it"). Today's consultants are living in a world with very sophisticated clients who not only question the professionals' actions but also want to be involved in the professional service. Again, the parallels between organization development and medicine are compelling.

Professional liability insurance is not unlike medical malpractice insurance purchased by health care providers. The primary difference is that medical malpractice insurance (or social work malpractice insurance, or nursing malpractice insurance) is a prerequisite to entry into the field. For example, the second author has practiced both social work and law, and would not consider practicing either of those professions without carrying professional liability insurance.

National standards of practice could be incorporated into underwriting guidelines for professional liability insurance, providing a national (as opposed to a state-by-

Professionalization of Organization Development 363

state) measure of professional organization development practice. Thus, this approach to professionalizing organization development would represent a marshaling of market (clients', insurers', and practitioners') forces to achieve what neither group can achieve alone.

This approach would require marketing to managers (i.e., clients) the idea that organization development practitioners—for that matter, all consultants—should provide proof of professional liability insurance as a precondition to exploring an engagement. Such a requirement would lead to new questions, including:

1. working through how to resolve the experience/first job "catch-22" conundrum;
2. teaching the insurance industry where and how to underwrite risks and ensuring ongoing organization development/lay oversight of this process;
3. developing claims handling protocols that result in mediation (rather than litigation) of claims by clients;
4. determining how to deal with internal practitioners, as well as issues of company or individual coverage (e.g., would an employer be expected to pay for this coverage for an internal consultant?);
5. exploring development of a clearinghouse (like a credit-reporting agency or better business bureau) devoted to certifying consultant capabilities; and
6. exploring development of a disciplinary commission, a form of remedy hotline not unlike those established in several states for reporting unethical practice among members of the bar.

This line of thought and development leads to potential areas of convergence between various groups concerned with professional practice in organization development and consulting. Such groups include the ODN, the ODI, the Academy of Human Resource Development, the Academy of Management, the Institute of Management Consultants, and the Association of Management Consulting Firms.

This approval might have broad appeal for resolution in that the problem is not solely an organization development problem. Rather, this approach would solve a problem inherent in consultative relationships in and around organizations. Organization development may possess a unique articulation of its values, but those values have been shown to be highly consistent with those of other helping professions (Wason, 1998; Weidner, 1998).

Several sources (Maister, 1997; O'Shea & Madigan, 1997; Stewart, 1998) suggest that businesses are often dissatisfied with the performance of consultants, and would likely be highly motivated to better manage their consultants in a variety of fields, such as information technology, human resources, organization development, compensation, or strategic planning. Business organizations have a motivation to better manage consultative relationships that merely needs to be tapped. For example, in the wake of the decision against Microsoft (Bernstein, 1998), companies immediately began to delineate more clearly who were employees versus who was a contractor or consultant. Such changes were made fairly quickly and resulted in policies being adopted in many companies that precluded hiring of contractors who did not meet the standards established by the Internal Revenue Service for contractors. Those standards included indemnification from risk by contractors, and explicit statement (if not proof) of the contractor's possession of professional liability insurance.

Strengths. One of the strengths of this approach is that it could be put in place relatively quickly. One set of standards could be easily applied without legislative action in fifty states. A draft of these standards already exists in the form of the practices identified by the ODN's professional liability insurance carrier in its policy application.

Weaknesses. As with the "body of knowledge" approach described above, the "marshaled forces" approach possesses some limitations as well. One of these is the development of standards: Who will develop them? Will control of these standards reside within organization development or inside the insurance industry? How can we insure that control of those standards' development and maintenance/renewal will occur by those with content experience or expertise? Perhaps an advisory panel speaking for the "profession," combining the multiple perspectives of practitioners, researchers, educators, thought-leading authors, and leaders might be an effective approach to answering these questions. Expanding on this notion to be more inclusive, perhaps we can apply the emerging knowledge about the use of large group interventions (Bunker & Alban, 1997; Dannemiller & Jacobs, 1992; Dannemiller-Tyson Associates, 1994) to capitalize on the talent and wisdom of many more people in a condensed time frame. Professional liability insurance, in and of itself, does not address values, raising two questions that might be tackled in such a forum.

1. What terms, if any, of professional liability insurance now offered through ODN is inconsistent with "OD values"?
2. What organization development values are not represented in the professional liability application process?

Some limitations exist to such an approach. First, small, independent practices on the cusp of financial viability might be hard-pressed to afford initial premiums. Second, while depending on arms-length contracting to establish external consultative relationships as subcontractors, internal consultants would present a set of issues for companies everywhere: Who is an internal consultant? Which employees are professional liability insurance needed for? By way of protecting "the profession," individuals between jobs who call themselves "consultants" might be harder pressed to find "interim" consulting work. It is our belief that such a discourse would not only be good for the profession of organization development, but also would greatly benefit its practice.

Licensing: A Worthy Ideal, But Not an End-All

In light of the above discussion, a logical question becomes, "What of licensure?" There are several potential answers. One potential answer is certification in lieu of licensure. To accomplish certification, a certifying body (e.g., ODI or ODN) could serve as the certifying organization, much as how SHRM conveys different levels of certification on HR professionals. This approach has some limitations, primary among them being risk of "compliance" with continuing education requirements that can turn associations into continuing education "money mills" powered by individuals' ongoing need for certification compliance.

Licensure will require state action, and this may be beyond the influence of the existing institutional mechanisms in the field. The states do a fine job of managing the continuing education process for many professions, often in conjunction or in close cooperation with their state's professional association, without creating a "money mill" of continuing education. As a field, we might think through what fields continuing education models appear most effective at advancing the professionalism, both individually and collectively, of their members.

Although most licensure acts (and many certification programs) deal with topical and ethical knowledge, they do not necessarily include the preventative structures that prevent poor practice from occurring in the first place. Thus, additional mechanisms would need to be designed and implemented to deal with breaches of ethical practice. Our observation is that a substantial improvement in the practice of consulting to and within organizations would take place if only consultants were expected to carry professional liability insurance as a condition of gaining entry to the market or client.

Licensure as established and managed by the state would provide the consumer (client) an additional layer of confidence, but is alone insufficient—and may not be necessary—for professional practice in the presence of a robust liability insurance underwriting structure. With such norms in place, organizations which do not require professional liability insurance of their consultants are more likely to work with a consultant or firm who may not conduct themselves professionally. These clients are more likely to sue their consultants; sued firms are more likely to be large practices with financial resources available for a settlement (O'Shea & Madigan, 1997). Small firms are less likely to be targets of suits (no deep pockets available to sue for), but would likely be scrutinized by underwriters for proof of procedures, practices, and the like. Thus, while large firms might consider professional liability insurance as a true *insurance* policy, small firms may come to regard the insurance and the practices associated with the insurance as a *"cost of doing business"* or cost of market entry. Our view is that the field would benefit from the motivation regardless of its source if it improves the practice of consulting professionals

(Schellhardt, MacDonald, & Narisetti, 1998), particularly those in organization development.

Please do not interpret our idea as a suggestion that professional liability is the end-all to what ails organization development. Rather, it might be thought of as a highly productive, readily implemented "next step" on the road to professionalizing the practice of organization development along the journey to recognition of organization development as a profession. We believe the organization development profession "mountain," such that it is, has seen attempts to scale it from the harder (licensure) side of the summit. Perhaps such idealism is organization development's enduring legacy. However, now may well be the time to push forward to the summit from a more gradual approach than from the licensure side of the cliff.

Summary

Where might organization development go from here in its journey-in-progress to become a profession? In this section we outline some of the choices the field will need to make in order to realize this goal.

First, obtaining licensure would appear from the above discussion to be dependent on different organizations coalescing around an agreed-upon body of knowledge, practice standards, and code of ethics. This convergence of energy flies in the face of the spirit of plurality and participation that organization development practitioners value so much. There exists a natural tension, with broad participation and inclusion at one end and a small group "getting things done" at the other. There will never be participation of everyone interested in organization development in such an effort (e.g., the American Medical Association today counts only 35 percent of all physicians among its members, and is perceived by some nonmembers as both exclusive and exclusionary). Our view is that a failure to resolve this natural tension has likely been central to the limited success organization development has had in becoming recognized as a profession.

In the absence of licensure, the use of professional liability insurance guidelines in the interim—until licensure—would guide practitioner behavior toward greater professionalism. Both structures for licensure and insurance are currently in place. Insurance appears more available and open to organization development practitioners' direct influence; working where we have direct rather than indirect influence is just good OD.

The Bottom Line: No Easy Shortcut to Professionalism

The first question to ask regarding organization development becoming a profession is, "Is it important?" We believe it is, for reasons that can be traced to Freidson's arguments about professional autonomy and trust. We know from the study of organization development, better goals are an essential ingredient for better results. Thus, rather than an exclusive focus on licensure, we think a multifaceted approach—including professional liability insurance—will both enliven the field and advance organization development toward more rapid recognition as a profession. Furthermore, if Benveniste (1987) is correct, and the independence of thought, trust, and competence found in professionalism is important for organizational adaptability, these same qualities should be important for organizational development practitioners because they are responsible for helping organizations deal with change. Thus, it seems logical that organization development practitioners could also benefit from professionalization.

"How can organization development become a profession?" The professionalization and professionalism of organization development will require hard, sustained work by individuals and organizations over time in order to become more than a future-tense "ideal." Part of this work will be done by individuals as they attend to their own practice on a daily basis to strive for Maister's description of professionalism. Another portion of this work is related to the leaders of firms, practices, and departments to guide and monitor the professional behavior of their employees.

A third set of work involves bringing normative and "marshaled market forces" to bear on the need for consultants of all types to possess professional liability insurance. This might resemble the kind of marketing done by the Software Publishers Alliance on the issue of pirating software, after which CIO's in major companies put in place vigorous procedures to ensure software licensure compliance. In organization development consulting, such a campaign may require formation of an alliance of the associations described earlier, as well as the Association of Management Consulting Firms and the Institute for Management Consultants. This might not be as difficult as it seems, as associations and individuals with an interest in professional practice would have similar motivations for supporting such an initiative.

A fourth set of work surrounds the structures necessary to precipitate demand for professional recognition by law. This will require leaders of a current professional organization to rise above their group's individual identities and act for the greater good of the field. It will require a desire and commitment for influencing political entities and process and power—a subject organization practitioners should be well familiar with at this stage in our field's development. Unfortunately, the proliferation of such professional groups and their orientation toward establishing their own unique identity may be further exacerbating the emergence of organization development into a profession.

Conclusion

We would suggest organization development attempt to emulate the profession of law, including the notion of peer review, and the recognition of the responsibility of the profession to provide its services to those unable to afford them. We urge organization development practitioners to use the words *pro bono* when providing those donated services.

Pro bono organization development work has occurred for some time within the field. For example, in Chicago, the ODN/Chicago conducts a *pro bono* services program, providing organization development support to nonprofit and community organizations who otherwise would not be able to take advantage of members' expertise. The community service program also provides an apprenticeship program of sorts in that new entrants to organization development gain an opportunity to work with an experienced practitioner in a field setting. Also in Chicago, students from Loyola University's Center for Organization Development work on a wide range of corporate, public sector, and nonprofit settings, engaged in organization development interventions as part of their graduate level coursework, and supervised by the Center's faculty. Other academic programs often provide similar experiences, above and beyond those available to working students in their own workplaces. While donated services may help a profession's image, such activities do not create or make a profession. For example, social work began as the work of unpaid volunteers, but such efforts did little for the profession's development. To become a profession, social work had to meet Wilensky's criteria discussed earlier.

If a professional services firm needs a management and enforcement procedure in place to help employees act consistently with the firm's values as Maister (1997) suggests, what can be said about an entire profession that lacks the same structure? How are organization development professionals be expected to live out their values in light of market pressures to abandon them? Many organization development professionals do not even work in firms. How can the field help all of them to behave in a consistently value-driven manner when they continually face competing free market demands?

In order to become a profession, organization development must first closely resemble a profession. To progress beyond an occupation, organization development must consistently "walk" the profession "talk." It would be our hope that the questions raised in this chapter will help organization development to more resemble recognized, legitimate professions.

Should organization development have a future? We think so, but the question is one that is beyond us to answer. The agenda put forward in this paper contains both an action agenda and a research agenda. Such a description fits well with our sense of the field of organization development.

ACKNOWLEDGMENTS

The authors would like to acknowledge Mariela Adams and Susan Black of Loyola University-Chicago, who provided invaluable research assistance in preparation of this chapter. The manuscript benefited greatly from supportive comments from our colleagues John Furcon, Leanne Hunt, and Barbara Weidner.

REFERENCES

Abbott, A. (1988). *The system of professions: An essay on the division of expert labor.* Chicago: University of Chicago Press.
Academy of Human Resource Development (AHRD). (1998). *Academy of Human Resource Development: 1998 Conference Proceedings.* Baton Rouge, LA: Author.
Adams, S. (1996). *The Dilbert principle: A cubicles'-eye view of bosses, meetings, management fads & other workplace afflictions.* New York: HarperBusiness.
Adizes, I. (1988). *Corporate lifecycles: How and why corporations grow and die and what to do about it.* Englewood Cliffs, NJ: Prentice-Hall.
Adizes, I. (1997). *The pursuit of prime.* Santa Monica, CA: Knowledge Exchange.
Anderson, L.A., & Dedrick, R.F. (1990). Development of the trust in physician scale: A measure to assess interpersonal trust in patient-physician relationships. *Psychological Reports, 67,* 1091–1100.
Beckhard, R. (1969). *Organization development: Strategies and models.* Reading, MA: Addison-Wesley.
Beckhard, R., & Harris, R.T. (1977). *Organizational transitions: Managing complex change.* Reading, MA: Addison-Wesley.
Benveniste, G. (1987). *Professionalizing the organization: Reducing bureaucracy to enhance effectiveness.* San Francisco, CA: Jossey-Bass.
Bernstein, A. (1998, December 7). When is a temp not a temp? *Business Week,* pp. 90–92.
Berner, R. (1999, January 28). Pharmacists are starting to move in on doctors' turf. *The Wall Street Journal,* pp. B1, B15.
Bunker, B.B., & Alban, B.T. (1997). *Large group interventions: Engaging the whole system for rapid change.* San Francisco, CA: Jossey-Bass.
Burnham, J.C. (1996). Garrison lecture: How the concept of profession evolved in the work of historians of medicine. *Bulletin of the History of Medicine, 70*(1), 1–24.
Burke, W.W. (1997). The new agenda for organization development. *Organizational Dynamics, 26*(1), 6–20.
Burke, W.W., & Goodstein, L.D. (1980). Organization development today: A retrospective applied to the present and future. In W.W. Burke & L.D. Goodstein (Eds.), *Trends and issues in organization development: Current theory and practice* (pp. 3–11). San Diego, CA: University Associates.
Church, A.H. (1999). The future of O. D. *Organization Development Journal, 17*(1), 2–3.
Church, A.H., & Burke, W.W. (1995). Practitioner attitudes about the field of organization development. In W.A. Pasmore & R.W. Woodman (Eds.), *Research in organizational change and development* (Vol. 8, pp. 1–46). Greenwich, CT: JAI Press.
Church, A.H., Burke, W.W., & Van Eynde, D.F. (1994). Values, motives, and interventions of organization development practitioners. *Group & Organization Management, 19,* 5–50.
Church, A.H., Hurley, R.F, & Burke, W.W. (1992). Evolution or revolution in the values of organization development: Commentary on the state of the field. *Journal of Organizational Change Management, 5*(4), 6–23.
Cummings, T.G., & Worley, C.G. (1997). *Organization development and change* (6th ed.). Cincinnati, OH: South-Western.
Curry, L., Wergin, J.F., & Associates. (1993). *Educating professionals: Responding to new expectations for competence and accountability.* San Francisco, CA: Jossey-Bass.
Dannemiller, K., & Jacobs, R.W. (1992). Changing the way organizations change: A revolution in common sense. *Journal of Applied Behavioral Science, 28,* 480–498.
Dannemiller-Tyson Associates. (1994). *Real time strategic change: A consultant's guide to large scale meetings.* Ann Arbor, MI: Author.
Deaner, C.M.D. & Miller, K.J. (1988). Organization development: An evolving practice. *Organization Development Journal, 16*(3), 11–18.

DuBois, B., & Miley, K.K. (1999). *Social work: An empowering profession.* Boston: Allyn and Bacon.
Foltz-Gray, D. (1999, January 1). Medical breakthrough: Advanced practice nurses, the up-and-comers in the health-care field, can do (almost) anything your doctor does. *American Way, 32*(1), 80.
Franklin, S., & Japsen, B. (1999, February 7). Union idea may be just what doctors ordered. *Chicago Tribune,* Sec. 5, pp. 1, 8.
Freidson, E. (1991). Nourishing professionalism. In E.D. Pellegrino, R.M. Veatch, & J.P. Langan (Eds.), *Ethics, trust, and the professions: Philosophical and cultural aspects* (pp. 193-220). Washington, DC: Georgetown University Press.
Freidson, E. (1994). *Professionalism reborn: Theory, prophecy, and policy.* Chicago: University of Chicago Press.
French, W.L., & Bell, C.H. (1995). *Organization development: Behavioral science interventions for organization development* (5th ed.). Englewood Cliffs, NJ: Prentice-Hall.
Gellerman, W., Frankel. M.S., & Ladenson. (1990). *Values and ethics in organization and human systems development: Responding to dilemmas in professional life.* San Francisco, CA: Jossey-Bass.
Gorman, C. (1999, February 8). Bleak days for doctors. *Time,* p. 53.
Jelinek, M., & Litterer, J.A. (1988). Why OD must become more strategic. In W.A. Pasmore & R.W. Woodman (Eds.), *Research in organizational change and development* (Vol. 2, pp. 135–162). Greenwich, CT: JAI Press.
Jensen, B. (1999, February 17). Ex-editor blasts AMA interference. *Chicago Tribune,* Sec. 1, pp. 1, 16.
Kimball, B.A. (1992). *The "true professional ideal" in America: A history.* Cambridge, MA: Blackwell.
Krause, E.A. (1996). *Death of the guilds: Professions, states, and the advance of capitalism, 1930 to the present.* New Haven, CT: Yale University Press.
Larson, M.S. (1977). *The rise of professionalism.* Berkeley: University of California Press.
Maister, D.H. (1997). *True professionalism: The courage to care about your people, your clients, and your career.* New York: Free Press.
McArthur J.H., & Moore, F.D. (1997). The two cultures and the health care revolution: commerce and professionalism in medical care. *Journal of the American Medical Association, 277*(12), 985–990.
McMahan, G.C., & Woodman, R.W. (1992). The current practice of organization development within the firm. *Group & Organization Management, 17,* 117–134.
Miller, E. (1993). Organizational consultation: A craft or a profession? *Leadership & Organization Development Journal, 14*(4), 31–32.
Mitroff, I.I., & Pauchant, T.C. (1990). *We're so big and powerful nothing bad can happen to us: An investigation of America's crisis prone corporations.* Seacaucus, NJ: Carol Publishing.
Morin, W.J. (1990). *Trust me.* New York: Drake Beam Morin.
National Society of Professional Engineers (NSPE). (1999). *Engineering licensure laws: Summary and analysis* (1999). Alexandria, VA: Author.
Organization Development Institute (ODI). (1996). *The international registry of organization development professionals and organization development handbook.* Cleveland, OH: Author.
Organization Development Institute (ODI). (1999, February 15). *The Organization Development Institute: International O.D. code of ethics.* [WWW document]. URL: http://members.aol.com/odinst/
Organization Development Network (ODN). (1999, February 15). *Organization and human systems development credo.* [WWW document]. URL: http://www.odnet.org/credo.html
Olthuis, R., & Hiebert, M. (1995). *High performing internal consultants: Their customers' view.* Paper presented to the American Society for Training and Development Annual Conference & Exposition, Dallas, TX.
Ozar, D.T. (1995). Professions and professional ethics. In W.T. Reich (Ed.), *Encyclopedia of bioethics* (rev. ed., Vol. 4, pp. 2103–2112). New York: Macmillan Library Reference.
PMI Standards Committee. (1996). *A guide to the project management body of knowledge.* Upper Darby, PA: Project Management Institute.

Porras, J.I., & Robertson, P.J. (1987). Organization development theory: A typology and evaluation. In R.W. Woodman & W.A. Pasmore (Eds.), *Research in organizational change and development* (Vol. 1, pp. 1–57). Greenwich, CT: JAI Press.

Schellhardt, T.D., MacDonald, E., & Narisetti, R. (1998, October 20). Consulting firms get an unexpected taste of their own medicine. *The Wall Street Journal*, No. 78 EE/PR, pp. A1, A10.

Schein, E.H. (1992). *Organizational culture and leadership* (2nd ed.). San Francisco, CA: Jossey-Bass.

Smith, R.D., Houston, J.M., & McIntire, S.D. (1996). *Organization development: Strategies for changing environments*. New York: HarperCollins.

Stewart, T.A. (1997, April 28). An amazing, all-purpose executive show. *Fortune*, pp. 371–372.

Stewart, T.A. (1998, Marchs 16). Gray flannel suit? Moi? *Fortune* pp. 76–82.

Torres, D.L. (1991). What, if anything, is professionalism?: Institutions and the problem of change. In S.B. Bacharach (Ed.), *Research in the sociology of organizations* (Vol. 8, pp. 43–68). Greenwich, CT: JAI Press.

Varney, G.H., & Darrow, A.L. (1996). Market position of master level graduate programs in organization development. *Organization Development Practitioner*, 27(2), 39–43.

Walters, C., Jr. (1971). *Unforgiven: The biography of an idea*. Kansas City, MO: Economics Library.

Wason, K.D. (1998). The physician-patient metaphor: A practitioner's viewpoint. *Organization Development Practitioner*, 30(3), 41–44.

Weidner, C.K., II. (1995). *Innovation and the research of extremes: Mastering the possibilities*. Paper presented to the Third Annual Interdisciplinary Students of Organizations Conference Chapel Hill, NC (September).

Weidner, C.K., II. (1998). The physician-patient metaphor reconsidered: Reassessing the nature and practice of organization development. *Organization Development Practitioner*, 30(2), 51–60.

Weimer, D. (1999, February 8). The doctor is in—a union meeting. *Business Week*, p. 6.

Wilensky, H.L. (1964). The professionalization of everyone? *The American Journal of Sociology*, 70, 137–151.

Wilkins, D.B. (1995). Redefining the "professional" in professional ethics: An interdisciplinary approach to teaching professionalism. *Law and Contemporary Problems*, 58, 241–258.

Woodman, R.W. (1989). Organizational change and development. New arenas for inquiry and action. *Journal of Management*, 15, 205–228.

Woodman, R.W. (1993). Observations on the field of organizational change and development from the lunatic fringe. *Organization Development Journal*, 11(2), 71–74.

Worley, C.G., Hitchin, D.E., & Ross, W.L. (1996). *Integrated strategic change: How OD builds competitive advantage*. Reading, MA: Addison-Wesley.

Zussman, R. (1985). *Mechanics of the middle class*. Berkeley: University of California Press.

ABOUT THE CONTRIBUTORS

Achilles A. Armenakis is the James T. Pursell, Sr. Eminent Scholar in Management Ethics at Auburn University. He completed his doctorate in business administration at Mississippi State University. His teaching, research, and consulting interests include diagnosing, planning, implementing, and evaluating organizational change.

Sheri J. Bischoff is an Assistant Professor in the Department of Organizational Leadership and Strategy at Brigham Young University. She received her Ph.D. in Organizational Studies at the University of Oregon. Her research interests include stress and job strain, work/life balance, and cross-cultural management.

Jacqueline A-M. Coyle-Shapiro is a Lecturer in Industrial Relations at the London School of Economics and Political Science. She was formerly a Lecturer in Management Studies at the University of Oxford. Her research focuses on evaluation of organizational change interventions, the psychological contract, organizational citizenship behavior, and retaliatory behavior. Her work is published in several journals including *The Journal of Applied Behavioral Science* and the *Journal of Management Studies*.

Robert B. Denhardt is Professor of Public Affairs at Arizona State University and Distinguished Visiting Scholar at the University of Delaware. He is a past president of the American Society for Public Administration, a nationwide organization of academics and practitioners in the field of public administration at all levels of government. He was the founder and first chair of ASPA's National Campaign for Public Service, an effort to assert the dignity and worth of public service across the nation. He is also a member of the prestigious National Academy of Public Administration and a Fellow of the Canadian Centre for Management Development. Dr. Denhardt has published 13 books, including *In the Shadow of Organization, Theories of Public Organization, Public Administration: An Action Orientation, Executive Leadership in the Public Service, The Revitalization of the Public Service,* and *Pollution and Public Policy.* He has published over 70 articles in professional journals, primarily in the areas of public administration theory and organizational behavior, especially leadership and organizational change. His most recent research has focused on leading change in American local governments and has been supported by a grant from the Price Waterhouse Endowment for the Business of Government.

Hubert S. Feild is Torchmark Professor of Management at Auburn University. He received his Ph.D. from the University of Georgia. His professional interests include human resource selection and research methods in human resource management.

Jeffrey D. Ford is Associate Professor of Management in the Max M. Fisher College of Business at The Ohio State University. Prior to joining Ohio State, Jeff was on the faculty at Indiana University and Rutgers University. His current research focuses on management in general, and the management of change in particular, as phenomena in and of language. He has also been working with his wife Laurie in the development of an original approach to the design and management of organizations called "link management."

Joseph W. Grubbs is an Assistant Professor of Public and Nonprofit Administration at Grand Valley State University. His research examines the cultural implications of interorganizational relationships. In addition to numerous conference presentations, his published work has appeared in the *American Review of Public Administration,* the *International Journal of Public Administration,* the online *Journal of Public Administration and Management,* and forthcoming in the *Journal of Organizational Change Management.* He is also a contributing author in Robert B. Denhardt's book, *Public Administration: An Action Orientation* (1999, Harcourt Brace).

La Verne H. Higgins is an Assistant Professor in the Department of Industrial Relations and Human Resource Management at Le Moyne College, where she also teaches in the undergraduate and graduate business programs. Her research interests include human resource management in highly competitive industries, strategic international human resource management, the impact of intercultural dynamics on human resource practices, and the pedagogy of international management education. Dr. Higgins received her B.A. and M.B.A. from the University of Minnesota. After more than a decade as a manager, she continued her education at the

About the Contributors

University of Oregon where she completed a doctorate in human resource management and international management.

Stanley G. Harris is Associate Professor of Management at Auburn University. He received his Ph.D. from The University of Michigan. His research interests and consulting activities revolve around the domains created by the overlap of several topics: organizational change and transformation, organizational culture, strategic human resource management practices, and individual cognition and sense-making.

Gurudev S. Khalsa is a member of the Center for Social Innovations in Global Management (SIGMA) at Case Western Reserve University, where he is earning his doctorate in Organizational Behavior. He specializes in consulting with and writing about people and organizations involved in transformational change at multiple levels. His research interests include the role of personal transformation in social change organizing, large group processes for transboundary organizational alliance building; and the integration of spirit and soul into organizational life. His dissertation on "Organizing as Pilgrimage" builds on all of these themes and SIGMA's work with the United Religions Initiative. His published work has appeared in *The Academy of Management Review*, *Organization and Environment*, *The Journal of Management Inquiry*, and *The Journal of Management Systems*.

Orisha A. Kulick is an organization development practitioner in private practice who advocates adopting systems theory as the disciplinary base for organization development research and practice. She recently received her M.S.O.D. from the Center for Organization Development at Loyola University Chicago. In addition, she holds a M.A. from the University of Chicago School of Social Service Administration and a J.D. from the DePaul University School of Law. As an Illinois Assistant Attorney General, she argued appellate briefs, including cases before the Illinois Supreme Court. She also prepared a continuing legal education program for the Illinois Department of Employment Security in which she taught lawyers how to conduct professionally sound administrative hearings. As a psychotherapist she supervised clinicians in a research setting and participated in a high volume multidisciplinary private practice. She welcomes the exchange of ideas at rishkulick@aol.com

Long W. Lam is an Assistant Professor of Management at the University of Houston, Clear Lake. He received his Ph.D. in corporate strategy from the University of Oregon. His research interests center on the role of context in shaping organizational strategy, structure, and process. He is currently doing research on human resource competencies, internationalization process, cross-cultural influences on top management values, and null modeling of corporate performance.

Craig Lundberg is the Blanchard Professor of Human Resources Management in the School of Hotel Administration, Cornell University. His recent scholarship focuses on promoting and enacting alternative inquiry strategies, new forms of organization design, development and consultancy, messy variations in field experiments, and the tactics and strategies of human systems development. A Fellow of the Academy of Management, he is active in several professional societies. He has published extensively, facilitates organizational and personal development, and is currently an Associate Editor of *Group and Organization Management*. He is a rancher, skier, and fly fisherman concerned with ecological stewardship.

D. Lynne Persing earned her M.B.A. and Ph.D. degrees in management from the Graduate School of Management at the University of Oregon. She is an associate research professor in strategy at the Ecole Superieure de Commerce de Toulouse in France. Her research interests include the management of intellectually intensive work, particularly in R&D, the management of innovation, temporal processes in organizations, the role of risk in strategic decision making, and international management. Her industry experience includes several years as a program management analyst for Northrop in Washington, DC.

Peter J. Robertson is Associate Professor in the School of Policy, Planning, and Development at the University of Southern California. His research focuses on the requirements for successful collaboration within and between organizations; the process and outcomes of organizational change oriented toward empowering employees and improving their quality of work life; the impact of organizational contexts on participants' attitudes and behavior; and reform efforts in public school systems. His research has been published in a variety of journals and books, including the *Academy of Management Journal*, *Public Administration Review*, the *Journal of Public Administration Research and Theory*, *Educational Administration Quarterly*, the *Handbook of Industrial and Organizational Psychology*, and *Research in Organizational Change and Development*. He has done consulting and training for a variety of organizations, including Lockheed Missiles and Space, Apple Computers, Sun Microsystems, the National Credit Union Administration, the United States Air Force, and the Federal Aviation Administration. He is a member of the Academy of Management, the American Society for Public Administration, and the Institute of Noetic Sciences.

David S. Steingard is an Assistant Professor of Management at St. Joseph's University's Erivan K. Haub School of Business. His current research and teaching interests include: corporate spirituality and transformation; new paradigm and stakeholder strategic management; consciousness-based organizations; the social and environmental responsibilities of business; alternative research methodologies; and the integration of science and spirituality in business and academe. He has a variety of publications and conference presentations in the fields of organizational studies and management education. In business, he has held positions in organizational development as Director of Human Resources, Internal Consultant for Culture Development, and Training Specialist. He received his Ph.D. in Organizational Behavior from Case Western Reserve University.

C. Ken Weidner, II, Ph.D., is Assistant Professor of Organization Development at Loyola University's Center for Organization Development, and is President and Director of Practice of ValueWorks, Inc., which provides consultation and applied research in organization culture, change and communication, and leadership development. Previously, Ken was responsible for leadership development and organization development at the University of Chicago Hospitals' UCH Academy, which received a "Best Practices" award from the American Hospital Association for implementing a corporate university in health care. His dissertation research, "Trust and Distrust at Work," received the 1997 American Society for Training and Development's Dissertation Award. Ken is author of the forthcoming book, *Trust and Distrust at Work* (tentative title, Davies-Black Publishing, 2000).